T0323524

REPRODUCIBILITY IN BIOMEDICAL RESEARCH

REPRODUCIBILITY IN BIOMEDICAL RESEARCH

EPISTEMOLOGICAL AND STATISTICAL PROBLEMS

ERWIN B. MONTGOMERY JR.

Medical Director, Greenville Neuromodulation Center, Greenville, PA, United States

Professor of Neurology, Department of Medicine, Michael G. DeGroote School of Medicine at McMaster University, Hamiton, ON, Canada

ACADEMIC PRESS

An imprint of Elsevier

Academic Press is an imprint of Elsevier
125 London Wall, London EC2Y 5AS, United Kingdom
525 B Street, Suite 1650, San Diego, CA 92101, United States
50 Hampshire Street, 5th Floor, Cambridge, MA 02139, United States
The Boulevard, Langford Lane, Kidlington, Oxford OX5 1GB, United Kingdom

British Library Cataloguing-in-Publication Data
A catalogue record for this book is available from the British Library

Library of Congress Cataloging-in-Publication Data
A catalog record for this book is available from the Library of Congress

ISBN: 978-0-12-817443-2

For Information on all Academic Press publications
visit our website at https://www.elsevier.com/books-and-journals

Working together
to grow libraries in
ELSEVIER   Book Aid   developing countries
            International

www.elsevier.com • www.bookaid.org

Publisher: Andre Wolff
Acquisition Editor: Erin Hill-Parks
Editorial Project Manager: Timothy Bennett
Production Project Manager: Maria Bernard
Cover Designer: Miles Hitchen

Typeset by MPS Limited, Chennai, India

Dedication

To Lyn Turkstra for everything...

And to the "Saints Thomas," Hobbes and Kuhn, our scientific consciences

An experiment is never a failure solely because it fails to achieve predicted results. An experiment is a failure only when it also fails adequately to test the hypothesis in question, when the data it produces don't prove anything one way or another.

— *Robert M. Pirsig,* **Zen and the Art of Motorcycle Maintenance: An Inquiry into Values** *(1974)*

Contents

Preface

There is a problem in biomedical research. Whether it is a crisis and whether it is a new problem or an old endemic problem newly recognized are open questions. Whether it represents an ominous turn of events or merely a hiccup in the self-correcting process of biomedical research also is an open question. At the very minimum, it may be just a crisis of confidence. But it seems to have captured the imagination and concern of journal editors and administrators of research-granting institutions and, consequently, should be of concern to everyone involved in biomedical research.

The problem is the failure of reproducibility of biomedical research. Indeed, as will be discussed in this book, there is an appreciation of local reproducibility in that virtually every experiment involves more than one observation or trial. Just as virtually every experiment, at least implicitly, appreciates the importance of replication within an experiment, the same issues and concerns apply to the larger issues of reproducibility across different experiments by different researchers. Even when the failure of reproducibility is defined narrowly, such as a failure to achieve the same results when other researchers independently replicate the experiment—the narrow sense of irreproducibility—there are concerns. (Note that narrow reproducibility is different from local reproducibility, described previously, and broad reproducibility, described later.) This narrow sense often focuses on issues of fraud, transparency, reagents, materials, methods, and statistical analyses for which better "policing" would solve the problem. Perhaps these may be the major factors, but they may not be the only factors. None of this is to deny the importance of fraud, transparency, reagents, materials, methods, and statistical analyses, but at the same time it is possible that some causes of irreproducibility may be in the logic inherent in the research studies. Indeed, numerous examples will be presented and, thus, there is an obligation to carefully consider the logical basis of any experiment. Also, absent from the discussion is the fact that irreproducibility in a productive sense is fundamental to scientific progress. Such consideration is the central theme of this book.

Generally, results of studies are dichotomized into "positive" and "negative" studies. Irreproducibility affects positive and negative studies differently. Most of the debate centers on positive claims later demonstrated to be false, examples of a type I error—claiming as true what is false. In statistics, a type I error is occasioned by rejecting the null hypothesis inappropriately, which posits no difference in some measured phenomenon between the experimental and the control samples. Understandably, type I errors shake confidence and risk misdirecting subsequent research. Perhaps even more of a problem are type II errors, where negative studies claiming, falsely, that there is no change in a phenomenon as a result of an experimental manipulation or no differences between phenomena just because the null hypothesis could not be rejected. It is just possible that the experiment, by design or statistical circumstance, would have only a low probability of being able to reject the null hypothesis. A better term for these studies are "null studies," as using the term "negative" implies some inference with a degree of confidence even if undefined. The null study contrasts to situations of a negative study where the null hypothesis is rejected in a setting of sufficient statistical power and thus a positive claim of no difference can be made with confidence (equivalence, noninferiority, or nonsuperiority studies). The latter is termed a negative study and provides confidence just as the modus tollens form of the Scientific Method provides certainty, as will be discussed in the text. The extent of the problem of null studies is magnified by the bias toward not publishing null studies and, when published, are typically confused as negative studies. Type II errors in null studies may result in lost opportunities by discouraging further investigations.

Replicability, the narrow sense of reproducibility, is of primary importance as it is the first requirement. The author agrees with those who are concerned about irreproducibility in the narrow sense and indeed supports proposals for the policing of fraud; transparency; reporting of reagents, materials, and methods; and robust statistical analyses—but it does not and should not stop there. There is the temptation to dismiss concerns about irreproducibility in the narrow sense, believing it acceptably low and little more than the cost of doing scientific business. Perhaps so, if these irreproducible studies were merely outliers or flukes and that the vetting process for awarding grants and accepting papers for publication is otherwise effective. However, one must remind oneself that many of the papers describing research subsequently found to be irreproducible were vetted by experts in the field. What does this say about such expertise or the process?

There may be another view, which is the broad sense of irreproducibility, that does not focus on the exact replication of a specific experiment, but rather the failure to generalize or translate. These may be called conceptually irreproducible. Of particular concern is the failure to generalize or translate from nonhuman studies to the human condition; after all, is that not the *raison d'être* of the National Institutes of Health? But even further, is it not a founding pillar of modern science's

Reductionism where there is faith in the ability to generalize and translate from the reduced and simplified? When considered in the broader sense, there is far more evidence for and concern about these forms of irreproducibility. One need look no further than the postmarket withdrawal of drugs, biologics, and devices by organizations such as the US Federal Food and Drug Administration (FDA), whose preapproval positive clinical trials were supposedly vetted by experts and found valid. Only later were the efficacy and safety conclusions found irreproducible. This is not without consequences for patients.

The lives of biomedical researchers would be made much easier if irreproducibility was merely the result of fraud; improper use of statistics; lack of transparency; or failure to report reagents, materials, and methods. But what if there are other causes of irreproducibility fundamental to the paradigms of biomedical research, such as the Scientific Method (hypothesis-driven research) and statistics? As will be argued in this text, much of scientific progress requires the use of logical fallacies in order to gain new knowledge. Strict deduction while providing the greatest certainty in the conclusions in an important sense does not create new knowledge and induction to new knowledge is problematic. Traditional valid logical deduction is the logic of certainty, it is not the logic of discovery. Indeed, the Scientific Method is an example of the logical Fallacy of Confirming the Consequence (or Consequent), also known as the Fallacy of Affirming the Consequence (or Consequent). Also, the logic of the Scientific Method is referred to as abduction. This claim itself is not controversial as both scientists and philosophers have pointed this out for decades. What is novel here will be the demonstration that this fallacy is a cause of irreproducibility in biomedical research. Fortunately, if recognized, there are methods to blunt the effect of the fallacy, thereby reducing the risk of unproductive irreproducibility. What is needed, it will be argued, is the judicious use of logical fallacies.

Perhaps novel and counterintuitive, at least from the perspective of some scientists, is that the discipline of logic, typically in the domain of philosophy, could be considered relevant to empirical biomedical research. Scientists from the beginning of modern science, as seen in the founders of the Royal Society, rejected philosophy by taking aim against scholastic metaphysician natural philosophers. While scientists as late as the early 1900s invoked past philosophers in scientific discussions as Sir Charles Sherrington did with Rene Descartes in his text "The Integrative Action of the Nervous System" in 1906, such discussions are virtually absent in any literature typically read by today's biomedical researcher. Thus, it is understandable that biomedical researchers would be skeptical of any argument that proceeds from anything that smacks of philosophy, such as logic. But lack of experience in logic, epistemology (concerning how knowledge is gained rather than the content of knowledge), more generally, is a poor basis for skepticism. All this author can do is ask for forbearance, as the author is confident that such patience will be rewarded.

This author's ambitions for any role of logic in this particular text are circumscribed. It is critical to appreciate that logic alone does not create biomedical scientific knowledge. Science is fundamentally empiric and its success or failure ultimately relies on observation, data, and demonstration. All logic can do is provide some degree of certainty to the experimental design and analytical methods that drive claims of new scientific knowledge. Yet, the issues of reproducibility fundamentally involve issues of certainty, as reproducibility is at its core a testament to certainty. On this basis alone, logic has a role to play in concerns about scientific reproducibility. The discussions in this text are intended to strengthen, perhaps by just a bit, the already strong and important position of empirical biomedical research.

It may well be that an experiment is so obvious that there are no concerns as to the underlying logic raised. However, the success of such an experiment is not evidence that logic is not operating. For example, consider the Human Genome Project, which has been called a descriptive research program in contrast to a hypothesis-driven program. Perhaps it could be argued that the domain of the research was clearly defined and marked, that is, the human genome, thereby obviating any inductive ambiguity. Essentially, the Human Genome Project consisted of "turning the crank." Interestingly, the project set the stage for subsequent hypothesis-driven research (Verma, 2002), for example, that identified *gene A* causes *disease B* such that affecting *gene A* cures *disease B*. In that regard, hypothesis-driven research has not fared well given only two FDA-approved gene therapies (as opposed to genetic testing), despite the estimated $3 billion spent on the Human Genome Project. It is important to note that this is not a criticism of the Human Genome Project and that there is every reason to believe that the results will change medical therapies dramatically, but it will take time because it is difficult to go from data collection to cause and effect required of a hypothetico-deductive approach critical to biomedical research.

Further attesting to the potential contributions of logic is the fundamental fallacy inherent in statistics as used in biomedical research, which is the Fallacy of Four Terms. The experimental designs typically involve hypothesis testing on a sample thought representative of the population of concern with inferences from the findings on the sample transferred to the population through a syllogistic deduction. The sample is the entities studied directly, for example, a group of patients, in an effort to understand all patients, which is the population. There are many reasons why all patients cannot be studied and thus scientists have little choice but to study a sample. Consider the example: *disease A in a sample of patients is cured by treatment B, the population of those with disease A is the same as the sample of patients; therefore, the population of patients with disease A will be cured with treatment B.* The syllogism is extended to *my patient is the same as the population of patients with disease A and therefore my patient will be cured with treatment B.* However, it is clear there are many examples where *my patient* was not *cured with treatment B.* Indeed, this would be an example of irreproducibility in the broad sense, and one needs only to look to the drugs, biologics, and devices recalled by the FDA to see this is true. Something must be amiss. The majority of pivotal phase 3 trials that initially garnered FDA approval and later abandoned were not likely to have been type I errors based on fraud; lack of transparency; failure to report reagents, materials, or methods; or statistical flaws within the studies.

There are a number of other fallacies to which the scientific enterprise is heir to. These include the Fallacy of Pseudotransitivity that affects the formulation of hypotheses critical to the Scientific Method, the Fallacy of Affirming a Disjunctive, the Fallacy of Limited Alternatives, and the Gambler's Fallacy. Each will be explored in detail. However, it is very important to note that just because an experiment may have committed a fallacy, it does not mean that the results of the experiment are false, but only one cannot be certain. In that uncertainty lies the risk of irreproducibility.

Scientists should not take umbrage when a risk for a logical fallacy is demonstrated or, more generally, when it is pointed out that the Scientific Method and the scientific enterprise are prone to fallacies. It only means that one cannot have absolute confidence in the result and in that doubt, there is the risk of irreproducibility. Indeed, circumstances exist where a greater risk means a greater chance of new knowledge. Thus, the optimal science may well require a judicious use of logical fallacies, as will be explained in this book.

It should not come as a surprise that the scientific enterprise may be liable to fallacies. Gaining new knowledge is difficult and complicated, particularly in biomedical research. Indeed, it will be demonstrated that many fallacies are critical to the advancement of science, as little progress would be made without them. This may seem counterintuitive but patience in coming to understand the utility of fallacies will be rewarded. The Scientific Method developed for a reason. Statistics evolved for a reason. Both are responses to the fundamental uncertainties of gaining knowledge—any knowledge, but particularly scientific knowledge. Thus, misuse of the Scientific Method or statistics, in many cases, may be symptoms of failures in understanding the challenges of gaining new knowledge—epistemology—that led to the Scientific Method and statistics in the first place. It is not the fault of the Scientific

Method or statistics if its practitioners fail to understand the fundamentals that lead to practitioners misusing the tools. One cannot blame the hammer if a carpenter chooses to use it to saw a piece of wood.

There is a reason why philosophers avoided logical fallacies—they risk irreproducibility. Indeed, one of the most powerful philosophical methods of analysis is to demonstrate that the consequence of an argument results in a contradiction or absurdity, which could be considered an example of irreproducibility in philosophical analyses. Biomedical research experimentation, being a form of argumentation, thus is inherently at risk for irreproducibility by virtue of its necessary trading in logical fallacies. Indeed, the experiment cannot be immunized by statistics against the effects of the inherent logical fallacies. As will be demonstrated in this book, statistics is derivable conceptually from an extension of syllogistic deduction, a logical form, to the partial syllogism—an invalid (not guaranteeing certainty), but useful (creating the possibility for new knowledge), form of logical reasoning.

This book focuses on the epistemic, particularly, logical foundations for biomedical research reproducibility. It is not a primer on statistical analyses. Rather, it examines the implications of the necessary judicious use of logical fallacies on biomedical research experimental designs, including statistical design. The implications are many. Cases are presented that demonstrate a proper use of logical fallacies that decrease the risks of irreproducibilities, yet judicious irreproducibility, such as in a negative study, can provide a contribution that is certain—an example of productive irreproducibility.

The potential for an irreproducible experiment contributing positively to biomedical research is not only helpful, it is fundamental to the advancement of biomedical research. Indeed, productive irreproducibility falsifies the hypotheses, which conveys the greatest, perhaps only, certainty. However, from the reactions voiced in editorials and policy statements, it would appear that irreproducibility is an anathema, a plague that must be avoided at all costs. When "vaccinations" (education) do not work, "quarantine" (rejection of papers and grants) is required. But is this really the case? At least some form of irreproducibility is critical to the success of biomedical research. Perhaps what is critically needed and should be supported is productive irreproducibility with only unproductive irreproducibility avoided. Indeed, understanding what constitutes productive and unproductive irreproducibility is a central theme of this book. In contrast, cases are presented that demonstrate irreproducibilities resulting from poor logic in the experimental design, even if statistically sound, that produce results that are *indeterminant* and of little or no use—an unproductive irreproducibility. Approaches used to mitigate the increased risk of unproductive irreproducibility from the necessary use of logical fallacies are presented.

The epistemic status of statistics is necessarily self-referential and hence internal. Indeed, statistical claims, in themselves, are not ontological (a claim as to reality). Key assumptions are required to be met to maintain the epistemic safety of the logical fallacies that are the basis for the statistical method used. Chief among those assumptions is the Large Numbers Theorem, which holds that with increasing sample size, the measure of the Central Tendency, as measured by the mean (average), approaches a constant (unwavering) number—itself a form of reproducibility. In other words, the Central Tendency, such as the average or mean, becomes stable as the number of measurements included in the calculation of the Central Tendency increases. The convergence onto a stable measure of Central Tendency is thought to provide at least some credibility to the notion that the mean has some status in reality (ontological status). Another important concept is ergodicity, meaning the degree to which measures of the Central Tendency of samples are the same as the Central Tendency of the entire population. For example, if one were to measure the blood pressure from several groups of patients thought to have the same disorder, one would have little confidence if the Central Tendency, for example, the average blood pressure, was different among all the groups. At the very least, the question arises whether one or more groups of patients did not actually have the disorder. Alternatively, it just may mean that the population of all patients with the disorder is not distributed uniformly so that selecting different samples results in very different measures of Central Tendency. If the fish in a lake were distributed evenly (randomly), it would not matter where in the lake one fished. Violations of these assumptions have been demonstrated in the mathematics and physics of Chaotic and Complex Systems, and it is probable that these key assumptions, the Large Numbers Theorem and ergodicity, are not operable in biological Complex or Chaotic systems.

There is increasing evidence that virtually any living biological organism operates in the domain of Complexity and Chaos, in part due to the fact that living biological systems operate far from thermodynamic equilibrium. Typically, an equilibrium is achieved when opposing forces balance each other. In such Chaotic or Complex systems, ergodicity may not apply and thus an application of standard statistics may not be appropriate. Their use nonetheless increases the risk of appearance of irreproducibility, regardless of whether the experiments actually are irreproducible. New statistical approaches are needed. A sketch of what the new statistics may look like is provided.

None of the concerns expressed here should be construed as devaluing biomedical research. This author has been a biomedical researcher and physician long enough to have seen changes in medical treatments that would qualify as miraculous by virtually any standard. This author has great optimism that the pace of new miracles will only increase in the future. Further, the large majority of other biomedical researchers encountered by this author are dedicated to advancing biomedical research and are honest and honorable. But biomedical research is difficult and challenging at nearly every level. It is easy to trip and fall, particularly over a stone that is not seen, such as an injudicious use of a necessary logical fallacy. It is hoped that this book will help illuminate the path, at least as it concerns the obstacles to the judicious use of logical fallacies.

Thanks to Melissa Revell, whose editing spared the reader the gibberish of my dyslexia. Also, many thanks to Mr. Fred Haer, President of the Greenville Neuromodulation Center, for the continuing support of the scholarship this book required.

Erwin B. Montgomery, Jr., MD

Prologue

Is not reproducibility a significant concern, as evidenced by the remarkable responses by journal editors and grant reviewers? How can any biomedical researcher be sure their research won't be found irreproducible or that their work has not been compromised by the potentially irreproducible work of other scientists on which she or he depends?

Are there a number of types of reproducibility? Certainly, published concerns relate to repeatability of the same types of experiments by different scientists, but has there not also been concerns about reproducibility from animal research to analogous human research, for example, successful therapeutics in animal models of human diseases only to fail in humans? Perhaps one could consider repeatability as reproducibility in the narrow sense and consider reproducibility across different species or experiments as broad or conceptual reproducibility.

Is not the individual biomedical experimenter also concerned about reproducibility even within the single experiment being conducted? Is that not the reason for repeating runs of the same experiment or using multiple subjects or observations? Could this be considered the narrow type of reproducibility, which some may refer to replicability?

If there are multiple types of reproducibility, do they share similar risks for irreproducibility? Could comparing and contrasting these different types provide general insights into all?

Contributing factors to irreproducibility largely have been confined to considerations of fraud, transparency, reagents, materials, methods, or statistical rigor; however, is there not clear evidence of at least some irreproducibility not due to fraud, transparency, reagents, materials, methods, or statistical rigor? As it is likely true there must be other factors, but what are they?

Some experts attribute irreproducibility to experimental design. But what is experimental design? Does not experimental design actually mean the logic within the design of the experiment? If so, then is not logic or its misapplication also a source of irreproducibility? What is the nature of the logic inherent in experimentation?

Can we understand at least some instances of irreproducibility by analyzing the logic, recognizing logic in its widest connotation? What is the logic in biomedical research such that appropriate use results in reproducibility while its misapplication results in unproductive irreproducibility?

What is logic? Logic is not a set of rules; rules are only the outcome of the genesis of logic. Indeed, rules derive from whatever is necessary to assure a true and certain conclusion. Through the millennia, thinkers, such as, but not limited to, philosophers and logicians, have tried to construct general methods to assure certainty and, thus, assure that the results of the application of those methods are true (reproducible). In this not the same as what every biomedical researcher does in the course of his or her experimentation?

Current discussions view irreproducibility solely in a negative sense, but is there a positive sense? For example, a hypothesis demonstrated clearly and soundly irreproducible means that the hypothesis can be discarded. In this sense, irreproducibility can be productive. Indeed, this is the basis of experimental designs termed futility studies, many sponsored by the National Institutes of Health. How does one distinguish productive from unproductive irreproducibility?

If logic is involved in at least some forms of irreproducibility, then is there some form of logic that leads to reproducibility and productive irreproducibility while other forms of logic lead to unproductive irreproducibility? Should not we encourage the use of logical forms that lead to reproducibility and productive irreproducibility and discourage those that lead to unproductive irreproducibility? Doesn't this not mean that we have to be able to recognize those different forms of logic?

Isn't reproducibility a question of certainty? Does not logic bred certainty, which increases reproducibility? If so, the whole history of logic is to produce certainty. Could not the millennia of research and scholarship on deductive and inductive logic be relevant to the logic inherent in biomedical research and contribute to reproducibility and productive irreproducibility?

Is not the Scientific Method a logical method to help assure biomedical scientific progress? Is not the Scientific Method structured that hypotheses are translated to predictions and the predictions tested? Is this not the same as the hypothetico-deductive method of a general logic (abduction)? If application of the Scientific Method results in irreproducibility without evidence of fraud; lack of transparency; misuse of reagents, materials, or methods; or statistical rigor, could the logic of the Scientific Method be a source of unproductive irreproducibility?

Are not the hypotheses tested in the Scientific Method important? If application of the Scientific Method in a specific study results in unproductive irreproducibility, could not the hypothesis be the cause? If so, what is it about the hypothesis that resulted in a reproducible or a productively irreproducible study compared to hypotheses that resulted in unproductive irreproducibility?

Is not scientific reporting a form of argumentation where a scientist attempts to convince others in the scientific community of the truth and certainty of the experiments reported after first arguing the case within himself or herself? Is there not a structure or logic to the arguments? Can one identify a report that risks unproductive irreproducibility based on the logic inherent in the argument?

Is not statistics a means to assure certainty, and hence reproducibility, as much as possible?

Virtually every statistical measure or test requires certain rules, assumptions, and requirements, such as unbiased sampling and a sufficient number of subjects or observations. Yet, are not the rules, assumptions, and requirements derivative from and hence evidence of some inherent logic? Does this not mean that there is a necessary logic that must be followed as much as possible? What is that logic, fundamentally?

But what if at least some biological phenomena do not comport with the rules, assumptions, and requirements of typical or traditional statistical analyses? What does this mean for the ability to demonstrate reproducibility and productive irreproducibility in experiments involving these biological phenomena?

If we define reproducibility within experiments (local reproducibility) between the same experiments by different scientists (narrow reproducibility) as having *exactly* the same results, is it not highly unlikely that there would be *any* reproducibility? Therefore, how much irreproducibility is still allowed to be held reproducible?

As the vast number of experiments rely on digital computation for data acquisition, analysis, and testing, could not the use of digital computers in themselves be a source of irreproducibility?

This book seeks to answer these questions.

Introduction

SCIENCE AS ARGUMENTATION WITHIN THE EXPERIMENT AND WITHIN THE COMMUNITY

Biomedical research is fundamentally empirical. New scientific knowledge comes from observations, data, and demonstrations. Thus, it would seem that the reality of observations, data, and demonstrations should vouchsafe the knowledge claims of biomedical research. Perhaps that is why the concerns about relative widespread instances of irreproducibility are so jarring. The presumption is that "data don't lie" and thus the sin lies within the researcher, whether it is overt fraud, covert fraud in the sense of failed transparency, or carelessness in the use of reagents, materials, methods, or statistics. The words of Mark Twain have humor and edge: "Figures often beguile me, particularly when I have the arranging of them myself; in which case the remark attributed to Disraeli would often apply with justice and force: 'There are three kinds of lies: lies, damned lies and statistics'" (Mark Twain's Own Autobiography: The Chapters from the North American Review). But, as will be demonstrated repeatedly, fraud; lack of transparency; misuse of reagents, materials, and methods; and statistical misadventures cannot account for all the cases of irreproducibility.

Logic is a tool for argumentation. Logic alone will not generate new knowledge. Logic is said to be truth (knowledge) preserving, not truth (knowledge) generating. However, scientific experimentation is a form of argumentation. The scientist first must argue to themselves that the findings, inferences, or conclusions are in accordance with the observations, data, and demonstrations. In running the same experiment multiple times, such as using multiple subjects, the experimenter must convince themselves of local reproducibility. Then, scientists must convince others of both narrow and broad reproducibility. Note that in

1

both cases, "convincing" derives from establishing certainty, and it is the forte of logic to test the certainty of any argument.

It is interesting in that a seminal study by John Ioannidis and colleague (Ioannidis, 2005), who examined reports of clinical trials in the most significant journals that were subsequently refuted, the point was made that one could not say which of the studies, the index or the refuting articles, were true based on the different outcomes. In these cases, there was no fraud; lack of transparency; misuse of reagents, materials, or methods; or statistical flaws that would point to the culprit. As will be seen in subsequent chapters, the implications and use of the knowledge offered by conflicting studies depend on which is more convincing, and the convincing ultimately will be played out in the logic within the studies rather than in the differences in results (see Chapter 6: Causation, Process Metaphor, and Reductionism).

Scientific experiments are structured arguments in themselves and adding claims of new knowledge to the scientific compendium requires argumentation to the scientific community. This is seen in the evolution of scientific reports. Four stages were described in an analysis by Charles Bazerman of scientific reports in the *Philosophical Transactions of the Royal Society* from 1665 to 1800 (Bazerman, 1997, pp. 169–186). During the period 1665–1700, scientific papers were uncontested reports of events. In a sense, the data presented was left to argue for itself. From 1700 to 1760, discussions were added and centered over the results. More theoretical aspects were addressed during the third period 1760–80; papers "explored the meaning of unusual events through discovery accounts" (Bazerman, 1997, p. 184). From approximately 1790 to 1800 (and arguably to this day), experiments were reported as claims for which the experiments were to constitute evidence with the intent of convincing the reader through argument. Consequently, it is at least possible that the methods of argumentation, following from experimental design, could be logically fallacious and, thus, a source of irreproducibility. These issues will be taken up in a rigorous manner in subsequent chapters.

Perhaps for most readers, the importance of logic to scientific demonstration (argumentation) would seem foreign. However, concerns regarding the logic of science and scientific argumentation for establishing new knowledge often have been explicitly expressed. For example, Galileo (1564–1642), early in the "Dialogues Concerning the Two Chief World Systems" (https://drive.google.com/file/d/0B9bX852JMJ__ODM4NjUyYzktZjE3OS00MmFlLTljNzktMmQzODM0ZDVmYWQy/view?hl = en), addressed the place of logic or reason in science and its relation to observation (evidence), having Salviati say:

> What you refer to is the method he [Aristotle] uses in writing his doctrine, but I do not believe it to be that with which he investigated it. *Rather, I think it certain that*

he first obtained it by means of the senses, experiments, and observations, to assure himself as much as possible of his conclusions. Afterward he sought means to make them demonstrable. That is what is done for the most part in the demonstrative sciences; this comes about because when the conclusion is true, one may by making use of analytical methods hit upon some proposition which is already demonstrated, or arrive at some axiomatic principle; but if the conclusion is false, one can go on forever without ever finding any known truth—if indeed one does not encounter some impossibility or manifest absurdity [emphasis added]. And you may be sure that Pythagoras, long before he discovered the proof for which he sacrificed a hecatomb, was sure that the square on the side opposite the right angle in a right triangle was equal to the squares on the other two sides. The certainty of a conclusion assists not a little in the discovery of its proof—meaning always in the demonstrative sciences. But however Aristotle may have proceeded, whether the reason *a priori* came before the sense perception *a posteriori* or the other way round, it is enough that Aristotle, as he said many times, preferred sensible experience to any argument. Besides, the strength of the arguments *a priori* has already been examined.

Clearly, Galileo is arguing for a logic in science. The process starts when Aristotle (as a scientist) "first obtained it by means of the senses, experiments, and observations, to assure himself as much as possible of his conclusions." This can reasonably be ascribed to the logic of induction, generating general principles from specific individual observations. But Galileo also left open the possibility that it was prior theory that may allow organization of observations in a deductive logical manner, reasoning from a general principle to a specific expectation, to subsequently create an induction. Galileo wrote "But however Aristotle may have proceeded, whether the reason *a priori* came before the sense perception a posteriori or the other way round, it is enough that Aristotle, as he said many times, preferred sensible experience to any argument." In any event, Galileo does not discount the importance of demonstration, that is, proof through argumentation.

The issue of logical structure in science has an extensive history intervening in the years from Galileo to today. Arguably, the Enlightenment extended the movement to empirical experimental science begun formally with the establishment of the Royal Society in 1660. The Enlightenment went well beyond the strict empiricism of the Royal Society to include a rationalism suggesting a type of logic that transcends the specifics to general principles. Perhaps, Isaac Newton's "Philosophiæ Naturalis Principia Mathematica" ("Mathematical Principles of Natural Philosophy") in 1687 is a prime example. Isaac Newton answered Edmund Hailey's question about how Newton knew the orbit of the moon was elliptical as empirically demonstrated by Kepler by replying that he had derived it (by applying his newly developed calculus to the two-body problem in physics). The following is Abraham de Moivre's account:

In 1684 Dr Halley came to visit him at Cambridge. After they had been some time together, the Dr asked him what he thought the curve would be that would be described by the planets supposing the force of attraction towards the sun to be reciprocal to the square of their distance from it. Sir Isaac replied immediately that it would be an ellipse. The Doctor, struck with joy and amazement, asked him how he knew it. Why, saith he, I have calculated it. Whereupon Dr Halley asked him for his calculation without any farther delay. Sir Isaac looked among his papers but could not find it, but he promised him to renew it and then to send it him (http://www.mathpages.com/home/kmath658/kmath658.htm).

The logical (rational) basis for science likely continued to be so generally accepted that little explicit analysis or commentary by scientists was necessary. To be sure, philosophers took up the question, if only to try to grasp the remarkable advances in science, not only in terms of knowledge, but in view of the remarkable technological translations. It is not inconceivable that the philosophers felt they had to "catch up." The philosophy of science turned sharply to considerations of logic in the early- to mid-1900s with the development of Logical Empiricism. This philosophical movement was based on a set of fundamental observations (scientific facts) and logical rules to infer from the fundamental observations that all scientific knowledge could be constructed. Indeed, philosophers such as Bertrand Russell looked to empirical science as a means of refining or reconstructing philosophy, particularly in matters of epistemology and specifically logic.

The logical nature of science did not become a major concern for scientists until the development of postmodernism, particularly following Thomas Kuhn's *The Structure of Scientific Revolutions* (1962). Kuhn's historical analysis belied the notion of a strict incremental rational basis for scientific advancement. Interestingly, Kuhn's work would be somewhat problematic as it relates to irreproducibility. Indeed, Kuhn would appear to suggest that virtually all scientific work is inherently irreproducible as it is tied to the specific paradigm (a particular perspective or approach) in which the scientific work takes place and because of Kuhn's notion of incommensurability between scientists operating in different paradigms. He held that a scientist operating in one paradigm just could not fully understand or appreciate scientists operating in a different paradigm. Kuhn goes on to suggest that new paradigms must ignore contrary observations (at least at first), hence ignoring irreproducibility, in order to survive long enough to contribute.

Kuhn's work seems to precipitate the postmodern attach on rational science. The postmodernists, particularly Paul Feyerabend (1975), dispute the privileged position of the Scientific Method and offer historical records as evidence. It seems most of the concern in the book by Gross and colleagues, and elsewhere, in response to the postmodernists relates particularly to the method and the historical successes and failures of modern science. The drivers of scientific progress were sociological or psychological rather than logical or rational. However, is should be noted that these critiques were spawned by commentators on science and rarely, if ever, from leading scientists. Indeed, the reaction to postmodernists reflects a continued faith in the logic or rationality of science, including biomedical research (Gross et al., 1997), and perhaps with a bit of hyperbole, suggesting a "nerve was struck" (Sokal and Bricmont, 1998).

Scientific rationality has also been inherent in the school of medicine, known as the Rationalists, since the time of the ancient Greeks. This school posited that all manifestations of human disease could be understood as some combination of general principles. Archetypical of this approach is Galen's conception of imbalances among four humors as they affect particular organs (Arikha, 2007). The Rationalists claimed to practice scientific medicine and this was a cornerstone in the Code of Ethics of the American Medical Association in 1847 (Montgomery, 2018). Current-day allopathic physicians are descendants of the Rationalist physicians.

Interestingly, Logical Empiricism, like its breatharian Logical Positivism, holds that from a set of axiomatic self-evident axions and a limited number of rules of inference, all human knowledge could be constructed. Logical Empiricism was the version where the axioms are derived from empirical science. Both efforts collapsed because of the demonstration of Incompleteness, particularly Gödel's theorems of Incompleteness. Essentially, Gödel's Incompleteness theorems demonstrate that their logical and mathematical statements generally were regarded as true, but cannot be proven, that is, derived from a set of self-evident axioms and rules of inference. It turns out that there are a number of different forms of Incompleteness, including computational, measurement (instrumental), and informational. The implications of these forms of Incompleteness on reproducibility are discussed in Chapter 10, Chaotic and Complex Systems, Statistics, and Far-From-Equilibrium Thermodynamics. The Incompleteness theorems do not obviate the value of logic, but rather demonstrate that the absolute certainty in logic is not possible. But given the aforementioned discussions, this should not be surprising.

SCIENTIFIC ARGUMENTATION AND LOGIC

As biomedical research is a form of argumentation within the experiment and among the scientific community, every argument entails a logic, in the widest connotation of the term "logic." However, as already intimated, logic, in its traditional sense, alone will not provide new scientific knowledge. Yet, there must be some means by which the absolute certainty derived from valid uses of logical deduction can be "borrowed" from to help with generation of new knowledge. To be sure, these alternative logical forms will not provide absolute certainty but nonetheless allow the generation of new knowledge. Further, the degree of uncertainty can be estimated by how these alternative forms differ from the valid traditional forms of deductive logic. In the traditional notion of logic to provide knowledge with absolute certainty, the actual forms (rules) of traditional logic are derivative of that purpose. If one is willing to loosen certainty from absolute, then these alternative forms, the fallacies, can serve the purpose of logic to provide knowledge that is reasonably certain, and importantly, to determine how reasonable is that certainty. Thus, an argument could be made that for the sake of biomedical research, the judicious use of logical fallacies should be combined with traditional approaches to create a new corpus of logic suited to the advancement of biomedical research.

The expanded corpus of logical structures in the scientific argument can only attest to the degree, not absolute, of certainty of the experimental argument. This unique corpus of logical forms that aids certainty in the scientific discovery of new knowledge can be called "evolutionary logic" (Montgomery, 2018). Evolutionary logic includes the traditional forms of logic, for example, deduction, but also includes logical fallacies derived from the traditional forms. This type of logic is called evolutionary, as it is necessary to evolve new knowledge in a sense that mutations (fallacies) are necessary to create new species (new knowledge). This logic is evolutionary as it is responding to the needs of scientists to gain new knowledge. It is organic in that it is in response to human needs. It is synthetic in that it is a creation of humans faced with the epistemic challenges in understanding the universe.

As will be demonstrated, the unique logic important to scientific discovery actually is logical fallacies, and the certainty of any experiment depends on the judicious use of logical fallacies. Further, the degree of certainty in any experimental claim can be gauged by how far the fallacy necessarily used departs from the traditional logical form that affords the greatest certainty. For example, deduction provides absolute certainty. A fallacious form of deduction important to biomedical scientific progress is the Fallacy of Confirming the Consequence, otherwise known as the Scientific Method. Again, it is important to note that claims to new knowledge based on empirical observations, data, and demonstrations are not falsified by the necessary use of the logical fallacy but only that the certainty of the knowledge claim cannot be absolute.

A central theme of this book is that logical fallacies are fundamental to the Scientific Method and the scientific enterprise. A judicious use of logical fallacies underlies the hypothetico-deductive approach in biomedical experimentation, the creation of the hypotheses for experimentation, and the statistical means to arrive at and justify inferences. For example, it has been recognized for decades and much ink has been spilt on describing how the Scientific Method, in the form typically used, is the Fallacy of Confirming the Consequence (or Consequent), also known as the Fallacy of Affirming the Consequence (or Consequent) (Lee and Hubona, 2009; Poldrack, 2008; Schmidt, 1956; Sjoberg, 2017). Perhaps what is novel in this text is the argument that the injudicious use of logical fallacies may be a source of unproductive irreproducibility, while the judicious use helps ensure reproducibility and productive irreproducibility.

THE MULTIFACETED NOTION OF IRREPRODUCIBILITY

First, it is important to come to grips with the notion of irreproducibility. Arguably, science could not advance without irreproducibility. Indeed, it is when flies could not be reproduced on rotting flesh in jars covered with gauze in the experiments by Francesco Redi (1626–97) were concepts such as spontaneous generation abandoned. In the aforementioned quote, Galileo, through Salviati, says *"if the conclusion is false, one can go on forever without ever finding any known truth* [emphasis added]—if indeed one does not encounter some impossibility or manifest absurdity."* The notion of "absurdity" is conceptually a case of irreproducibility. The quote clearly demonstrates the importance of productive irreproducibility in avoiding going on forever from a false conclusion. Unless one thinks that the current state of knowledge is complete, then it is inevitable, and indeed it is to be hoped for, that there will be many instances of productive irreproducibility in the future, as per Galileo. The onus is to ensure that future irreproducibility is productive—through the judicious use of logical fallacies (as will be discussed later).

To be sure, the notion of productive irreproducibility in biomedical research may seem strange. However, it recently has been formalized in what are called futility studies. For example, the National Institutes of Health and others have conducted futility studies whose purpose is to confidently eliminate potential treatments specifically by demonstrating their failure to result in anticipated outcomes, again a form of irreproducibility (e.g., see Koch et al., 2015). When a study fails but the nature of the study is one that does not lend confidence, this study would be an example of "unproductive" irreproducibility. There are statistical as well as logical reasons why it is easier to disprove something than to provide something.

There are at least three notions relative to the domain over which concerns of irreproducibility range. There is local reproducibility within an experiment. Another is the narrow sense of irreproducibility, which relates to the exact replication of specific experiments down to the regents, specimens, materials, instruments, and statistical analyses used. Given this view, then the issues of narrow reproducibility concern readily boil down to fraud, transparency, and the mechanics of the experimentation and statistical analyses. Indeed, journal editors and grant administrators have addressed issues at this level (Landis et al., 2012; Steward and Balice-Gordon, 2014). There seems little in this view that would be considered unreasonable.

To be sure, there are instances of narrow irreproducibility, and addressing the risk factors is very important in biomedical research as evidenced by the policies being set by journal editors and grant administrators. However, this is not necessarily the case as discussed in Chapter 8, Case Studies in Basic Biomedical Research. Replicability, at least in the simple sense, may not be possible in chaotic and complex biological systems. The violation of simple replicability needs not argue for irreproducibility, but rather requires some changes in how replicability is understood. It just may be inherent in data analysis through the use of digital computers (Chapter 9: Case Studies in Computational Biomedical Research).

There is a broad sense of irreproducibility that includes concerns of the narrow sense but goes beyond the specific experiments to generalize or translate to other actual and potential experiences. Consider the example of a clinical trial for a new treatment. Typically, the treatment is applied to a sample of the population of concern (e.g., a group of 100 patients with Parkinson's disease is studied and the results considered applicable to the many thousands of patients with Parkinson's disease). Typically, these clinical trials are vetted by governmental organizations, such as the US Federal Food and Drug Administration, and by peer-reviewed articles to be published describing the clinical trial. The validity of the clinical trial is vouchsafed by market approval and publication. Yet, the extent of postapproval withdrawal of drugs, biologics, and devices most assuredly must be a form of irreproducibility. Further, it cannot be an issue of reagents, as the drug, biologic, or device is no different preapproval versus postapproval. It is not likely that there are methodological errors, as it is highly likely that if the same subjects were tested again, the results would be the same. Similarly, it is unlikely, or at least one cannot assume, that the statistical analyses were at fault. There likely is something more fundamental. As will be discussed elsewhere, it is likely that the problem lies in the Fallacy of Four Terms, the risk for which is fundamental to clinical trials (Chapter 5: The Logic of Probability and Statistics and Chapter 7: Case Studies in Clinical Biomedical Research).

The broad view of reproducibility, with its emphasis on generalization and transfer, is critical to the very nature of biomedical research by virtue of its reductionism. In vitro (e.g., test tube, slice, or tissue culture studies) experiments are valuable in their generalizability to in vivo (e.g., whole animals) experiments; experiments in lower species are critical to understanding higher species, such as humans. Biologically inspired in silico (computational or mathematical modeling and simulations) experiments are important as they relate to in vitro and in vivo experiments. The ability to generalize from the specifics of an experiment is critical to science, as this ability makes science a general discipline of knowledge rather than a database. Particular knowledge can be derived from an economical set of principles rather than as a lookup in a database. Irreproducibility across specific methods, specimens, and reagents strikes at the nature of science.

The notion of reproducibility limited to replication would be very narrow and consequently not much use to biomedical science. Indeed, it would be a form of Solipsism of the Present Moment, which means that what is true is only what is true in the exact circumstances of the present moment in time. This form of solipsism would not provide any understanding of the past or any predictions of the future. This is the antithesis of biomedical research, which appeals to generalizable knowledge that transcends the individual experiment. An experiment on roundworms, *Caenorhabditis elegans*, is important to those concerned with the impact of roundworms, but it also has value to biological science on a much larger scale. That experiments conducted on worms can be reproduced in other species, such as humans, not only expands the utility of such research, but also lends further credence to concepts demonstrated in the roundworm when similar concepts are found to hold in a number of other species. Reproducibility when concepts vindicated with experiments in nonhuman species are demonstrated in humans, for example, would be conceptual reproducibility. Further, there may be evolutionary reasons based on the Second Law of Thermodynamics (a physical law that describes changes or dynamics originally derived from studies of steam locomotion) for generation across biological species, particularly conservation of genes throughout phylogeny (discussed in Chapter 10: Chaotic and Complex Systems, Statistics, and Far-From-Equilibrium Thermodynamics). Clearly, the broad sense of irreproducibility is important.

PRODUCTIVE IRREPRODUCIBILITY

In the search for new biomedical knowledge, the problem of experiments not being reproducible has become a major issue for the biomedical research enterprise. At the very least, it may be that the recognition of a lack of reproducibility has become a problem. Irreproducibility is seen as reducing confidence in all of biomedical research as it is not clear which study eventually will be found irreproducible. After all, many studies later found irreproducible were vetted by experts in the field through peer review. The concern is clearly evident, as shown by leaders in the biomedical research enterprise, such as journal editors and officials of sponsoring organizations and governmental programs. It clearly is a concern to the practicing scientist who surely will be affected by policies instituted by journals and granting agencies. It would appear from the steps recently undertaken or being entertained that irreproducibility is like a plague that must be prevented if possible, quarantined if not prevented, and avoided at all costs. But is this truly the case? These concerns regarding irreproducibility seem to reflect a notion that science is progressive and incremental. Each new advance is built on top of a previous advance, giving the impression that rarely does old knowledge have to be reversed. But this does not accurately reflect the history of science, as pointed out by Thomas Kuhn (1962). The extent of the problem of the failure of reproducibility is discussed later.

Irreproducibility is critically necessary for the advancement of science—productive irreproducibility should be embraced. Productive irreproducibility provides confidence that the experiment and the inferences generated are wrong and safely dismissed. In addition, productive irreproducibility importantly targets for revision the presumptions and assumptions entailed in the failed experiment and in the process by which the inferences were made. Indeed, productive irreproducibility has been formalized in biomedical research as futility studies as described earlier. The injudicious use of logical fallacy results in uncertainty, consequently providing little direction for further research. Again, referring to Galileo's writings, "if the conclusion is false, one can go on forever without ever finding any known truth—if indeed one does not encounter *some impossibility or manifest absurdity* [emphasis added]." It is the impossibility or manifest absurdity that constitutes productive irreproducibility and prevents the scientist from continuing from a false conclusion, the likely consequence if an unproductive irreproducibility was encountered.

Analyses by philosophers of science, such as Karl Popper (2002), present a cogent argument for the centrality of irreproducibility in science. Popper held that the very possibility of a claim to knowledge being demonstrated irreproducible (falsifiability) is what makes science and not pseudoscience. Lest this not be seen as only of interest to hoary (and younger) philosophers, leading scientists used the same notion of the potential to demonstrate irreproducibility to argue against admitting Creationism and Intelligent Design into the science curriculum. It would seem then, conceptually, wholesale aversion to irreproducibility, at the very least, is tantamount to "throwing the baby out with the bathwater." Perhaps there is some good use of irreproducibility.

Unproductive irreproducibility is to be avoided. This type of irreproducibility contributes little but confusion. One cannot reject the conclusion found unproductively irreproducible because such rejection might eliminate further consideration of the hypotheses that subsequent experiments might have demonstrated as true. However, continued experimentation based on the hypotheses and experiments found unproductively irreproducible could be great wastes of time and resources. At the very least, a study found judiciously irreproducible has the positive effect of being able to eliminate the hypotheses.

Typically, unproductive irreproducibility results from technical issues around any experiment. For example, a poor experimental design or an improper use of statistics is likely to lead to an increased risk of unproductive irreproducibility should there be an attempt to reproduce the experiment. However, as will be seen, it also is possible that the sound uses of many experimental designs and statistical analyses still produce unproductive irreproducibility. The cause of subsequent irreproducibility lies in the logical errors in the framing of the experiment. Recognizing the logical errors at least points the scientist in a direction to inform future experiments.

As will be demonstrated, productive irreproducibility frequently involves a form of logic called modus tollens. It is of the form *if a implies b* [hypothesis] *is true and b* [experimental prediction] *is false, then a must be false*. As it is certain that *a* is false, it can be abandoned from the considerations of what might make *b* true. If the prediction is false, that is, not demonstrated in the experiment, hypothesis *a* can be rejected with certainty. This would be a form of productive irreproducibility.

ONTOLOGY VERSUS EPISTEMOLOGY

Philosophers make a distinction between ontology, the theory of what reality is like (e.g., there is just such a thing as pneumonia that is caused by just such a thing as *streptococcal pneumonia*), and epistemology, the method by which a knowledge of what reality is like is obtained (e.g., Koch's Postulates for determining a bacterial cause of a disease). There may be a presumption among many scientists of an equivalency between the nature of reality of concern to scientists (ontology) and the manner by which the ontology is discovered, that is, the scientific method in experimentation (epistemology). Note that the equivalency is not presumed to be circular or self-referential (as some postmodernist critics of science may claim). In other words, it is not a self-fulfilling prophesy or tautology (true by definition) that the hypothesis necessarily means that the prediction will be found true. A tautology would be if the definition of the hypothesis is exactly the same as the prediction, which would be of the logical form *if a implies a is true and a is true, then a is true*, how could it be otherwise? The first presumption in a successful experiment, particularly one that demonstrated reproducibility, is that the results are not hallucinations or illusions. The results of the experiment are not some contingent proxy for what is real. The results are presumed to be really real.

Certainly, the statement just made may seem categorical and thus extravagant, if not exuberant. Certainly, no scientist would admit to such a wholesale claim that his or her results are not contingent or tentative. However, the actual operations of the scientific enterprise reflected in its history would suggest otherwise, at least implicitly, as suggested by Thomas Kuhn (1962). Lest one be dismissive of Kuhn, it should be noted that the prestigious journals *Science* and *Nature* appreciatively marked the 50th anniversary of the publication of Kuhn's "The Structure of Scientific Revolutions."

Perhaps it is the conflation of ontology and epistemology with the presumption of their equivalence that makes the published concerns regarding irreproducibility so jarring. Unappreciated methodological errors resulting in irreproducibility are seen as challenging the ontological status of the science. The issue may lie in biomedical science's epistemology rather than its ontology (discussed in greater detail in Chapter 2: The Problem of Irreproducibility). At the very least, the demonstrations of scientific irreproducibility should cause a reevaluation of the epistemology, the route to scientific knowledge. The epistemic analysis is likely to uncover causes other than fraud; lack of transparency; misuse of reagents, materials, and methods; and statistical rigor.

Perhaps another indication of the conflation of ontology and epistemology lies in the habit of scientific reports to default to the measure of Central Tendency (the mean, median, or mode) rather than the variance of the actual measurements, as will be discussed in detail in Chapter 4, The Logic of Certainty Versus the Logic of Discovery. Further, the Central Tendency measure becomes the real ontology, even if the measure of central tendency was never found in the actual measurements. An interesting example of this faith is seen in S. M. Stigler's *The Seven Pillars of Statistical Wisdom* (2016) and others (Stahel, 2016), although doubt was expressed by the early founders of modern statistics (Gigerenzer et al., 1989). The Central Tendency is further installed as the reality by demonstration; the results are dichotomized as true (real) or false (unreal) based on statistical significance, such as by rejecting the null hypothesis by a $P \geq .05$. The results are reported as statistically significant (a true or positive finding) or as statistically not significant (interpreted as a false or negative finding). When considered uncritically, such a dichotomization may provide a false sense of certainty. Rarely, if one askes what the mass of a Higgs boson is, the answer is "there is a 95% chance that the mass is $125.3 \pm 0.6\,\mathrm{GeV}/c^2$" or "the mass is $125.3 \pm 0.6\,\mathrm{GeV}/c^2$." Measures of statistical significance typically are continuous variables that are dichotomized to a yes or no, possibly leading to false confidence in the reality of the measure of Central Tendency. How can something that weighs $x \pm y$ be real? At the very least, this discussion points out that statistics is an epistemic device that does not on its own vouchsafe a claim to what is real.

Those with a narrow view of reproducibility likely would have little disagreement with the description just given regarding statistics. But that suggests a concept of statistics as a tool that can be misused. The argument made elsewhere is that statistics is a logical reaction to the condition of human knowledge and depends on logical fallacies. The misuse of statistics may be more likely if the underlying logical fallacy is not recognized and, consequently, not appreciated.

THE USE OF LOGICAL FORMALISM

An example of a logical formalism is the modus tollens form of deduction, which is *if a implies b is true and b is false, then a is false.* Conceptually it is no different from mathematical formalisms such as $a + b = b + a$ or *if $a = b$ and $b = c$ then $a = c$.* The logical form contains logical variables such as *a* and *b* that stand for some value, for example, an object such as Parkinson's disease or another proposition such as Parkinson's disease and dopamine depletion. Thus, it is possible to create large and nested arguments by embedding other logical arguments as variables in the main argument. The nested arguments also would contain logical variables. The logical variables are linked by logical operators such as *if, and, or, not,* and *implies.* Logical variables are linked to form propositions and propositions are linked to form arguments. This is not unlike mathematical variables that might contain numbers and are linked by operators such as addition, subtraction, division, and multiplication.

Further, the logical variables, propositions, and arguments in propositional and syllogistic deduction can have a value of true or false to comply with the Principle of the Excluded Middle, also call the Principle of Bivalence. The principle holds that any variable, proposition, or argument is either true or false, not both true and false, and not neither true or false. As will be discussed in Chapter 4, The Logic of Certainty Versus the Logic of Discovery, the Principle of the Excluded Middle is what gives propositional and syllogistic deduction its certainty but also is the source of their limitations, as most arguments in biomedical research are not totally and only true or only false. There are other forms of logic that attempt to escape the Principle of the Excluded Middle, such as predicate and fuzzy logic. However, these other forms are not necessary for the purpose of this book. An introduction to logic is given in Appendix A for those readers who wish, or need, to understand propositional and syllogistic deductive logic better.

The great value of framing scientific arguments, such as experimental design and analyses, as logical propositions is that the validity of the scientific argument can be assessed quickly by just the logical form. For example, if the rational for an experiment is *if dopamine neuron lesions by toxin* (*a* or a hypothesis) *results in parkinsonism* (*b* or prediction) *and an animal given the toxin* (*a* or a hypothesis) *develops parkinsonism* (*b* or prediction), *then the toxin* (*a* or a hypothesis) *produces dopamine neuron lesions* (*b* or prediction). Such an experimental design can be reduced to the logical form of *if a implies b is true and b is true, then a is true*. This logical form is invalid, which means any and every value put into the logical variables, such as the toxin for *a* and parkinsonism for *b*, is invalid. Indeed, one does not even have to ask what values are being put into the variables. Note that the phrase in the proposition, dopamine lesions by toxin, is not necessarily true because of the conditional *if* that is applied to it. The conclusion being invalid does not mean the conclusion, such as the toxin produces parkinsonism, is false; it may be true but one cannot be certain that it is true or that it is false.

Virtually every general rule is analogous to the logical form. For example, in many courts, hearsay evidence is not admissible. Hearsay evidence is of the form "What did X (not in the court room) tell you about what Y said to X about what Y did?" One does not need to know who X or Y is or what Y did to still know that such evidence is not admissible in court. Likewise, the argument used very frequently in biomedical research is of the form:

ARGUMENT 1.1

Major premise: *Patients with X* (bridging term) *improve with Y* (major term)

Minor premise: *Patient Z* (minor term) *has X** (bridging term)

Conclusion: *Patient Z* (minor term) *should improve with Y* (major term)

The argument is invalid if X and X* are not exactly the same. It does not matter what Z, Y, X, or X* is.

The use of logical formalisms may strike one as "theoretical," perhaps leading to the temptation of dismissing the logical formalism as being irrelevant to biomedical research. This would be unfortunate because the value of such use would be lost. However, no one would argue that $F = m^*a$, which is the formal statement of Newton's Second

Law of Motion where F is the force applied to an object, a is the resulting acceleration, and m is the mass of the object, is "theoretical." Being theoretical has the connotation of being tentative or unproven. There is nothing tentative or unproven about $F = m^*a$ (at least at nonrelativistic speeds). Likewise, there is nothing tentative about the logical form *if a implies b is true and a is true, then b is true* (the modus ponens form in propositional logic) or *if a implies b is true and b is false, then a is false* (the modus tollens form in propositional logic). These arguments are absolutely certain and not theoretical.

SUMMARY

There is a crisis of irreproducibility in biomedical research, at least as judged by recent exposures of the problem in published scientific reports and the response by editors of journals and administrators of granting agencies. Whether this is new or newly discovered is not the issue, as the lack of reproducibility strikes at the confidence in biomedical research generally. The problem is magnified manyfold if irreproducibility is defined in a broad sense beyond whether the exact same experiment can be conducted another time with exactly the same results (narrow sense of reproducibility). To hold such a narrow sense of reproducibility is tantamount to Solipsism of the Present Moment and may be a disservice to biomedical research in terms of its goals and mission.

It is likely that not all instances of irreproducibility are the result of fraud; lack of transparency; variations in reagents, materials, and methods; or lack of statistical rigor. In those cases, what accounts for the irreproducibility? The argument has been advanced, and will be pursued throughout the book, that many of these studies were logically flawed. The possibility of logical flaws stems from the necessity of using logical fallacies in order to gain new knowledge. Both the risks and the potential for gaining new knowledge depend on the distance between the logical fallacy used and its valid counterpart. Together, the judicious use of logical fallacies, combined with the traditional valid counterparts, constitutes evolutionary logic.

The Problem of Irreproducibility

Biomedical research—particularly in the area of preclinical animal-based research, which is supposed to be the basis for future human treatments—has been shown in a series of contemporary articles to have a problem, which is irreproducibility (Begley and Ellis, 2012; Prinz et al., 2011). In the narrow sense of irreproducibility, investigators were unable to later reproduce an estimated 20%−25% of findings from previous research. A series of prospective attempts to reproduce five important studies in cancer research in laboratory animals were undertaken (Aird et al., 2017; Horrigan, 2017; Horrigan et al., 2017; Kandela and Aird, 2017; Mantis et al., 2017). Of the five, two replicative studies were indeterminant, one was not reproducible, and the last two were reproducible, except that neither study reached statistical significance. The latter may be reproducible in the narrow sense but not in the broad sense. If one is disciplined, zero out of three achieved reproducibility. Only 61% of psychology studies could be replicated. In neuroscience, in a meta-analysis that included 730 published studies, the median power in the statistical analyses was 21% (Button et al., 2013). This means that the probability of demonstrating a true effect was only 21% in these studies. In a study of 70 agents initially reporting increased survival in a rodent model of amyotrophic lateral sclerosis, none had significant effects in 221 replication experiments (Scott et al., 2008). Statistical power generally relates to the probability of being able to detect a difference. Factors determining power are the magnitude of the effect and the variability of the measures of the effect. For example, consider a situation where a set of persons weights, in pounds, is measured and gives the following results: 100, 125, 110, 190, 150, and 150 with an average of 156. The subjects are given a treatment and the weights are remeasured. The hypothesis is to detect a 5-pound difference in the average weight after treatment with confidence. However, if the predicted change in average weight is 20 pounds, then it should be relatively easy to have confidence if, in fact, a 20-pound change occurs.

Reproducibility in Biomedical Research
DOI: https://doi.org/10.1016/B978-0-12-817443-2.00002-1

Similarly, if the weights were 156, 156, 156, 156, 156, and 156, it may be possible to detect a true 5-pound average weight loss. An example may be stated of what is the probability (from 0 to 1, or 0% to 100%) of detecting an effect, x, with a standard deviation (measure of the variability) to a confidence of $P < .05$ (meaning less than a 5% change that the change is just due to chance) with y number of subjects?

The problem with irreproducibility in the broader sense is reflected in the governmental postapproval recall of drugs, biologics, and devices as evidence of irreproducibility. In the first 8 months of 2014, the FDA recalled 836 drugs; in 2013, 1225 drugs were recalled (www.raps.org/Regulatory-Focus/News/2014/08/11/20005/Number-of-Drug-Recalls-Surges-at-FDA-Led-by-Mid-Level-Concerns/). With respect to medical devices, 91.5% of devices were not recalled within 5 years of approval. However, this means that 8.5% were recalled. The Institute of Medicine wrote: "Device design, that is, failure of a device to perform as intended despite the product's meeting all its design specifications, accounted for 28.4% of recalls" (Institute of Medicine, 2011). However, it must be noted that FDA recalls could suffer ascertainment problems and actually reflect only the most egregious examples. It is more likely that physicians and healthcare professionals simply stop using drugs, biologics, and devices that post-FDA approval experience demonstrate to not have reproduced the benefits that the FDA-approved studies would suggest, or to have greater adverse effects. In post-FDA approval studies, only 4% and 25% of patients responded to the 10 greatest selling prescription medications (Schork, 2015).

One may argue that recalls simply indicate the success of the FDA and other governmental agencies in conducting ongoing postapproval surveillance in order to discover problems and to mount effective responses to them. Yet, such an argument rings hollow in the ears of anyone who pauses to consider the number of humans injured by recalled drugs and devices. The nonsteroidal anti-inflammatory drug and pain reliever Vioxx (Rofecoxib), for example, won FDA approval in 1999 and was recalled in 2004. The FDA estimates that between 1999 and 2003 Vioxx may have contributed to 27,785 heart attacks and sudden cardiac deaths (Graham, 2004).

TYPE I AND II ERRORS

Type I errors are defined as claiming a difference (rejecting the null hypothesis) when no difference actually exists. This may be finding *drug X* effective in *disease Y* when in fact *drug X* is not effective as demonstrated subsequently. Most of the studies of irreproducibility are those of type I errors. Type II errors are defined as claiming there is no difference (accepting the null hypothesis) when a difference actually exists. Type II errors are more likely to occur in what are called "negative studies." However, one has to be careful with definitions, and one suggestion has been to use terms "positive," "negative," and "null" studies (Scott and Wasserman, 1997). A positive study is one in which the hypothesis claims to be true when its predictions are confidently found to be true. A negative test is a study where the hypothesis is thought to be true is tested with the intent of demonstrating it to be false and then the predictions are found to be false with confidence, as in futility studies (such as an experiment with sufficient statistical power). A null study is one in which the hypothesis is claimed to be true and the null hypothesis is rejected, but with insufficient certainty. Similarly, a null study may be one in which there is a claim of no effect, the null hypothesis is accepted, but there is insufficient certainty. Most often, null studies result because the experiments did not have sufficient statistical power as described previously. The result is not negative, rather it is indeterminant. To claim an actual null study to be a negative study would be to commit a type II error. Unfortunately, it appears that the reasonable suggestion of describing positive, negative, and null studies has not gained traction, as most of the literature continues to conflate null studies with negative studies.

Type II errors are not inconsequential. For example, early trials of Levodopa, which later revolutionized the treatment of Parkinson's disease and served as a model for other neurotransmitter-replacement therapies, were nearly abandoned when early studies did not demonstrate benefits (discussed in greater detail in Chapter 8: Case Studies in Basic Biomedical Research). The same was true for thymectomy for myasthenia gravis. Thalidomide created a generation of deformed children because rodents did not demonstrate toxicity in preclinical trials. Given the prejudice against publishing null and negative studies (Matosin et al., 2014), the full extent of loss due to type II errors can only be guessed at, but likely it is not small. A frequent example of type II errors is seen in randomized clinical trials of new treatments, for example, using effectiveness as the primary outcome measure. In such cases, statistical analyses, such as power analyses, are appropriate for the primary endpoint. Power analyses estimate the probability of finding a statistically significant effect given the magnitude of the effect and the variability. The publications in question often go on to discuss adverse effects, often arguing no statistically significant difference (null hypothesis not rejected) in adverse effects between the experimental group and the control group. Many of such studies are treated as negative studies when in reality they are null studies. Although these claims are often taken at face value, studies have shown that power relevant to the experience of adverse effects is frequently insufficient to exclude a type II error and that clinically significant differences may exist in the adverse effects (Tsang et al., 2009). Indeed, it is the insufficient power to validate negative claims as to adverse effects that is likely the greatest cause of postmarket withdrawal of drugs and devices, or the application of FDA "black box warnings" if there is no withdrawal. In response to largely serious adverse effects, the FDA recalled 113 previously approved medical devices between 2005 and 2009 (Zuckerman et al., 2011).

The general presumption is that statistical methods are robust in excluding type I errors, although this is contradicted by the demonstrated irreproducibility in the manner of a type I error. Sample sizes insufficient to confidently accept the null hypotheses to result in a valid negative study nevertheless have status as being true (negative study). Specific approaches have been developed when a negative outcome is to be claimed as true (Walker and Nowacki, 2011). For example, *treatment A* is considered equal to *treatment B* because no statistically significant differences were identified on standard hypothesis testing. This would be a null, not a negative, study. Paraphrasing Carl Sagan, "the absence of evidence is not evidence of absence."

It is clear that insufficient sample sizes also increase the risk of type I errors. This means there is even a greater chance of type I errors that would be appreciated just based on the incidence of studies demonstrated irreproducible. Indeed, studies have shown that over 90% of examined neuroscience studies are underpowered (Button et al., 2013) and, therefore, are at a high risk of type I errors and irreproducibility.

CONTROLLING THE INEVITABILITY OF IRREPRODUCIBILITY RISK

The centrality of falsifiability, another name for irreproducibility, to the progress of science is seen in its history as pointed out by scientists and philosophers, such as Karl Popper (2002). Indeed, in many situations, the scientific rewards are directly proportional to the risk of irreproducibility (see, e.g., the discussion of Epistemic Risk in Chapter 4: The Logic of Certainty Versus the Logic of Discovery). It may be optimal, scientifically, to aim for a certain level of inevitability of productive irreproducibility, analogous to the acceptable false positive rate in appendectomies. For example, a surgeon preforming appendectomies, with 15% of surgeries resulting in the removal of a normal appendix would be considered acceptable (Mariadason et al., 2012). A record of removing fewer normal appendixes suggests that the criteria for surgery are too strict, resulting in an unacceptable number of patients at risk for rupture of their appendix. Similarly, insisting on no irreproducibility likely means that promising research that could make significant contributions may be removed from consideration.

The challenge is to identify those potential studies while at a relatively higher risk of irreproducibility, also have a high probability of scientific contribution. To be sure, the value of the statement just made is not new. The National Institutes of Health (NIH) has specific programs to invite high-risk—high-reward research proposals (www.nih.gov/news-events/news-releases/nih-common-fund-announces-2016-high-risk-high-reward-research-awards). The challenge is how to identify those research proposals. An approach has been developed based on Epistemic Risk, which is composed of Epistemic Distance and Epistemic Degrees of Freedom between the conceptual and the testable hypotheses (discussed in detail in Chapter 4: The Logic of Certainty Versus the Logic of Discovery) based on the logic and logical fallacies necessary for discovery.

If a degree of irreproducibility is inevitable, then the question becomes how one may distinguish, whether prospectively or retrospectively, reproducible and productively irreproducible from unproductively irreproducible studies? Does the answer lie solely in conducting every study twice by the originating scientist first and a different, independent scientist second (Russell, 2013)? Were this to be done, current resources would have to be doubled or original research halved if current rates of scientific support are only maintained. The alternative of selectively attempting to reproduce certain high-value studies immediately raises the question as to what constitutes a high-value study (Baker, 2012). Recall that both the transistor and the personal computer initially were greeted with a great deal of skepticism as to their worth.

Irreproducibility of results is no mere academic concern. The budget of the NIH is over $30 billion a year. The rate of unproductive irreproducibility of research study results suggests that many billions of dollars have been wasted over the years because they were spent on producing results in which no one may have confidence. This is not to say that many, if not most, funded research studies failed to produce real knowledge; it is to say, rather, that one may not know confidently which among them did, and which did not. Uncertainty extends to human clinical research trials that have been based on suspect preclinical research. How many patients have been exposed to experimental treatments based on irreproducible preclinical research? How many human research subjects have been placed at risk for injury, despite the high rate of false claims in preclinical animal studies? The success rate for phase 2 clinical trials is estimated at 18% (Arrowsmith, 2011). A bit of humorous, if cynical, wisdom in medical and research circles holds that biomedical science cures cancer all the time—in mice—but humans are less cooperative. However, how many promising treatments have been abandoned because they were beset by type II errors? Methods that may help predict the risk of irreproducibility will be discussed later in this chapter.

INSTITUTIONAL RESPONSES

These and other questions are being raised in the professional literature and by leaders of institutional science—which demonstrates that these questions are important and still open (Collins et al., 2014)—and by editorial boards of leading scientific journals (McNutt, 2014). Further, governmental agencies and editors are instituting significant changes in policies that could compromise biomedical research if not addressed by all stakeholders.

Suggestions for remedying this problem have included greater rigor in the description of methodologies, particularly statistical methods, as well as the creation of repositories for data (McNutt, 2014). But these are not new, and the revisiting of the issues suggests they have difficulty gaining traction. Interestingly, the NIH instituted a policy for data sharing in 2010:

> NIH believes that data sharing is essential for expedited translation of research results into knowledge, products, and procedures to improve human health. NIH endorses the sharing of final research data to serve these and other important scientific goals and expects and supports the timely release and sharing of final research data from NIH-supported studies for use by other researchers (http://grants.nih. gov/grants/policy/nihgps_2010/nihgps_ch8.htm#_Toc271264947).

Similarly, the NIH has a policy for the sharing of research materials:

> Investigators conducting biomedical research frequently develop unique research resources. NIH considers the sharing of such unique research resources (also called research tools) an important means to enhance the value of NIH-sponsored research. Restricting the availability of unique resources can impede the advancement of further research. Therefore, when these resources are developed with NIH funds and the associated research findings have been accepted for publication, or after they have been provided to NIH, it is important that they be made readily available for research purposes to qualified individuals within the scientific community (http:// grants.nih.gov/grants/policy/nihgps_2010/nihgps_ch8.htm#_Toc271264947).

Notable is the absence of any suggestion of facilitating the replication of research studies.

What problems are the institutional responses intended to address? There are a number of potential causes of irreproducibility, which include: (1) intentionally misleading (fraud); (2) lack of sufficient information in the original publication to allow replication; (3) variations in reagents, materials, and methods; and (4) technical and statistical errors.

Fortunately, fraud is thought to be infrequent, although often headline-grabbing. Fraud accounts for approximately 26.6% of irreproducible results (Steen, 2011); however, it is likely an overestimate when the full extent of irreproducible studies is considered. Issues of transparency and insufficient reporting are being addressed, as many journals use Internet-based repositories for detailed information to be available as supplements to the original reports. It remains to be seen whether increasing transparency and provision of greater technical details will reduce the risk of unproductive irreproducibility.

The degree to which statistical errors contribute to irreproducibility is unknown and difficult to assess. The availability of "turnkey" statistical software packages appears to invite less rigorous considerations by users, particularly prior assessment whether the assumptions required in the appropriate use of a test have been met. However, it is important to note that many of the studies later demonstrated to be irreproducible were vetted by senior and respected scientific referees. The likelihood of an egregious misuse of statistics escaping notice would not appear to be great, although one cannot know this with confidence. Many journals specifically ask reviewers about concerns for statistical errors and many offer independent review by statisticians.

At best, statistical validity is internal (see Chapter 5: The Logic of Probability and Statistics). Statistical analyses can be considered as a formal (logical) system with specific assumptions (axioms) and rules of manipulation and inference. An example of a formal system is modus ponens of classical propositional logic, *if a implies b is true and a is true, then b is true* and modus tollens *if a implies b is true and b is false, then a is false.* It does not matter what *a* or *b* specifically are. This is a great aid as many experiments demonstrated to be in the form of modus ponens or modus tollens can be taken as certain independently of what *a* or *b* specifies. Similarly, statistics are like a formal system and actually independent of specific circumstances. For example, one can demonstrate that the median weight of duck-billed platypuses is statistically significantly different than the median weight of an adult human. Note that the nature of statistical analyses does not determine the nature of the argument to which statistical analysis is to be applied. As will be demonstrated later, statistics are a derivative from the failure of logical syllogistic deduction and the subsequent necessity to use the logical fallacy version of the syllogistic deduction (see Chapter 5: The Logic of Probability and Statistics).

REPRODUCIBILITY AND IRREPRODUCIBILITY

In important, but subtle ways, there is a difference between reproducibility and irreproducibility. Indeed, the notion of reproducibility is more complex and extensive in its role in science than the notion of merely replicating specific experiments, as discussed later. Irreproducibility is very different as it follows directly from attempting to replicate an exact same experiment within the same experiment (local), the same experiment by different investigators (narrow sense), and to replicate the concept over a series of different but relevant experiments to establish conceptual reproducibility (the broad sense). It is important to note that the effort of this text focuses on irreproducibility, which is far narrower in its implications than the notion of reproducibility. The concern here is that the reader not conflate the notions of reproducibility and irreproducibility and thus judge the discussion of irreproducibility in the context of a much larger and complex domain as science. Notions of irreproducibility are not just the inverse or denial of reproducibility. An analogy to medical reasoning is helpful. The Positive Predictive Value of a test (a form of reproducibility where the test result confirms the disease) is different than the Negative Predictive Value (a form of productive irreproducibility where the test result confirms the disease is not present). These are no mere reciprocals of each other and examples are ubiquitous in medicine.

Reproducibility is more complex and at the heart of several issues related to science and the scientific enterprise. Some scientists and philosophers have argued that reproducibility is central to the scientific enterprise and the historical successes that science has achieved. Indeed, one function of the Royal Society of London, founded in 1660 and given a royal charter in 1662, was to conduct public experiments (essentially testing reproducibility) as a means for vouchsafing scientific claims. Indeed, this is reflected by the motto of the Royal Society, *Nullius in verba*, "Take no man's word for it." However, the notion of public demonstrations should be taken with a grain of salt and, more typically, the public were like-minded persons (Shapin and Schaffer, 1985).

Yet, public demonstration was not necessarily an exercise in reproducibility. Few biomedical experiments are purposely replicated, arguing that the centrality of reproducibility is a qualified notion. Often the rationality or eloquence of the scientific claims and reports or their authors was sufficient (Feyerabend, 1975; Kuhn, 1962; Steinle, 2016). At times, the mathematical derivation of the scientific claim was sufficient. Further, there are any number of claims, such as the Big Bang or String Theory, thought to be scientific but are not subjectable to experimentation, as replication of the phenomena, and therefore not reproducible. There may be situations where reproducing experiments is not practically feasible.

One interpretation of the fact that many scientific claims accepted as truly reflecting nature is the rationality and eloquence in themselves argue for a faith of reproducibility rather than empiric demonstration of reproducibility. Indeed, Galileo essentially praised Copernicus for ignoring the "data" as the heliocentric model of the solar system left many unexplained contrary "facts" until they could be resolved with Galileo's demonstration of inertia and Kepler's identification of elliptical orbits based on precise astronomical data received from Tycho Brahe. Apparently, greater confidence is afforded when mathematics provides the rationale and eloquence.

To be sure, many claims not subjected to specific tests or reproducibility are quarantined from accepted fact by calling them theories, but often they are controlling of the scientific enterprise as though they were fact. The postmodernist notion of science argues that science is an enterprise. In other words, science is what scientists do, a striking departure from the origin of the term "scientist" by William Whewell in 1834 who defined the term as those who do science (in an anonymous review of Mary Somerville's "On the Connexion of the Physical Sciences" published in the Quarterly Review). The Strong Social Program suggests that reproducibility is just what scientists will agree to. Science as psychology suggests that reproducibility is what scientists are psychologically predisposed to in forming consensus—the tightness of that consensus notwithstanding (Atamspacher and Massen, 2016). Science as technology, to invert the typical conception, holds that science is what the technology that created it says science is and issues of reproducibility become ones of production technology (Zimmerli, 2016). It is a form of Operationalism but without any anchor on some self-event measure or thing measured. As Zimmerli writes, "The proof of reproducibility is reproduction" (Zimmerli, 2016).

Yet, hardly any experiment is exactly reproducible and, thus, reproducibility is better understood as a limited irreproducibility (Collins, 2016). Thus, the question becomes how much irreproducibility is allowed for there to be reproducibility? In this sense, the issue of how much irreproducibility is allowed with the experiments still taken as reproducible is analogous to the Positive and Negative Predictive Values. In part, differences between the original experiment could be considered reproduced when the difference between it and the reproducing experiment is less than the minimum scientific or clinical meaningful difference (see Chapter 7: Case Studies in Clinical Biomedical Research).

To mitigate the uncertainties within any particular experiment, an appeal is made to some external standard, assuming there could be one (for a contrary position, see Quine and Ullian, 1978). However, this is not certain as the external standard is likely the result of some experiment and now the same question can be asked of the experiments that generated the external standard. This process can go on without end in what Collins called the Experimenter's Regress (Collins, 2016).

While reproducibility is an important element in virtually every discipline considered science, as opposed to the humanities, each has somewhat different guidelines thought to increase the chances of reproducibility. In physics, notions of conservation, as in conservation laws, invariance, and symmetry, are thought hallmarks of studies likely to be reproducible. In mathematics, reduction to tautology and the avoidance of self-contradiction or absurdity are the hallmarks. The situation is likely to be quite different in biomedical research.

CONTRIBUTIONS OF LOGIC TO BIOMEDICAL RESEARCH

A great many scientists, philosophers, mathematicians, statisticians, computer scientists, engineers, editors, and administrators are wrestling with these issues. However, the effort in this book is much narrower but, as hopefully will be appreciated, very powerful. The object of analyses is logic of the kind often taught in undergraduate courses in philosophy. The themes are deductive logic, specifically propositional and syllogistic deduction, and induction. The overarching theme is that at least *some* forms of both productive and unproductive irreproducibilities arise from logical fallacies, which are necessary departures from the valid form of propositional and syllogistic deduction, and induction. Perhaps, as yet counterintuitive, biomedical research requires the use of these logical fallacies (see Chapter 3: Validity of Biomedical Science, Reproducibility, and Irreproducibility). Thus, it is critically important that the logical fallacies be used in a judicious manner.

However, it is not so simple because any robust formal system (logic) will run up against Gödel's Incompleteness Theorem where true statements cannot be derived from the axioms and rules of inference. If so, how can one have certainty? Gödel's Incompleteness Theorem is a Number-Theoretic Incompleteness that will extend to Information-Theoretic Incompleteness, particularly in complex and chaotic biological systems (discussed in more detail in Chapter 10: Chaotic and Complex Systems, Statistics, and Far-From-Equilibrium Thermodynamics). As biomedical research is about information, the limits of formal logical systems will affect reproducibility or at least the demonstration of it. As will be demonstrated, a new approach to reproducibility in complex and chaotic nonlinear systems will be necessary (Chapter 10: Chaotic and Complex Systems, Statistics, and Far-From-Equilibrium Thermodynamics).

REDUCTIONISM AND CONCEPTUAL (BROAD) IRREPRODUCIBILITY

As discussed further in Chapter 5, The Logic of Probability and Statistics, Reductionism is a form of the Fallacy of Four Terms, a fallacious form of the certain syllogistic deduction, as illustrated schematically in Argument 2.1. The typical syllogistic deduction that provides certainty is shown in Argument 2.1 and is offered as a review of what constitutes a certain deduction:

ARGUMENT 2.1

Major premise: *All humans* (bridging term) *are mortal* (major term)

Minor premise: *Socrates* (minor term) *is human* (bridging term)

Conclusion: *Socrates* (minor term) *is mortal* (major term)

In this argument, *all humans* are isomorphic to being *mortal* in the major premise and human is isomorphic to *Socrates* in the minor premise. The conclusion is based on the Principle of Transitivity, where *if $a = b$ is true and $b = c$ is true, then $a = c$ is true*. Thus, *Socrates* (c) "equals" *mortal* (a). The critical issue is that the term *humans* (b) in the major premise is the same as *human* (b). As can be appreciated, there are three terms: the major term, *mortal* (a); minor term, *Socrates* (c); and bridging term, *human* (b). Another manner to explicate the deduction is that the set of *all humans* is a wholly contained subset within the set of all things *mortal*. *Socrates* is a wholly contained subset within the set of *all things human*. Therefore, *Socrates* is contained within the set of all things *mortal*.

But what if *Socrates* is just a different kind of human (b') from the humans (b) referred to in the major premise? In this case the Principle of Transitivity does not hold, *if $a = b'$ is true and $b = c$ is true* it cannot be asserted that *$a = c$ is true*. Note that this does not mean that a cannot equal c, but only that it cannot be claimed with certainty that $a = c$. It may then be a matter of additional experimentation to demonstrate $a = c$. The syllogism where *Socrates* is a different type of *human* compared to *human* in the major premise is the Fallacy of Four terms: a, b, b', and c. Similarly, *Socrates* may be not wholly contained in the set of all *humans* and, consequently, cannot be assured of being contained in the set of all *mortal* things.

Reductionism in biomedical research necessarily invokes the Fallacy of Four Terms as represented in Argument 2.2:

ARGUMENT 2.2

Major premise: *Epilepsy in rodent* (bridging term) *entails mechanism A* (major term)

Minor premise: *Humans* (minor term) *have epilepsy* (bridging term)

Conclusion: *Humans* (minor term) *entail mechanism A* (major term)

To have certainty, *epilepsy in rodent* in the major premise must be the same as *epilepsy* in the minor premise so as to maintain the three terms. However, if *epilepsy in rodent* in the major premise is not the same as *epilepsy* in the minor premise, then there are four terms and hence a fallacy. Reductionism requires relating epilepsy in the rodent to epilepsy in humans, thus necessitating the risk of the Fallacy of Four Terms. Again, this is not to say that the conclusions from the use of the Fallacy of Four Terms are true or false but only that the certainty of the conclusion cannot be established. One can substitute tissue culture or computational simulations for rodent and the Fallacy of Four Terms still applies.

In the example just given, *epilepsy in rodent* "stands in" for *epilepsy* in humans, thus generating the risks associated with the Fallacy of Four Terms. Specific examples are described elsewhere. There may be different ways in which the object of experimental manipulation, such as the rodent epilepsy model, can stand in for the object for any generalized knowledge claim, such as epilepsy in humans. For example, a part of the brain that is of interest can be isolated from the rest of the brain (Methodological Reductionism), such as a tissue slice preparation, in order to render experiments more feasible and inferences more tractable. Success in doing so is often associated with a simplified or reduced hypothesis allowing the now demonstrated hypothesis in the experimental manipulation to "stand in" for hypotheses in humans (Theoretical Reductionism). Yet, the concepts gleaned from the tissue slice are thought applicable to the whole brain; consequently, there is no Conceptual Reductionism—an interesting critique of the relevance to tissue slice research to whole brain understanding is presented by one of the pioneers in tissue slice research, Mircea Steriade, in his "The Intact and Sliced Brain" (2001). Otherwise, what is the utility of Methodological and Theoretical Reductionisms? This is the *modus operandi* of Scientific Reductionism.

It would seem that there is another aspect of Scientific Reductionism, that is, Scientific Reconstructionism, in which the object of more general interest is reconstructed from the reduced or simplified experimental preparations. Failure to do so represents irreproducibility in the broad sense and is exemplified by all the times cancer has been cured in mice, but the same manipulations are unable to cure cancer in humans. The failure then to reconstruct the original phenomenon from the reduced preparation or replicate findings from lower and less complex species in higher and more complex species can be considered a problem in reproducibility—in the broad sense. Clearly, if the intent of Methodological and Theoretical Reduction is to understand higher or more complex phenomena, failure to do so represents conceptual irreproducibility.

There has been increasing concern that Reductionism is not meeting the expectations of science for a complete understanding of the larger scale phenomena to which the Reductionist methods were applied. There is no question that the technology associated with or enabling of Reductionism has been remarkable where the phenomenon of life can be dissected down to the molecular level. The question now is whether the whole phenomenon can be reconstructed from its parts. Indeed, this very concern about reductionism motivates the discipline of Systems Biology (Bickle, 2007; Bizzarri et al., 2013; Cornish-Bowden et al., 2007; Gilbert and Sarkar, 2000; Joyner, 2011; Joyner and Proctor, 1999; Li et al., 2009; MacDougall-Shackleton, 2011; Martinez-Lavin et al., 2008; Moalem and Percy, 2002; Neugebauer et al., 2001; Rose, 1998; Saks et al., 2009; Schadt et al., 2005; Strange, 2005; Van Regenmortel, 2001; Villoslada et al., 2009; Wolkenhauer and Green, 2013). If it is true that reconstruction from the increasingly smaller and finer bits of phenomena resulting from the dissecting technology is increasingly problematic, is it at least possible that Reductionism paradigms may increase the risk of irreproducibility?

VARIABILITY, CENTRAL TENDENCY, CHAOS, AND COMPLEXITY

One cannot exclude the possibility that the very methods that enable empirical hypothetico-deductive and inductive biomedical research are not, at least in some cases, counterproductive to the quest for reproducibility. Particularly of concern are classical statistics. As will be demonstrated in Chapter 5, The Logic of Probability and Statistics, statistics is a means of providing certainty to probability and probability is the means to provide utility to a syllogistic argument, that being the Partial Syllogism. The point here, to be expanded on later, is that statistical errors that risk irreproducibility may in fact stem from the logical structure of the Partial Syllogism translated to probability and statistics. Indeed, many of the necessary assumptions for the proper application of statistics relate to the Fallacy of Four Terms, such as how samples are collected to "stand in" for the population that is of ultimate interest.

Consider the ubiquitous use of the notion of Central Tendency (the mean, median, or mode) in statistics. Most scientists still believe that the mean or the median accurately describes the ontology (or reality) of the phenomenon. However, this was controversial even in the early development of statistics (Gigerenzer et al., 1989) and remains so. The famous physiologist Claude Bernard was hostile to the idea that a statistical approach emphasizing the mean (average) was appropriate. At that time, there was the concept of *l'homme moyen* (or average man) where the statistical descriptor was sufficient to characterize what it is to be a man. Bernard wrote:

> If we collect a man's urine during twenty-four hours and mix all this urine to analyze the average, we get an analysis of a urine which simply does not exist; for urine, when fasting, is different from urine during digestion. A startling instance of this kind was invented by a physiologist who took urine from a railroad station urinal where people of all nations passed, and who believed he could thus present an analysis of the average European urine! (Gigerenzer et al., 1989).

One might qualify the mean or median by adding standard deviations or percentile ranges, but it still is an open question how well these measures reflect reality. For example, even given the mean, standard deviation, and sample size, it is impossible to recreate the original sample, as virtually an infinite number of distributions can be constructed from the same mean, standard deviation, and sample size. The descriptive statistics are abstractions. To the degree information is lost, the abstractions become less representative of the data the researcher is trying to describe. Yet, it is questionable from an epistemological basis whether one should present actual distribution of the data, as how would one generalize from that specific instance or experiment? In other words, the true generalization could be difficult to extract as it would be in the "weeds." What is clear is that information is irretrievably lost when reducing the sample to a statistical descriptor (an example of Methodological and Theoretical reductionism). The loss is irretrievable because the original sample cannot be recreated (Orcutt et al., 1968).

This has led many philosophers of science to apply the notion of entropy and the Second Law of Thermodynamics as applied to Information to understanding the epistemic basis of science, as will be explained more fully in Chapter 10, Chaotic and Complex Systems, Statistics, and Far-From-Equilibrium Thermodynamics. The Second Law of Thermodynamics holds that in any closed system, entropy cannot decrease. Entropy is a measure of randomness and hence can be considered the converse of information. Information can be considered as non-random state changes in sequential or in parallel states. For example, the order of letters, spaces, and punctuation marks in sensible English is not random. Illumination of the pixels (essentially parallel states) making up a normal computer screen presentation is not random, when used as intended. Thus, in any closed system, as entropy cannot decrease, information cannot increase. Any irreversible process in the closed system will result in a loss of information that cannot be regained. As described previously, using the mean as the descriptor of phenomena risks an irretrievable loss of information. It is possible that should the experiments be repeated, the loss of information reflected in the use of the Central Tendency increases the risk of irreproducibility.

These issues are discussed in greater detail in Chapter 10, Chaotic and Complex Systems, Statistics, and Far-From-Equilibrium Thermodynamics.

Classical or standard statistics depend on the Large Numbers Theorem. This theory holds that the mean (average) of a sample will converge onto a single value as the number of items in the sample increases—a form of reproducibility. In this sense, convergence onto a single value gives the impression of validity, but actually only in the sense of internal consistency. The internal consistency seems to afford the mean result of an experiment some semblance of reality, but not proof. Further, the convergence provides a sense of predictability. For example, the mean of a sample with $n + 1$ observations is likely to be the same or very (arbitrarily) close to the mean of n observations, provided n is sufficiently large. Hence, an experiment with a sample of $n + 1$ observations will reproduce a mean result of experiments that necessarily is close to the mean of a sample of n observations. Any increments to the sum of observations contributing to the single observation are divided by n and become vanishingly small as n increases toward infinity. Thus, in the limit of n approaching infinity, the difference between the mean of n observations and $n + 1$ observations becomes zero.

The situation is analogous to the distinction between precision (variance around the Central Tendency) and accuracy (the degree the Central Tendency reflects reality). Precision relates to the reproducibility of the outcomes following repeated application of the process—the narrow sense of reproducibility. Accuracy relates to the reproducibility of actually being in the target such that the same target can be achieved with different samples and different experiments—the broad sense of reproducibility. However, within the experiment, there is no way of knowing whether the Central Tendency or differences in Central Tendencies in controlled studies reflect reality. Rather, the scientist looks to the variance for some confidence that the Central Tendency comports with reality. This distinction is appreciated readily by the fact that experiments can be precisely inaccurate.

Precision is defined internally and thus is an epistemic question, while accuracy is defined by an external target and thus is an ontological question. There may be biological systems where the Large Numbers Theorem may not hold; consequently, the meaning of a Central Tendency may be a difficult and elusive notion, even though the Central Tendency can be determined easily. For example, manipulations of some complex biological systems may result in one of several possible states, called meta-stable states (discussed in greater detail in Chapter 10: Chaotic and Complex Systems, Statistics, and Far-From-Equilibrium Thermodynamics). An analog is shown in Fig. 2.1. The shaded surface of the three-dimensional image of the Necker Cube normally can be seen as in either the front or the back of the cube. It is not seen as both or

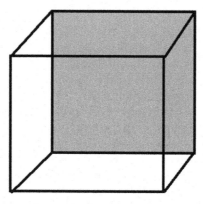

FIGURE 2.1   The Necker Cube Illusion and an example of bifurcation between two states given the same initial conditions. The cube appears as a three-dimensional cube with transparent sides. Some will view the shaded surface to the plane in the back. Others will see it as the plane in the front. Many will see the position appear to shift instantaneously between the front and the back; in other words, the perception of the shaded surface bifurcates between a perception of the shaded surface on the front or on the back. The physical appearance of the image is exactly the same regardless of whether the shaded surface is in the front or the back.

neither. The key point is that the same exact data, the visual sensations induced by viewing the image, are exactly the same for either perception—the inference from the data. More importantly, it generally is not possible to predict which perception will supervene on the same data. These systems may display Chaos and Complexity (see Chapter 10: Chaotic and Complex Systems, Statistics, and Far-From-Equilibrium Thermodynamics). The relevance to the Large Numbers Theorem is that it is possible that adding more data, such as staring at the Necker Cube, may cause the perception (shaded area in the front) to jump to the alternative (shaded area in the back). The perception will not converge onto a single Central Tendency—a single unchanging perception.

Another important concept is ergodicity, which means, forgiven a great deal of simplification, that a descriptor, such as the Central Tendency, is the same over a region of the entire population (discussed in greater detail in Chapter 5: The Logic of Probability and Statistics). Thus, one can sample from anywhere in the population and the Central Tendencies will be the same as the other samples and, importantly, the same as the population. Indeed, much of statistical sampling methodology assumes and requires ergodicity. For example, fishing in a lake is a very different proposition if the fish are all distributed evenly in the lake compared to when the fish are hiding in only one or a few spots (Fig. 2.2). Absence of an independent measure, a fish finder, for example, moving from spot to spot to find the average number of fish, may be problematic. Fishing in a

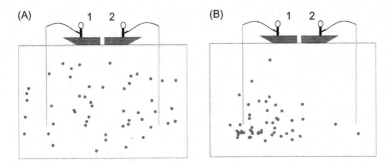

FIGURE 2.2 Schematic representation of the effects of ergodicity. In condition A, the fish are distributed evenly in the lake—high ergodicity. In condition B, the fish are concentrated under fisherperson 1. The experience of the two fisherpersons in condition A will be the same and considered reproducible. Their experiences will be different under condition B and the experiments considered irreproducible. *Reproduced from Montgomery Jr., E.B., 2019. Medical Reasoning: The Nature and Use of Medical Knowledge. By permission of Oxford University Press, USA (In Press).*

single spot is like collecting a sample with the hope that the sampling reflects the population, that is, fish everywhere in the lake. If this is not the case and the sample is to "stand in" for the population, then there will be the risk associated with the Fallacy of Four Terms.

Statistical methods unique to Chaotic and Complex Systems are needed. For example, correlational analyses often make assumptions as to the relationship between the dependent and the independent variables, such as linear (Fig. 2.3A), and utilize appropriate models. Logistic regression analysis often assumes a sigmoidal relationship (Fig. 2.3B). The relationships or models (analogous to linear or sigmoidal) also can be derived from the Chaos and Complex Systems theory such as supercritical pitchfork bifurcations (Fig. 2.3C), where a process beginning from an initial state and then bifurcating to more than one subsequent state can be used for regression modeling, much like the shifting visual perceptions of the Necker Cube (Lipovetsky, 2010). The problem is that typical regression analyses applied in the condition depicted by Fig. 2.3C are likely to result in the conclusion that the x and y variables (along the horizontal and vertical axes, respectively) are not related (in addition to violating the necessary assumption of homoscedasticity, which is a measure of how close each of the data points are to the model, applying a straight horizontal line through the data in Fig. 2.3C would show wide variability in the distance of the data points to the left and the straight line). But as can be seen, the results are consistent with a pitchfork bifurcation.

There are a number of areas in biomedical research where the systems under analysis display Chaos and Complexity, such as intracellular

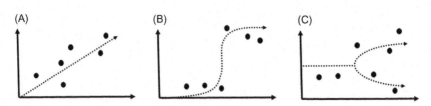

FIGURE 2.3   Schematic representations of correlations demonstrating true relationships between two conditions represented by the horizontal and vertical axes. (A) The relationships are presumed to be linear, and the goodness of fit of the data, presented by points, to a strike line can be computed. (B) The relationship is sigmoidal and is used in logistic regressions where a continuous variable on the horizontal axis, such as a diagnostic test, differentiates between two conditions in the vertical axis, for example, health or disease. (C) The dynamics between the conditions represented by the horizontal and vertical axes display Chaos or Complexity. As the value of the condition represented in the horizontal axis increases, the value represented by the vertical axis bifurcates into two conditions. Note that traditional regression models would not capture the relationship between the values of the dynamic represented in the horizontal and vertical axes and the potential inference that there is no relationship, hence a type II error.

signaling (Feldherr and Akin, 1999; Houart et al., 1999; Legenstein and Maass, 2007; Pham and Ichikawa, 2013; Shen and Larter, 1995; Sigalov, 2011; Matrosov and Kazantsev, 2011; Zhu et al., 2007). Also, borrowing from physics, evidence shows that some Chaotic and Complex Systems do not follow theorems fundamental to standard statistics such as the Large Numbers Theorem. Thus, it is at least possible that the standard statistical approaches to biomedical research that involve Chaotic and Complex Systems may not be applicable (Kaneko, 1990) and their use results in apparent, although not actual, irreproducibility.

One way to avoid false conclusions as to irreproducibility is to first determine whether the system under investigation is Chaotic or Complex. This is no different than first determining whether the outcomes are distributed normally, as a Bell curve, so as to select the appropriate statistical method, such as parametric rather than nonparametric methods. Parametric methods depend on the data being distributed normally, whereas nonparametric methods do not. One might ask why not use just nonparametric methods? However, nonparametric methods are less powerful. One can ask if the variance appears to increase and the mean changes with increasing sample size, in contrast to the Large Numbers Theorem, which would suggest a dependence on initial conditions indicative of a Chaotic or Complex System (Kaneko, 1990). Also, analysis techniques analogous to classical or standard statistics but better suited to Chaotic and Complex systems need to be used. Various means to do so will be discussed in Chapter 10, Chaotic and Complex Systems, Statistics, and Far-From-Equilibrium Thermodynamics.

CONCEPTUAL REPRODUCIBILITY AND THE IMPORTANCE OF HYPOTHESIS GENERATION

When the experiment attempting to reproduce another experiment in which the species, preparation, or sample is different from the original, reproducibility more often involves issues of conceptual reproducibility (the broad sense) rather than technical or statistical reproducibility (the narrow sense). This is evident when animal studies fail to predict a human benefit or when a human sample fails to predict what happens in the population of concern, as evidenced by the removal or disuse of previously FDA-approved drugs, devices, and biologics. Interestingly, the large majority (perhaps all) of institutional responses to the crisis of reproducibility appear not to address issues of conceptual reproducibility.

Conceptual errors that increase the risk of failure to reproduce can be appreciated in the discussion given previously of the Fallacy of Four Terms when there is conceptual confusion as to the bridging term. In some ways, the bridging term is fundamental to the hypothesis that an experiment is attempting to support or refute. If the hypothesis is not well formed, the derivative bridging term risks confusion, placing the experiment at the risks associated with the Fallacy of Four Terms.

The importance of well-formed hypotheses goes much further. Virtually every experimental hypothesis is derivative of a long list of explicit assumptions and implicit presumptions (Ullian and Quine, 1978). A schematic representation can be demonstrated in Argument 2.3 given here:

ARGUMENT 2.3

Logical form

Premise:	*If a implies b is true and*
Premise:	*If b implies c is true and*
Premise:	*If c implies d is true and*
Premise:	*If d implies e is true and*
Conclusion:	*if a is true, then e is true*

Finding: *e* is false, then one or more of the premises (*a*, *b*, *c*, or *d*) is false or propositions (the implications) invalid as the conclusion is invalid (by modus tollens)—but which one, or all?

Example

Premise:	*If increased neurometabolic activity on functional magnetic resonance imaging implies increased neuronal activity is true*
Premise:	*Refractory depression implies brain area 25g (subgenu cingulum) has increased neurometabolic imaging is true*

(Continued)

ARGUMENT 2.3   (Continued)

Premise:	*High-frequency Deep Brain Stimulation (DBS) reduces neuronal activity is true*
Conclusion:	*Then high-frequency DBS should improve depression*
Finding:	*Thus far in controlled trials DBS does not appear to improve depression*

So, one of the premises must be false or propositions invalid, but which one? According to the Duhem–Quine thesis, it is not possible just from the argument alone to know which premise is false or which implication is invalid. In the case of DBS, from the fact that DBS does not appear to help depression demonstrates that at least one of the premises must be false or one implication is invalid but the finding contrary to the logical conclusion does not demonstrate which premise is false or implication invalid. The premise *Refractory depression implies brain area 25g (subgenu cingulum) has increased neurometabolic imaging is true* is an empirically demonstrated fact and must be true (assuming no methodological errors). However, the premise *If increased neurometabolic activity on functional magnetic resonance imaging implies increased neuronal activity is true* is not necessarily true as exceptions have been demonstrated. Regarding the premise *High-frequency Deep Brain Stimulation (DBS) reduces neuronal activity is true*, the very large percentages of studies demonstrate the premise false (Montgomery and Gale, 2008). However, demonstrating the falsehood of the premise *High-frequency Deep Brain Stimulation (DBS) reduces neuronal activity is true* has no bearing on whether the other premises are true or false and, thus, also are suspect.

In the absence of external knowledge, how could one assess the relative risk of any particular premise being false or implication being invalid? Clearly, such an assessment would help the experimenter prioritize which premise or implication would be the hypothesis of the next experiment. Typically, premises are held to be facts of the matter, usually via direct observations. However, it is important to distinguish facts from quasi-facts, which actually are inferences but used as though they were facts. Perhaps it is the generation of the implications—the hypothesis and the prediction—that are most important for ensuring reproducibility (Chapter 6: Causation, Process Metaphor, and Reductionism); if irreproducibility does occur, then it is hoped it is of the modus tollens type and a positive knowledge claim can be made with confidence.

Most hypotheses are generated from analogies or by interpolation or extrapolation from prior theories. Analogies are metaphors, which actually are the Fallacy of Pseudotransitivity (discussed in greater detail in Chapter 4: The Logic of Certainty Versus the Logic of Discovery).

An example of a very productive analogy was by Mary Broadfoot Walker (1888–1974), who noted that patients with myasthenia gravis were similar in their weakness to patients receiving curare during surgery. Reasoning from the analogy, she used anticholinesterases that were effective in reversing curare-induced weakness to improve the weakness of myasthenia gravis. The Fallacy of Pseudotransitivity is of the form *if a implies b and c implies b, then a implies c*, where *a* is myasthenia gravis, *b* is weakness, and *c* is curare. The analogy or metaphor establishes a form of synonymy between myasthenia gravis and curare that allows the treatment of curare to migrate to the treatment of myasthenia gravis. This is a fallacious form of the Principle of Transitivity, which is of the form *if a implies b and b implies c, then a implies c*. Thus, when assessing a hypothesis derived by analogy, the critical question is how great is the departure of the instantiation of the Fallacy of Pseudotransitivity from the Principle of Transitivity? These issues will be addressed in detail in Chapter 6, Causation, Process Metaphor, and Reductionism.

SUMMARY

It would appear from much of the current discussion that irreproducibility is to be avoided at all costs. Yet, the progress of biomedical research, and science in general, depends critically on irreproducibility. It is a historical fact that most scientific breakthroughs come with the overthrow of prior understandings, typically because these prior understandings were "irreproducible" in the light of new experiments. Thus, irreproducibility cannot just be the "bad guy." How then to best understand irreproducibility?

There are two forms of irreproducibility. One is productive of new knowledge, even if only by eliminating reasonable theories and hypotheses, and is called productive irreproducibility. The other is unproductive, as the arguments posed by the experiments are invalid and lead to conclusions whose truth or falsity cannot be established—called unproductive irreproducibility. In these experiments, if the conclusions are true, it is only by chance or luck. If they are false and only to be detected later, if at all, then unproductive irreproducibility delays or retards biomedical research progress. As biomedical research necessarily involves the use of logical fallacies, the judicious use of logical fallacies will lead to productive irreproducibility instead of unproductive irreproducibility. Increasing productive and decreasing unproductive irreproducibility require a thorough understanding of the logical structures of the experiments that go beyond issues of statistics and transparency.

Validity of Biomedical Science, Reproducibility, and Irreproducibility

SCIENCE MUST BE DOING SOMETHING RIGHT AND THEREIN LIES REPRODUCIBILITY AND PRODUCTIVE IRREPRODUCIBILITY

Striking a balance in perspective regarding irreproducibility in biomedical research is often difficult. There can be no denying that biomedical research is a boon to humankind. Not only are humans living longer, but they are doing so with less suffering and disability. Yet, as will be discussed later, some of the very reasons for the success of biomedical research are the same reasons that put biomedical research at a risk of unproductive irreproducibility. Judicious attention to those reasons can be used to minimize the risk.

There would seem to be little need for defensiveness on the part of biomedical researchers. Some scientists acknowledge that studies have been irreproducible, but tend to diminish the impact on science (Fanelli, 2018). Some of these studies focus on the most obvious sources of irreproducibility such as fraud and bias, yet there maybe an ascertainment problem, as many cases go unexamined and therefore undetected. The problem is that one cannot readily determine, prior to a large investment in retesting, which study is immune from the possibility of being demonstrated unproductively irreproducible. In a survey of 1500 scientists from a wide range of disciplines, the results were

39

telling: "Although 52% of those surveyed agree that there is a significant 'crisis' of reproducibility, less than 31% think that failure to reproduce published results means that the result is probably wrong, and most say that they still trust the published literature" (Baker, 2016).

Perhaps a more productive perspective is that the current crisis in biomedical research is less an indictment of the past but more of a caution for the future. Certainly, there are those who judge themselves by what they have done and others by what they have yet to do. While biomedical research has done wonders, may be a bit more can be done.

Perhaps scientists are still smarting a bit following the attacks of the postmodernists on the unique and privileged position of science (discussed in this Chapter). It is not surprising that the crisis of confidence in reproducibility in biomedical research has been exploited by the severe critiques of science, such as the postmodernists (Saltelli and Funtowicz, 2017). It would seem that the significant rate of irreproducible experiments would play into the postmodernists' notion that science, in general, and biomedical research, in particular, do not enjoy any privileged epistemic position and, consequently, inferences to reality (ontology) are no better than anyone else's, even the most unscientific positions. This not the opinion held by this author. This critique of reproducibility and irreproducibility in biomedical research is not the story of "The Emperor's New Clothes" where the emperor, that being science, is found wanting of clothes and, thus, is exposed. Rather, the emperor's clothes are a patchwork of pieces with more or less clarity and confidence. Pieces will be replaced once recognized as needing replacement. Given the complexity of living organisms and the difficulties of understanding living things deeper than just as phenomena, biomedical science necessarily is a patchwork of knowledge. Critics of biomedical researchers should not be like the child of the original story of the "Emperor's New Clothes" who is silenced and taken away by the citizens—the unappreciated ending to Hans Christian Andersen's tale written in 1837. It is just that science is very difficult and challenges the capacities of humans to understand. Science is extraordinarily difficult at the very large—cosmology; the very small—quantum physics; and the very complex—biology. It is not unexpected that humans should stumble on occasion, but humans should avoid stumbling repeatedly over the same difficulty.

The premise that scientific discovery proceeds from the judicious use of logical fallacies and from the employment of productive irreproducibility may seem jarring, if not insulting, to science and hence to scientists. Given William Whewell's definition of a scientist as one who does science, then to say that science proceeds from logical fallacies is to suggest that scientists trade in logical fallacies. Scientists should not be insulted by this claim for reasons to be discussed later. However, a distinction has to be made between *what* is discovered and *how* it was discovered. A very great deal of the *what* discovered by biomedical science is sound even if the *how* is tentative. Alexander Fleming may have blundered onto the discovery that penicillin mold created a substance that destroyed bacteria, but that does not mean that the discovery of penicillin is in doubt. Louis Pasteur said, "In the fields of observation chance favors only the prepared mind" (Lecture, University of Lille, December 7, 1854). The history of science is replete with similar examples.

The key to the advancement of science is to demonstrate that inferences from observations—whether or not in the context of the laboratory or field—can be demonstrated to be true with certainty. The key test of scientific certainty, hence truth, is reproducibility, which can and does happen even if the inferences are generated by the use of logical fallacies. The incredible breadth and depth of corpus of biomedical knowledge attest to the capacity to generate hypotheses that become knowledge, even if hypothesis generation requires the use of logical fallacies. Science ultimately will be right, and increasingly so, even as it sheds its wrong reasons.

At this juncture, it is important to be clear as to the nature of the critique of biomedical research contained in this book. There maybe a tendency to see the term "critique" in a pejorative connotation. That is not the intent at all. Rather, the term "critique" as used here (in the philosophical sense) is value-free analysis and very much with a positive or constructive intent. The confidence of persons in science should not be shaken by any criticism as the success of science is unparalleled in any other domain of human knowledge. The success of science is evidenced by its pervasive influence in the human experience and to its contributions to the betterment of the human condition. Clearly, biomedical scientists are doing something right.

In many ways, the remarkable achievement of the Enlightenment, which touched every facet of the human experience since the 16th and 17th centuries, has its origins in the remarkable changes in the notion of science in the 17th century (Jordan, 2017). Rationality became the Zeitgeist and manifested itself in nearly every human endeavor. Increasingly, the rationality was a scientific one. Science became the basis for understanding even civil society and its institutions. Thomas Hobbes (1588−1679) advanced what then would be a radical idea, that the monarchs did not derive their authority by religious divine right. Hobbes was constantly hounded with accusations of being an atheist, which only showed that the grip of religion-based metaphysics was loosening. Regardless of one's opinion of Hobbes' final notions of the operations of civil society (the importance of an absolute sovereign even if not a monarch), he took inspiration from science. Hobbes' approach constituted an empirical assessment of the nature of human beings and then extrapolated from that scientific (or at least science inspired) assessment to what an optimal civil organization would be. While the ideals of the Enlightenment may have since faded, particularly in affairs among humans, it has not faded in the affairs of humans have with the universe and reality in the context of science.

William Whewell coined the term "scientist" in 1834 as those that do science, as an artist is one who does art. In Whewell's notion, the endeavors of scientists are shaped by science. The notion that nature's fundamental reality is independent of humans, constant and orderly even if the order is impossible for humans to fully understand, if true, vouchsafes the productivity and accomplishments of the scientific enterprise. Some postmodernist critics have attempted to invert the notion of science from that described by Whewell to science simply being what scientists do. The postmodernists' attempt is to disconnect the scientific enterprise from notions of an independent, constant, and orderly reality that drives humans to understanding. An independent reality does so by constraining what can be found true. Ultimately, reality is not artificial in the sense of a human psychological or social construct—only.

Extrapolating from a realist notion of an external, constant, and ordered universe, human understanding will be contained by reality as river banks constrain the river (even if sometimes the banks are breached). This is not to say that the ultimate understanding will result in crisp, certain, and unambiguous knowledge, but perhaps like in a painting by Claude Monet, when his cataracts diffused the images before his vision deteriorated too greatly, he was still able to grasp important perceptions, perhaps in a manner more revealing (Gruener, 2015). Indeed, according to Kant's Transcendental Aesthetic, humans are just so constructed as to respond to the reality in nature (Kant, 1781) for which there is increasing neuroscientific evidence. Thus, in fundamental ways, the accomplishments of human science resonate with and thus can reflect the structure of the universe, even if yet unrealized.

SCIENCE VERSUS HUMAN KNOWLEDGE OF IT

Yet, the scientific enterprise is conducted through humans and, consequently, subject to misadventures to which humans are heir, such as unproductive irreproducibility through the injudicious use of logical fallacies. Nature can only be seen through the human senses, even if augmented by technology. Humans, probably since the very beginning, appreciated, if not recognized, the unreliability of human sense perception, perhaps largely based on its irreproducibility. Yet, whatever consistency there was supported the faith of an external world, at least in the traditions of Western culture. The response was to disavow any veracity of human sensation; consequently, whatever consistency there was to experience was related to ideal forms, as in Plato's (The Republic c. 380 BCE), and which mathematics was the archetypical example. Human sensations only were of the shadows cast from the ideal forms. In contrast, the early empiricists, such as Aristotle (Leroi, 2014), recognized the limitations and precariousness of human sensation, but nonetheless saw human sensation as the route to natural knowledge.

In an interesting way, it could be argued that the ancient Greeks were more careful in their science. To be sure, ancient Greeks, from Aristotle to Galen, conducted experimentations by dissections of the dead and by vivisection of the living. Yet, the Greeks knew that inferences from observations of the dead or abnormal living could not fully inform knowledge about the living (Zaner, 1998). Yet, how else to unpack the complexity of living phenomena? Perhaps despite or unaware of the concerns of the ancient Greeks, biomedical researchers today seem untroubled by the limits of dissection and vivisection and perhaps that maybe a contributing factor to unproductive irreproducibility (see discussion of Reductionism).

Experimental designs to reduce the complexity (Methodological Reductionism) and to simplify the explanation (Theoretical Reductionism) were necessary and these approaches are continued to this day in countless biomedical research laboratories. Methodological Reductionism can be affected by reduced preparations, such as tissue slices, organotypic tissue cultures, lower animals, lesioned higher animals, computational modeling and simulation, sampling from the population, and the use of experimental controls to counterbalance complexity and confounds. Yet, the Greeks knew that knowledge gained from manipulated (vivisected) organisms was knowledge of vivisected organisms with generalization to the intact organisms being problematic or, at the very least, tentative. Doing so risks the Fallacy of Four Terms as introduced in Chapter 1, Introduction. However, Methodological Reductionism is almost the *sine qua non* (the essential ingredient) of biomedical research.

Science also strives to "work backward" from Reductionism to recreate the original whole phenomena. It maybe progressing from initial studies in simpler organisms or from reduced preparations to increasing complex platforms. It maybe mathematical and computational modeling and simulations. In situations where the phenomenon itself proves recalcitrant to the efforts of reductionism and subsequent reconstruction, reasoning from analogies, such as Chaotic and Complex Systems, maybe a way of "working backward." The difficulty inherent in humans trying to understand living things is a challenge, it is not debilitating.

THE NECESSITY OF ENABLING ASSUMPTIONS

The success of biomedical research is remarkable in its contributions to the human good. Biomedical research is even more remarkable in view of the challenges to understanding the complexity of living organisms. However, on the road to such success, a number of compromises were and are necessary. To denigrate biomedical research because of the necessary employment of enabling assumptions and presumptions and the judicious use of logical fallacies would be foolish and sophomoric. However, the enabling assumptions and presumptions and the judicious use of logical fallacies limit the conduct of science and, therefore, what can be known to science. But this is no different than the assumptions of normally distributed data in order to apply certain statistical measures and tests (see Glossary of Concepts and Appendix B: Introduction to the Logic of Probability and Statistics for further explanations). The successful biomedical researcher needs to be aware of the enabling assumptions and presumptions just as she would the assumptions about the distribution of data prior to the application of statistical tests. In the case of non-normally distributed data, alternative statistical measures and tests must be applied. A dependency on the enabling assumptions and presumptions and the judicious use of logical fallacies does not mean that biomedical research is unfounded. Consequently, there should be no embarrassment on the part of biomedical researchers to acknowledge these enabling assumptions and presumptions and judicious use of logical fallacies. Rather, by acknowledging them, these enabling assumptions and presumptions and judicious use of logical fallacies can be used productively. As importantly, when these enabling assumptions and presumptions and use of logical fallacies prove counterproductive, they are replaced and biomedical research continues.

The Centrality of Reproducibility

It is important to appreciate the centrality of reproducibility in the scientific enterprise. It is not just replicability of the exact same experiments by other researchers, narrow reproducibility, or conceptual or broad reproducibility in generalizing the results of any particular experiment to the larger question, whether from species to humans or from a sample of subjects to a population. Reproducibility is an issue within every experiment, at least in the local sense of replicability, even if not generally appreciated. This is local reproducibility. Try to publish a research project that runs an experiment on a single subject, slice, tissue culture, or simulation. The truth, value, or utility (depending on one's metaphysical position) is tied to reproducibility over multiple subjects within or iterations of the same experiment. Thus, the concerns voiced in journals and public pronouncements reviewed in Chapter 1, Introduction, with respect to narrow reproducibility are fundamental to what a biomedical researcher does nearly every day in terms of local reproducibility. Interestingly, the exact same experiment run multiple times will lead to multiple results. Indeed, any experiment that has exactly the same result (zero variance) over multiple runs or iterations is suspect. Consequently, there is a paradox in that a certain degree of irreproducibility is required to have a reproducible result.

The Central Tendency

The enabling presumption is that within or reflected in the multiple different outcomes within a single experiment lies the true result. Generally, it is presumed that under the proper conditions, the measure of Central Tendency—such as the mean, median, or mode—reflects the true state of affairs even if none of the results equal the Central Tendency. This is evident in the writing of statisticians, for example,

> The basic task of data analysis is to determine the parameter(s) that make the model the best description of the data in some clearly defined sense. The core business of statistics is to supplement such an "estimate" with a measure of precision, usually in the form of an interval in which the *"true value"* [italics added] should be contained with high probability. The probability model that is used in such a context is that of a random variable X with a supposed distribution, most commonly the normal distribution. It is given by a density f or, more generally, by the ("theoretical") cumulative distribution function $F(X) = P(X \leq x)$. The distribution is characterized by parameters: the expected value μ [the Central Tendency in this case the mean of the population] and the standard deviation σ [variance of the population] (in case of the normal distribution). For the posed problem, μ is the *parameter of interest* [italics added], and σ is often called a *nuisance parameter* [italics added], so is the problem would be easier if it were not needed for a reasonable description of the data (Stahel, 2016).

Another important enabling presumption is that the "true value" will be found among or reflected in the varying results of multiple executions within a single experiment, provided a sufficiently large number of executions were done. This is the basis of the Large Number Theorem and depends on the ergodicity, introduced in Chapter 1, Introduction and explained further in Chapter 5, The Logic of Probability and Statistics and Chapter 10, Chaotic and Complex Systems, Statistics, and Far-from-Equilibrium Thermodynamics. Yet, as discussed in Chapter 10, Chaotic and Complex Systems, Statistics, and Far-from-Equilibrium Thermodynamics the Large Number Theorem may not hold in Chaotic and Complex systems (those systems operating far-from-thermodynamic equilibrium). Atmanspacher and Demmel wrote

Basic assumptions in the statistical analysis of systems and traditional science are ergodicity, stationery, stability, and so on. All these assumptions can be inappropriate in complex systems. Particular examples show that, as a consequence, statistical limit theorems [such as the Large Number theorem] need to be applied with great care, for instance, with the help of large deviation statistics. Since limit theorems are a basic statistical backbone of reproducible results, it is evident that the concept of reproducibility must not be uncritically *assumed* [italics added] but carefully checked in each individual situation. Extremely long-living transients between stable behaviors [the notion of bifurcations to be discussed in Chapter 10] may be other indicators of intrinsically irreproducible behavior (Atmanspacher and Demmel, 2016).

The Scientific Method

As discussed frequently in this book, the Scientific Method is an enabling presumption. The archetype of the Scientific Method is as follows: (1) identify a problem or gap in knowledge; (2) gather as much information as possible relative to the gap or problem; (3) create a hypothesis to fill in the gap or resolve the problem (perhaps the weakest link in the process); (4) extrapolate or interpolate from the hypothesis to predictions; (5) translate predictions into experimentally testable predictions; (6) test for the presence or truth of the predictions; and (7) if the predictions are found true or present, accept the hypothesis (modus ponens-like form of the Scientific Method), or if the predictions are found false or absent, then reject (modus tollens form of the Scientific Method) and reframe the hypothesis (perhaps the second weakest link in the process). Yet, the Scientific Method in its modus ponens-like form is the Fallacy of Confirming the Consequence. But it is a necessity for scientific progress (see Chapter 4: The Logic of Certainty versus the Logic of Discovery). The remarkable progress of biomedical research specifically, and science in general, demonstrates the power and utility of this enabling presumption. But its power cuts both ways, inspiring experiments whose contribution of new knowledge have passed the test of time; but at the same time has led to unproductive irreproducibility.

The quality of the hypotheses is critical to the Scientific Method. Historically, the question of the origin of hypotheses has been perhaps the least studied critically (Montgomery, 2018). Nonetheless, as will be shown in this book, the quality of the hypothesis has great implications for the reproducibility, productive irreproducibility, and utility for gaining new knowledge. In large part, new hypotheses are a consequence of the narrative of scientific success. For example, the notion of causality is fundamental to biomedical research. Causality is the concept that things happen for a reason, that phenomena are caused by something, they do not just happen (this is the philosophical concept of the Principle of Sufficient Reason). Indeed, science is about finding and understanding all of the causes of phenomena. Notions of causality underlie interpolations and extrapolations among and beyond specific scientific knowledge claims to form the hypotheses subjected to the Scientific Method (see Chapter 5, The Logic of Probability and Statistics). However, it must be done wisely, which necessitates an understanding of its limitations.

As will be seen, metaphor is a powerful source of new hypotheses that, surviving testing, become new knowledge. Yet, metaphor is an example of the Fallacy of Pseudotransitivity, and designing experiments based on the metaphor risks the Fallacy of Four Terms. Again, the remarkable success of biomedical research attests to the power of the judicious use of the Fallacy of Pseudotransitivity and the Fallacy of Four Terms, but examples of unproductive irreproducibility also attest to its power to mislead. The successful and wise biomedical researcher would not deny the use of such enabling fallacies; indeed, cognizant of them, she uses the fallacies wisely.

The wise biomedical researcher recognizes that the use of enabling assumptions and presumptions and the judicious use of logical fallacies are necessary for discovery and carefully notes that their use alone does not produce certainty that justifies the resulting experimental claims as new knowledge. Yet recognizing the epistemic limits of the Scientific Method and then compensating for them provide greater certainty and justification for translation into ontology (what is real).

As the highly successful process from hypothesis to substantiated knowledge is not random, it must follow a logic of some kind. However, it cannot be logically valid. The reason is the very definition of logical validity. A valid logical argument cannot give rise to false conclusions. Demonstrations of irreproducibility, as discussed in Chapter 1, Introduction clearly showed that the logic utilized, at least in studies that are unproductively irreproducible, had to be invalid. Hence, the nonrandom progress of biomedical research must be predicated on invalid (in the traditional sense) logic. As will be seen in Chapter 4, The Logic of Certainty Versus the Logic of Discovery a novel notion of logic, evolutionary logic, is proposed that uses traditionally valid logical deductions and, at the same time, embraces the judicious use of logical fallacies.

One way the Scientific Method is enabling is in the generation of hypotheses that subsequently are tested to gain new knowledge. Consider the argument *if a is b is true and b is true, then a is true*. As evidence, consider the clause *if a is b is true*, then *a* can be substituted for *b* and the argument *if a is b is true and b is true, then a is true* becomes *if a is a is true* (Principle of Identity) and *a is true, then a is true*. It cannot be anything but true, but it is a tautology that provides no new knowledge other than *a* is identical to *b*. However, the Scientific Method is of the form *if a* (scientific hypothesis) *implies b* (scientific prediction or experimental outcome) *is true and b* (scientific prediction or experimental outcome) *is true, then a* (scientific hypothesis) *is true*. The notion of *implies* needs not be the same as the notion of *is* (a state-of-being verb). It maybe that *implies* means *causes* and the Scientific Method now becomes *if a causes b is true and b is true, then a causes b must be true*. A theory of cause can generate *a* (a hypothesis, for example) and *b* (a prediction from the hypothesis), allowing the construction of an experiment as *if a implies b* (from the causal theory) *is true and b is true, then a is true* and is counted as evidence for the causal theory. Although inherently the logical Fallacy of Confirming the Consequence, thereby voiding absolute certainty, the fallacy provides considerable means nonetheless to increase the certainty and thus its utility.

Reductionism

The complexity of biological systems generates an incredible variety of phenomena. Indeed, no two humans are exactly alike, for example. How can any explanatory theory and scientific approach deal with such variety? Two perspectives are possible. The first views variety as variations on central themes or principles. The second views variety as diversity where each instantiation is a new reality onto itself. Relative to utility in understanding reality, the latter position risks paralysis due to Solipsism of the Present Moment. Clearly, biomedical researchers adopted the former, holding that the great variety of phenomena can be explained or reconstituted from an economical set of fundamental principles and universals. This is the basis for Reductionism.

Perhaps one of the most remarkable facts since the founding of modern science with the chartering of the Royal Society in 1662 has been advances in technology and institutionalization of the scientific enterprise. The advancing technology has dramatically increased the varieties of phenomena that can be subjected to Reductionisms. For example, the development of neurometabolic imaging has now been brought to bear on human mental states, such as beliefs, emotions, and consciousness, among others, that heretofore were held beyond the physical or, at least, unexplainable in the physical. Neurometabolic imaging reduces these mental states to changes in the energy metabolism of specific parts of the brain—not the mind.

The power of Reductionism, however, comes from the faith that the original individual phenomenon, not just a class or average over the phenomena, can be reconstituted from the economical set of fundamentals. The mechanisms by which principles in the economical set interact to reconstitute the original phenomenon imply a form of causality that is analytically tractable. The notions of causality are one source of new hypothesis generation that feeds the scientific enterprise to create new knowledge.

The process of dissection, in the widest connotation of the term, and the subsequent reconstruction into an explanation or understanding have a long tradition in natural philosophy that carries over to modern science. An example is the group of Italian natural philosophers in the 16th century known as the School of Padua. They used the terms "resolution and composition" for dissection and reconstruction, respectively. It was a method used by Galileo, particularly in his examination of the trajectory of a projectile (Jesseph, 1996).

The power of the methods and approaches described previously also place biomedical research at risk for the Fallacy of Confirming the Consequence and the Fallacy of Four Terms. These risks will be examined throughout the book and methods will be provided for reducing the risks sufficiently to avoid unproductive irreproducibility while still gaining new knowledge. However, none of this should be taken as weakening biomedical science or impugning the methods and skills of the biomedical researcher.

Causality

As alluded to earlier, causality is critical to the generation of new hypotheses, some of which become new knowledge. However, it is a consequence of the logically invalid Practical Syllogism as discussed in Chapter 4, The Logic of Certainty Versus the Logic of Discovery and Chapter 6, Causation, Process Metaphor, and Reductionism. Invocation of causality is inherent in humans' reactions to the changing world around them. It is reflected in the explicit Principle of Sufficient Reasoning frequent in the conversations of the natural philosophers, the forbearers of today's scientists. The principle essentially held that everything happens for a reason. Although absent in conversations today, it only is because it is now taken for granted since the Age of Enlightenment. The centrality of causality is not diminished by being unspoken.

The remarkable utility of causality also has its risks and may underlie instances of unproductive irreproducibility. This risk was identified by the ancient Greeks in the Principle of Causational Synonymy, which held that whatever is in the cause that generates the effect must likewise be present in the effect. The electrons in the outer orbit of the atoms that make up the surface of the hand are what repel the electrons in the outer orbits of the atoms that make up water that allow the hand to stir the water. To be sure, the ideal end of Reduction is to reach a theory of ultimate causational synonymy, for example, the Theory of Everything in physics where all the fundamental forces of nature are unified into a single causal theory.

Whether a Theory of Everything in biomedical science will ever be achieved is an open question, but this does not mean it should not be pursued—quite the opposite. The remarkable success of biomedical research and Reductionism demonstrates that the trajectory is sound even if the destination is unachievable. A great deal of good will be done pursuing the destination. Indeed, biomedical science, in particular, and science, in general, have been very successful at creating interim destinations that are achievable and serve as a staging point to reach future destinations, ultimately to a complete theory with causational synonymy.

The danger lies with setting short-term destinations, that is, causal explanations that fall short of causational synonymy, and viewing them as sufficient. For example, the Scientific Method reasonably led to experiments based on *if gene A caused behavior B is true and behavior B is true, then gene A causes behavior B*. At the very least, successful experimental evidence of the argument has heuristic value and further research can show exactly how *gene A* causes *behavior B*. However, a gene is not identical with a behavior. Story Landis, then director of the National Institute of Neurological Disease and Stroke, together with Thomas R. Insel, then director at the National Institute of Mental Health, had to write an editorial in *Science* cautioning scientists, writing:

> One important lesson from neurogenetics is that genomic variations in regulatory regions can account not only for how much of a protein is made, or when it is expressed, but exactly where in the brain a protein is expressed. Because brain function is specified by precise regional circuits, even small differences in the location of the brain cells that produce a particular receptor or an enzyme can result in large differences in function. Importantly, the link between genomic sequence and behavior is the brain: We cannot hope to understand how genomic variation influences behavior without understanding how genomic variation influences neural circuitry (Landis and Insel, 2008).

Essentially, what Landis and Insel were insisting on is the Principle of Causational Synonymy. To be sure, the expectation is that someday there will be an explicit linkage between the expression of *gene A* and the consequent *behavior B* likely through a multitude of intervening states. Ultimately, there will be a linkage in that each state follows directly from the previous state. In that case, *gene A is intermediate step C, which is intermediate step D.... which is neural circuitry E... which is behavior B*. The Principle of Causational Synonymy will be fulfilled, thereby increasing certainty and reducing unproductive irreproducibility.

The risk is that these intermittent destinations are taken as sufficient. For example, these destinations become bridging terms in syllogisms for future experimentation. Because they are not fundamental, conforming to the Principle of Causational Synonymy, they expose the experiments to the Fallacy of Four Terms and the risk of unproductive irreproducibility as discussed in Chapter 4, The Logic of Certainty Versus the Logic of Discovery. Again, this risk is a challenge, not a limitation, as there are means to mitigate the risk, and biomedical researchers are resourceful.

SPECIAL CASES OF IRREPRODUCIBLE REPRODUCIBILITY

There maybe scientific claims where it is just not feasible in principle to test, let alone reproduce. It is unlikely that science will be able to replicate the Big Bang that gave rise to the universe. One approach is to substitute a theoretical analysis, typically in terms of mathematics, where the truth, validity, or utility of the mathematical/theoretical analysis is tested by its reproducibility in a mathematical sense of not resulting in a contradiction or absurdity (see Chapter 9, Case Studies in Computational Biomedical Research). The implication is that any experimental reproduction is not possible. However, this is not entirely true.

Another alternative is used frequently in science in which the process is reproducible even if the ultimate outcome is not. The reproducibility of the process involves the Process Metaphor. An excellent example is the demonstration of Galilean inertia. The scientific claim is that an object in motion remains in motion infinitely unless acted upon by an outside force. Clearly, no experiment will be able to follow an object in motion for an infinitely long time. How could Galileo substantiate the claim, thereby assuring reproducibility? He arranged two incline planes. He would release a ball from a specific height on one inclined plane, observe the ball rolling down one plane and up onto the other, and then measure the height on the second plane to which the ball rolled. He found them to be the same. He then reduced the angle of the second plane and found that while the ball rolled to the same height, the ball had to roll further. He repeated the process with decreasing angles of the second plane and found that the ball still would have to roll further and further to reach the same height. Galileo then extrapolated that if the angle of the second incline was zero, the ball would roll forever, thereby demonstrating a principle of inertia. The truth, validity, or utility of the claim of inertia is in the reproducibility of the phenomenon, albeit in slightly different circumstances, that is, the specific angle of the second incline plane. The regular and reproducible incremental increase in the distance the ball rolled up the second incline with an incremental decreasing angle of the incline is a kind of reproducibility that vouchsafed the generalization to the principle of inertia. This process is called a Process Metaphor because the truth, validity, or utility of the claim is established not by actual observation of the phenomenon claimed, but by the reproducibility of each condition (a larger angle of the second incline plane) serving as a metaphor to the next condition (a smaller angle of the second inclined plane).

The ubiquity of the Process Metaphor is evident every time differential calculus is used. For example, if velocity is defined as the change in distance over the change in time, how is it possible to measure instantaneous velocity where the change in time is zero (straight application of the definition would lead to division by zero, which is unallowable)? Yet the notion of an instantaneous velocity seems reasonable. Rather, a small segment of time is taken and the velocity is determined. The process is repeated as the small segment of time becomes less to the point of being infinitesimal small and consequently discounted as described previously. This is the basis of all differential calculus. To be sure, mathematicians have provided a sound footing for this extrapolation from the method (the Process metaphor) rather than the fact (the scientific claim) using limit theory. (For an amusing, but not without a point, counterargument, see George Berkeley, "The Analyst: A Discourse Addressed to an Infidel Mathematician wherein It Is Examined whether the Object, Principles, and Inferences of the Modern Analysis Are More Distinctly Conceived, or More Evidently Deduced, Than Religious Mysteries and Points of Faith," 1734).

The methods of Methodological and Theoretical Reductionisms are further examples of Process Metaphors. The presumption is that the process proceeds in step such that the difference between the reduced step and the just prior complex step is infinitesimal, and in the limit, the observations at the reduced step demonstrate the observations of the previous step, at least ideally. Note that this is not to suggest that the actual history of science proceeded in just such a manner, but it is illustrative of the reasoning (or faith, as Bishop George Berkeley would argue). While the exact circumstances of the Big Bang (the original experiment) may not be reproducible, scientists maybe able to get sequentially closer and, in the limit of that proximity, be said to reproduce the Big Bang. The reproducibility of the process or trend with each sequential closer approximation is the basis for certainty. The importance of reproducibility continues to be central to science.

SCIENCE AS INFERENCE TO THE BEST EXPLANATION

An alternative to strict reproducibility as evidence in support of a scientific claim is the model of Inference to the Best Explanation. The model is described as: "Its governing idea is that explanatory considerations are a guide to inference, that scientists infer from the available evidence to the hypothesis which would, if correct, best explain that evidence" (Lipton, 2000). Note that this formulation does not insist on reproducibility in the narrow sense. It could be considered reproducibility in a broad sense where the notion of "best explains the evidence" suggests alternatives and the best resonates more with the phenomenon to be understood, hence reproduces the phenomenon better. But this likely would be a stretch, as it is clear that factors other than the best explanation as the best reproducer of the phenomenon have been important in the history of the scientific enterprise. A full discussion giving the concept of Inference to the Best Explanation is beyond the scope of this text. Only briefly and partially, the discussion that does take place is to further the analysis of reproducibility and productive irreproducibility.

Consider the example of Copernicus's notion of a heliocentric solar system. There is reason to believe that Copernicus's heliocentric solar system was not the best explanation compared to Ptolemy, at least in terms of the ability to obviate the need for epicycles (orbits within orbits necessary to explain the path of planets observed in the sky). The epicycles were necessary, as some planets appeared to change direction as they moved across the sky, unlike the stars, and this is the reason why planets got their name from the Greek *planētēs*—meaning wanderer. It was not until Kepler's description of the orbits of the planets as ellipses that epicycles could be eliminated. Further, the Copernican theory could not explain how it is that objects do not fly off the earth as the earth spins or when one drops an object it falls at one's feet and not some distance away if the earth was rotating. Attempting to describe in a single metric necessary for the judgement of "best" requires a consolidation of the different components that are not comparable. For example, does a reduction in the number of epicycles from approximately 80 in the Ptolemaic system to 34 in the Copernican system outweigh the failure to explain why objects do not fly off the earth or dropped objects do not land elsewhere in order to argue that the Copernican system was the best explanation? There are many other examples, where at the time, what explanation subsequently would prove to be the victor was not the best explanation at the time.

The notion of the best explanation suggests a competition between alternatives that invite accusations of arbitrariness in the definition of what constitutes the best explanation. Further, subjective metrics, such as eloquence, parsimony, symmetry, and conservation, often enter into the judgment of what is the best explanation. Absent any deductive or logical calculation as rigorous to meet the Principle of the Excluded Middle (see Chapter 4, The Logic of Certainty Versus the Logic of Discovery), other factors were sought out, such as politics, psychology, or sociology, and were fodder for the postmodernists.

The larger question is the degree to which one can be certain that the best explanation is true, valid, or of use. Note that just an exclusion of alternatives does not ensure that what remains is true, valid, or useful. The rational remains the Fallacy of the Consequence—*if the best explanation implies the phenomenon is true, the phenomenon is true; therefore, the best explanation is true.* Even if one were to include alternatives, in the form of *if best explanation or second-best explanation or...implies the phenomenon is true, and the phenomenon is true, then the best explanation or second-best explanation or...is true.* If the second, third...best explanations are rejected, the result is still a fallacy. This is compounded by the fact that rejection of the second, third...explanations is often taken as evidence of the truth, validity, or utility of the remaining best explanation. This is the Fallacy of Limited Alternatives in logic and is the Gambler's Fallacy in Probability Theory (see Chapter 4, The Logic of Certainty Versus the Logic of Discovery and Chapter 5, The Logic of Probability and Statistics). The only way to be certain that the best explanation is certainly true, valid, or useful is to change the logic *if the best explanation implies the phenomenon is true, the phenomenon is true; therefore, the best explanation is true* to *if and only if the best explanation implies the phenomenon is true, the phenomenon is true; therefore, the best explanation is true.* This is difficult to assert in empiric biomedical research and argues that Inference to the Best Explanation may not be much comfort to scientists.

SUMMARY

Science has been remarkably successful and not by accident. To be sure, many scientific observations have been occasioned by accident or happenstance. But the transition from accidental observation to scientific principle is, for the most part, rigorous to the extent that any human enterprise constrained by the constant and universal laws of nature can be. Science ultimately will be right even as it sheds its wrong reasons. But the scientific enterprise is a human endeavor fraught with human frailties. Further, nearly every biological phenomenon, when viewed with sufficient resolution, is incredibly complex and maybe beyond complete explications. Consequently, enabling assumptions are necessary. Reducing the complexity by dissection or vivisection, where these terms are used in the widest sense, requires the enabling assumption that the reduced preparations are still generalizable to the whole phenomenon. The Scientific Method is an enabling assumption. However, enabling assumptions create a risk for unproductive irreproducibility.

Essentially, in biomedical science the surest gauge to true knowledge is reproducibility, yet rarely, if ever, are any experiments exactly reproducible. This results in the paradox that confronts biomedical research as what degree of irreproducibility still counts as reproducible? Reproducibility is central to the performance of individual experiments, but this form of local reproducibility is not the focus of this book. Rather, the focus is on situations where an individual experiment was conducted rigorously and reasonably, with transparency and without fraud. Yet when the same experiment is found irreproducible, in the narrow sense, in the hands of another experimenter or when the experiment is extrapolated to what is reasonably similar yet found to be irreproducible in the broad sense is the subject of this text.

The Logic of Certainty Versus the Logic of Discovery

CERTAINTY, REPRODUCIBILITY, AND LOGIC

The only sure way to absolute reproducibility of any experiment is to be absolutely certain that the experiment is absolutely true—good luck! However, it may be possible to "unpack" what it means to be absolutely certain and for the experiment to be absolutely true. The term "unpack" is just as when one unpacks a suitcase to find that this single suitcase contains a great many different things, each of which can be handled independently. One would not just know that by looking at the closed suitcase. It may be possible to find some things in the scientific suitcase that could be helpful in dealing with certainty and help in the utility of biomedical research in gaining new knowledge. In unpacking the epistemology (theory of what is knowledge and how it is obtained) suitcase, two potentially very useful tools are deductive logic and inductive logic.

It is important to make clear what logic is as used in this book. Logic is not facts or experimental observations—just as mathematics is not about facts or experimental observations, although facts and experiments may be couched in mathematical terms. For example, the mathematical statements $a + b = b + a$ and *if $a = b$ and $b = c$ then $a = c$* are true regardless of what facts or observations are inserted for a, b, or c. These statements are true and certain independent of the facts or observations. The statements are not about facts and observations, the statements are about the relations. Logic is about relations; how it is that a relates to b and c, for example. The great value of logic is that if an experimental argument, such as the rationale for the hypothesis and its prediction and the argument that seeks to test the prediction in light of the hypothesis, can be recast in its logical form, it is relatively straightforward to determine if the logic of the experimental argument is invalid and thus at risk for unproductive irreproducibility.

Valid logical relations between true logical variables guarantee true conclusions, no matter what facts are substituted for the variables *a*, *b*, or *c*. Similarly, the statement for any integer *a*, the statement $a \times a = 2$ is false no matter what integer value is substituted for *a*. The logical statement *a and not a* (*a*) is false no matter what is substituted for *a*. In this sense, logic and mathematics are equivalent. Any biomedical scientist using mathematics is using logic. The major difference is that some logical arguments are not necessarily expressed in mathematical terms. In those situations, it may be more difficult for the biomedical researcher to realize that she still is using logic even if not mathematics. For mathematical and logical systems that are complete, meaning that every theorem or valid logical argument can be derived from a set of axioms, certainty can be obtained (caution—not all mathematical and logical systems are complete).

Experiments are logical arguments even if not recognized as so. Consider an experiment where there is reason to believe that *hypothesis A* is true. Generally, *hypothesis A* is analyzed to arrive at *prediction B*. The experiment is constructed to determine whether *prediction B* is found to be true given *hypothesis A*. The experiment is conducted and *prediction B* is found to be false. One has little difficulty believing that *hypothesis A* must be false in some way. The experiment can be constructed as a logical statement *if hypothesis A implies prediction B and prediction B is found false, then hypothesis A is false*—with certainty (assuming that *prediction B* truly is false). This situation is logical deduction by modus tollens. *Hypothesis A* is false by modus tollens and is an example of productive irreproducibility (in the broad sense); *hypothesis A* can be discarded with certainty as a cause or necessity of *prediction B*. This would be an example of productive irreproducibility.

Rarely are experiments designed so that a prediction of the hypothesis is set up to be proven false, although this is the surest way to certainty. Examples of experiments set up in the manner of modus tollens include, for example, futility studies or negative studies (Snapinn et al., 2006), but these are few. Usually, the logic of most experiments is set up as *if hypothesis A implies prediction B and prediction B is found true, then hypothesis A is true*. Such a statement is not certain. *Hypothesis A* may or may not be true and the demonstration of *prediction B* to be true has no bearing on the certainty of *hypothesis A* as true or false. It does not matter what facts or observations are substituted for *hypothesis A* or *prediction B*, the conclusion cannot be certain. If the argument and conclusion cannot be certain, there will be a risk for unproductive irreproducibility. And it will be unproductive, unlike the case of modus tollens described earlier. The logical argument just described is the Fallacy of Confirming the Consequence (which is synonymous to other titles such as Fallacy of Confirming the Consequent, Fallacy of Affirming the Consequence, and Fallacy of Affirming the Consequent).

The theme pursued in this book is that biomedical experiments used to gain new knowledge nearly uniformly require the use of logical fallacies (the exceptions being futility studies). By recognizing the logical fallacies employed, biomedical research can take steps to reduce the uncertainty associated with the use of logical fallacies to increases the chance of reproducibility or productive irreproducibility. Important to note, this statement is not to say that the use of logical fallacies is the sole or even major reason for the crisis of irreproducibility in biomedical research. But substantial evidence will be brought to show that it is a cause of irreproducibility and that there are methods to reduce the risk of unproductive irreproducibility.

The logical forms and their associated fallacies are not difficult to appreciate. Indeed, these are the topics of undergraduate college introductory courses in logic. Perhaps what might be novel is the notion that logic at this level would have important things to say about biomedical research and reproducibility. While being novel may make it seem foreign to most biomedical researchers, being foreign should not be confused with being strange. The subsequent discussions demonstrate the reasonableness of the theme and that the logical structure of experimentation is indeed in front of every scientist. But as George Orwell said, "To see what is in front of one's nose needs a constant struggle."

There may be the temptation to dismiss the central theme that logic, simple enough for undergraduate college students, can be robust enough to address at least some of the issues of biomedical research and reproducibility. To be sure, there is considerable work underway at very complex systems of formal logic that are likely to make contributions to the issue of irreproducibility (Bell et al., 2001). However, this does not exclude the potential of relatively simple logic making a contribution, as noted in the *King James Bible*, Psalms 8:2, "Out of the mouth of babes and sucklings hast thou ordained strength."

DEDUCTIVE LOGIC—CERTAINTY AND LIMITATIONS

The only route to absolute reproducibility is absolute certainty. Perhaps the only route to absolute certainty is deductive logic. The paradox is that deductive logic is not a route to new knowledge other than discovering what one already knew but did not realize. As biomedical research is in the business of generating new knowledge, something other than deductive logic will be necessary. But there is a price to be paid for using logic other than valid deductive logic, which is uncertainty, and hence a risk of irreproducibility. What this means is that biomedical research must be as close as possible to pure valid deductive logic to increase reproducibility, yet different enough to gain new knowledge. As will be seen, any departure from deductive logic results in logical fallacies. However, the term fallacy must not be taken in a pejorative sense. That an argument is a logical fallacy does not mean that the results or conclusions are false, only that they are not certain. This means that biomedical research must embrace the use of logical fallacies, but in a judicious manner. Engaging in an injudicious use of logical fallacies likely contributes to unproductive irreproducibility, perhaps to a far greater degree than most realize.

There is an exception to the claims just made. There is one way to gain new knowledge with absolute certainty, if one accepts that learning that an idea, hypothesis, or theory is false constitutes new knowledge. The knowledge gained is that the hypothesis is false, hence the theory generating the hypothesis is irreproducible. In this case, it is productive irreproducibility. As described earlier, this involves a specific form of deduction called modus tollens that results in a falsification of the hypothesis. This is the basis for futility studies and negative studies that should not be confused with null studies, which are studies where no significant differences are found in an experiment but cannot be certain. The notion of null, from the null set, is a set that has no elements or members, nothing true or false. The philosopher Karl Popper used falsifiability as a way to demarcate science from pseudoscience in his "The Logic of Scientific Discovery" (2002). The necessity of a scientific theory to be falsifiable was the prevailing argument to keep theories of Intelligent Design out of the school science curriculum.

Scientific discovery as a "diagnosis of exclusion," where experiments are designed to refute candidate hypotheses via modus tollens, as opposed to directly vindicating preferred hypotheses, was never very attractive to philosophers or biomedical researchers alike. Over the many years, this author, as a reviewer of grants and papers, has yet to see a study proposed or executed specifically to refute a prevailing theory (the exceptions are the recent use of futility studies in clinical trials). Most studies that do end up refuting established theory do so more by accident. In fact, it would be difficult to predict how a study section of the National Institutes of Health or a panel at the National Science Foundation would greet a research proposal specifically aimed at refuting a prevailing hypothesis. While not impossible in principle, historical analysis suggests that it is nearly impossible in practice (see Kuhn, 1962). There are significant problems that would complicate any attempt at discovery by a falsification of alternatives—the "diagnosis of exclusion" approach. One obstacle is the Fallacy of Induction where one cannot be sure that every reasonable candidate hypothesis, save one, has been refuted, as will be discussed later.

There are many forms of deductive logic. The two principal forms discussed in this effort are propositional logic and syllogistic deduction. Most of the other forms are in reaction to the fundamental requirements in propositional logic and syllogistic deduction that provide certainty at the expense of utility. With fuller discussions to follow, these problematic requirements are the Principle of the Excluded Middle (also known as the Principle of Bivalence) and the necessity of the logical relations to be in the form of state-of-being verbs such as *is* or *are*, for example, *a is b*. The Principle of the Excluded Middle holds that any premise must be either true or false, not both true and false, and not neither true nor false. The nature of state-of-being logically linking verbs will be evident subsequently. As will be seen, an analysis of propositional logic and syllogistic deduction will serve the discussion of reproducibility quite well.

SYLLOGISTIC DEDUCTION

Classic syllogistic deduction is of the form:

ARGUMENT 4.1

Major premise:	*All humans* (bridging term) *are mortal* (major term)
Minor premise:	*Socrates* (minor term) *is a human* (bridging term)
Conclusion:	*Socrates* (minor term) *is mortal* (major term)

As can be readily appreciated, three terms are linked by the state-of-being verbs *is* and *are*. The syllogism can be interpreted as the set *all humans* is a subset or contained within the set *things that are mortal* (Fig. 4.1). Similarly, the set *Socrates* is a subset or contained within the set *things that are mortal*. As can be seen from Fig. 4.1, *Socrates* is contained within *all humans*, which is contained within *things that are mortal*. Consequently, *Socrates* is contained within the set *things that are mortal*. It cannot be anything else but that *Socrates is mortal*. The state-of-being verbs can be considered as evidence of membership in a set of entities.

The question can be raised whether anything new was learned as a consequence of the syllogism in Argument 4.1. It seems hardly surprising that Socrates would be mortal. In a sense, Socrates and humans are equivalent. What befalls humans would happen to Socrates as he is a member of the set of all humans. Thus, the conclusion that Socrates is mortal is a tautology; it follows from the definitions of mortality,

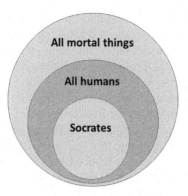

FIGURE 4.1   Venn diagram of the classic syllogistic deduction of Argument 4.1. As can be seen, the set of *humans* **is** a wholly contained subset within the set of *all things mortal*. Likewise, the set *Socrates* **is** a wholly contained subset within the set of *humans*. Thus, Socrates is contained within the set of *all things mortal*.

humans, and Socrates. While the syllogism is absolutely certain, it is not a great utility from the standpoint of gaining new knowledge.

The absolute certainty of the syllogism comes from the fact that the bridging term, *all humans*, is entirely true as there are no exceptions— depending on how a human is defined. Note that paramecia (single cell microorganisms) are not included within the set of *all humans* nor are there humans outside the set of *all humans*. The interesting question is whether the fact that there are no humans outside the set of *all humans* is a matter of definition. Whether there is just such an entity, otherwise the same in all relevant criteria as every human included in the set of *all humans*, considered not within the set of *all humans*; this is an epistemic question. Or is the set of *all humans* determined by the reality (ontology), the nature of the universe is just such that anything that has the reality of being human is contained within the set of *all humans*? Note that while appearing an academic and metaphysical (in the pejorative sense of the words) concern, it is serious enough to biomedical research. Definitions are human-made and hence contingent. The scientist hopes that reality is not contingent. Creating a correspondence between definition and reality underlies much of what is science.

Similarly, the set of *all things mortal* does not include anything that is not mortal and anything that is mortal is not outside the set of *all things mortal*. Thus, the premises conform to the Principle of the Excluded Middle. Similarly, the state-of-being linking verbs indicate class membership, for example, the set of *all humans* is in the class (set) of *all things mortal* and *Socrates* is in the class (set) of *all humans*.

The conclusion of Argument 4.1 obtains its validity and certainty by the bridging term that maps the minor term to the major term. The minor term is contained within the bridging term, which is within the major term. In doing so, the bridging term must stand in exactly the same relation to the major term as it does to the minor term. If not, the Fallacy of Four Terms may arise. For example, it might be possible that Socrates is just the type of human that is immortal. Thus, the humanity of Socrates in the minor premise is different from the sense of humanity in the major premise. Now there are two forms of the bridging term and one cannot have confidence in the conclusion that *Socrates is mortal.*

The argument of virtually every biomedical experiment is at risk for the Fallacy of Four Terms. The risk is inherent any time the bridging term in the major premise is different, in any way, from the bridging term in the minor premise. Yet, the nature of experimentation precisely is invoking different bridging terms. An example is shown in Argument 4.2:

ARGUMENT 4.2

Major premise:	*A sample of albino rodents with disease A* (bridging term) *respond to treatment B* (major term)
Minor premise:	*Humans* (minor term) *have disease A* (bridging term)
Conclusion:	*Humans should respond to treatment B*

History clearly has shown arguments of the form of Argument 4.2 often to be wrong, as in the case of human teratogenicity and thalidomide. A valid syllogistic deduction should provide absolute certainty and hence absolute reproducibility of the experience with albino rodents to the experiences with humans. Where does the syllogism go wrong? The truth or validity of the major premise is an experimental fact, give or take. The minor premise is a previously established fact. The problem lies in the bridging term. It just may be that *disease A* in this particular sample of albino rodents is different than *disease A* in humans. The consequence is the Fallacy of Four Terms.

The Fallacy of Four Terms also explains the importance of transparency in the reporting of the results of experiments, for example, the reagents, methods, species, ... used in the experiments. Consider Argument 4.3 that follows:

ARGUMENT 4.3

Major premise:	*Using reagents, methods, species...* (bridging term) *produced results A in experiment B* (major term).
Minor premise:	*Experiment C* (minor term) *used reagents, methods, species...* (bridging term)
Conclusion:	*Experiment C* (minor term) *will produce results A* (major term)

However, the actual conduct of *experiment C* did not produce *result A*; hence, *experiments B* and *C* are irreproducible. One possible cause is the Fallacy of Four Terms, where there are two versions of the bridging term. It may well be that the *reagents, methods, species...* in the major premise are different than the *reagents, methods, species...* in the minor premise. If there is a lack of transparency in the reporting of *experiments B* and *C*—in other words, a failure to specify the *reagents, methods, species...* used—there would be no way to know whether the irreproducibility was a consequence of two versions of the bridging term. The irreproducibility would be unproductive.

Extending this discussion, it may be that the pathoetiological (the cause of an abnormality) and pathophysiological (how the abnormality is expressed) mechanisms directly associated with *disease A* are identical in albino rodents and in humans with the single exception. A mechanism *X* in *disease A* in albino rodents is absent in humans, designated by *not X* or, logically, as *X̄*. Argument 4.2 can be reframed as in Argument 4.4:

ARGUMENT 4.4

Major premise:	*A sample of albino rodents with disease A and mechanism X* (bridging term) *respond to treatment B* (major term)
Minor premise:	*Humans* (minor term) *have disease A and mechanism X̄* (bridging term)
Conclusion:	*Humans should respond to treatment B*

In Argument 4.4, it is clear that there are two bridging terms containing *disease A and mechanism X* and the other *disease A and mechanism X̄*. Argument 4.3 is invalid because of the Fallacy of Four Terms; consequently, the conclusion is not guaranteed. Note that this is not to say that humans with *disease A and mechanism X̄* will not respond to *treatment B*, but only that this is not certain.

The fallacy in Arguments 4.2 and 4.4 allows introduction of the concept of Epistemic Risk. In these cases, Epistemic Risk translates to the risk of unproductive irreproducibility and depends on just how different the two versions of the bridging terms are. Argument 4.2 provides no assessment of how different the two versions are. Argument 4.4 suggests that the presence or absence of *mechanism X* is different in the two versions of the bridging terms. Some estimate, at least qualitatively, of the Epistemic Risk might be that the greater the involvement of *mechanism X* in the pathoetiological and pathophysiological mechanisms of *disease A*, the greater the Epistemic Risk because *mechanism X* is not present in humans. As will be discussed in a variety of different contexts, Epistemic Risk includes Epistemic Distance and Epistemic Degrees of Freedom. In the case here, Epistemic Distance is the length of the causal chain between the mechanisms of *disease A* and *mechanism X*, for example, the number of intermediary enzymatic events and their physical proximity, such as linked on a membrane or protein complex. Epistemic Degrees of Freedom relate to how different in kind are the mechanisms of *disease A* and *mechanism X*. For example, in an enzymatic chain of interactions that primarily involve phosphorylation or dephosphorylation, there are relatively little differences in kind and therefore relatively few Epistemic Degrees of Freedom. However, if the chain of interactions involves the translation of DNA to mRNA to protein synthesis to post-translation modification to intracellular trafficking, the Epistemic Degrees of Freedom could be quite large and, consequently, the Epistemic Risk correspondingly large.

It is hardly likely that every human with *disease A* will receive *treatment B* based solely on Argument 4.2. However, depending on the ethics involved, some humans with *disease A* will receive *treatment B* in the context of a human clinical trial. Thus, there will be costs and risks. The Epistemic Risk enters into the calculus, as does other ethical concerns, such as risks to receiving or not receiving *treatment B* in those humans with *disease A*. However, it is important to note that these considerations should be entertained prior to conducting the research that resulted in the major premise, that is, the animal experiments. The experiments involve cost, not only financially, but ethically, in terms of the treatment of laboratory animals. These costs should be considered an investment, but like all prudent investments, the (Epistemic) risks to a return on the investment, conceptual unproductive irreproducibility, should be considered prior to the investment.

Syllogistic Arguments 4.2 and 4.4 involve trans-species generalizations. In constructing such arguments, the biomedical researcher should weigh the costs, benefits, and Epistemic Risk. However, for purposes here, the notion of species should be expanded. For example, in vitro preparations such as cell or tissue culture, organotypic cocultures, and many others could be considered as different species. Consequently, biomedical research involving in vitro preparations would be subject to the same risks for the Fallacy of Four Terms as the in vivo trans-species generalization. Similarly, mathematical modeling and computational modeling and simulation, in silico preparations, can be considered as a species; thus, experimentation involving in silico species likewise is at risk for the Fallacy of Four Terms. Even confining the bridging term to within the same species can result in the Fallacy of Four Terms, as shown in Argument 4.5:

ARGUMENT 4.5

Major premise:	*A sample of humans with disease A and* (bridging term) *respond to treatment B* (major term)
Minor premise:	*All humans, or at least my patient* (minor term), *have disease A* (bridging term)
Conclusion:	*Humans, or at least my patient, should respond to treatment B*
Inference:	*My patient should receive treatment B*

If the argument is valid and the premises are true, then the conclusion can be guaranteed. The critical question relates to the two versions of the bridging term and the Epistemic Risk of assuming they are identical in order to avoid the Fallacy of Four Terms. Is the sample of humans with *disease A* who were studied for their response to *treatment B* (major premise) the same for all humans or at least my patient (minor premise) who may receive *treatment B*? If so, then there are only three terms; if not, then there are four terms and the argument is invalid. If the sample of humans included, for example, only Caucasian males age 20–30 years without any other medical or social problems and my patient is a 70-year-old non-Caucasian female with diabetes and heart failure, then there may be considerable Epistemic Distance and Epistemic Degrees for Freedom between the two versions of the bridging term. Indeed, it is likely that the Fallacy of Four Terms is a significant factor in the failure of medications, biologics, and devices abandoned or withdrawn after previously being approved by organizations such as the U.S. Food and Drug Administration (FDA) following what were thought rigorous clinical trials. For example, the sample of subjects treated in the clinical trials that resulted in FDA approval was different from the patients that were subsequently treated after more widespread clinical use. The difference between the subjects treated in the clinical trials results in the Fallacy of Four Terms and hence the broad unproductive irreproducibility that resulted in the postmarketing failures.

The risk for the Fallacy of Four Terms is nearly inescapable any time a sample rather than an entire population of interest or concern is the subject of the actual experiments and where a generalization to the entire population is the intent. In most cases, it is either infeasible or impractical to subject the entire population of concern to experimentation, although it is interesting to see an increasing number of multicenter studies enrolling thousands of patients. The perceived need to do so is evidence of the risk for the Fallacy of Four Terms. In this sense, as the sample size approaches the size of the population of concern, the Epistemic Distance between the sample and the population diminishes. The four terms move toward only three terms and the syllogism approaches the absolute certainty of a syllogistic deduction. However, despite enrolling very large numbers of subjects, from the Fallacy of Induction, discussed in detail later, it would be very difficult to be sure that everyone in the population of interest was examined. Further, there are disadvantages to very large-scale studies—demonstrating statistical significance but clinical meaninglessness (discussed further in Chapter 7: Case Studies in Clinical Biomedical Research).

Different bridging terms are inherent when the major term is a generalization, such as an attribution to an entire population or to a theory, and the minor term is a particular sample or testable hypothesis. In Argument 4.1, the major term, *all things mortal*, is a generalization and the minor term, *Socrates*, is a particular. Generalization of the major term forces the bridging term to take on the character of a generalization, as noted by the linking verb *are*, while the bridging term in the minor premise is singular, as noted by the linking verb *is*. Thus, the argument is at risk for the Fallacy of Four Terms. An alternative form of the syllogism can have a major premise that is a particular while the minor premise contains a generalization, as noted in Argument 4.4, but the concerns remain.

Locating the general and particular terms as major or minor terms serves different purposes. In induction, which is reasoning from particulars to a generalization, seeing many ravens that are black to the generalization that *all ravens are black*, the particular is in the major term, it is used to create a generalization, such as a principle or law, from a set of particulars. In the deductive approach, such as the Scientific Method, using the generalization or principle as the major term then sets up the minor term to be the particular prediction and the experimental methods as the bridging term.

JUDICIOUS USE OF THE FALLACY OF FOUR TERMS

As can be appreciated, nearly every experiment in biomedical research is a variation on the Fallacy of Four Terms. This is understandable as the valid form of the syllogistic deduction does not provide new knowledge other than the recognition of a tautology, which in itself is not without benefit. For example, a large nested syllogistic deduction can lead to new realizations, as shown in Argument 4.6:

ARGUMENT 4.6

Major premise$_1$:	*Bridging term$_1$ is major term$_1$*
Minor premise$_1$:	*Minor term$_1$ is bridging term$_1$*
Conclusion$_1$:	*Minor term$_1$ is major term$_1$*
Major premise$_2$:	*Bridging term$_2$ is minor term$_1$* (as major term in major premise$_2$)
Minor premise$_2$	*Minor term$_2$ is bridging term$_2$*
Conclusion$_2$:	*Minor term$_2$ is minor term$_1$, which is major term$_1$*

and so on. . .

Variant syllogisms, such as the Fallacy of Four Terms and practical and partial syllogisms to be discussed later, do not constitute proof. This is because they are inherently uncertain. The conclusions of such syllogisms require experimental vindication or refutation. The variants of the syllogistic deduction are useful when considered as templates for the construction of biomedical experiments, as well as suggesting research hypotheses. For an example, a syllogism whose main premise is a generality while the minor premise is a particular can be translated into an experiment where the minor term in the minor premise becomes the hypothesis, the bridging term becomes a means for operationalizing the hypothesis (converting the scientific hypothesis into a hypothesis that can be tested experimentally), and the major term becomes the informing theory. An example is shown here in Argument 4.7:

ARGUMENT 4.7

Major premise$_1$:	*Blocking dopamine reuptake of dopamine* (bridging term$_1$) *causes dopamine depletion* (major term$_1$)
Minor premise$_1$:	*Drug X blocks* (minor term$_1$) *dopamine reuptake* (bridging term$_1$)
Conclusion$_1$:	*Drug X leads to dopamine depletion*
Major premise$_2$:	*Dopamine depletion* (bridging term$_2$) *produces parkinsonism* (major term$_2$)
Minor premise$_2$:	*Drug X* (minor term$_2$) *produces dopamine depletion* (bridging term$_2$)
Conclusion$_2$:	*Drug X produces parkinsonism*

To the degree that dopamine depletion associated with parkinsonism in major premise$_2$ is not the same as dopamine depletion associated with *drug X*, the argument is open to the risk of the Fallacy of Four Terms. This variant of the syllogistic deduction is at risk because major premise$_2$ may not be entirely true, as there are forms of parkinsonism not associated with dopamine depletion, such as multiple infarctions in the globus pallidus externa. Nonetheless, this variant of the syllogistic deduction does provide a strong rationale for actually conducting the experiments to demonstrate whether *drug X* may produce parkinsonism. Such experiments could lead to research to determine the mechanisms by which *drug X* produces parkinsonism in the hopes that reversing or blocking the mechanisms invoked by *drug X* may lead to better disease management. This is precisely what happened in the development of levodopa for the treatment of Parkinson's disease (LeWitt and Fahn, 2016) and reviewed in detail in Chapter 8, Case Studies in Basic Biomedical Research.

Unlike the valid form of syllogistic deduction, which results in tautologies that often are of relatively little value, these variants are a critical source of research hypotheses. Indeed, one could argue that biomedical researchers should embrace and become experts at using these logical fallacies. Many may find such a claim counterintuitive, thus risking uncritical rejection of the claim. That would be unfortunate.

It is important to emphasize the fact that these variants do not constitute proof, although they often are taken as proving a particular hypothesis. For example, early in the development of Deep Brain Stimulation (DBS), it was held that high-frequency stimulation of a target reduces activity in the target, whereas low-frequency stimulation enhances neuronal activity. The arguments were as shown here:

ARGUMENT 4.8

Major premise:	*Surgical ablation* (destruction) *reduces neuronal activity* (major term) *and improves parkinsonism* (bridging term)
Minor premise:	*DBS* (minor term) *improves parkinsonism* (bridging term)
Conclusion:	*DBS reduces neuronal activity in the manner of ablation*

ARGUMENT 4.9

Major premise:	*DBS in the brain of animals* (bridging term) *results in reduced neuronal activities in the stimulated target* (major term)
Minor premise:	*DBS in humans* (minor term) *is DBS in animals* (bridging term)
Conclusion:	*DBS in humans* (minor term) *reduces neuronal activity* (major term)

Initial experiments were based on the presumptions contained in Arguments 4.8 and 4.9 and the conclusions were demonstrated. Later, these experiments were refuted; hence the experiments were irreproducible. In this case, the term irreproducible is applied in the larger context, that is, reproducibility of the concept as opposed to the actual experience (see Chapter 1: Introduction). Thus, these experiments constituted an unproductive irreproducibility, as the later demonstrated irreproducibility did not add to knowledge in their own right. In other words, there was no basis in the earlier and later experiments themselves to demonstrate which was in error and in what manner was the error made.

In Argument 4.8, there are at least two versions of the bridging term, *improves parkinsonism*, resulting in the Fallacy of Four Terms. There are many ways to improve parkinsonism, as attested to by the multiplicity of therapeutic actions, such as dopaminergic medications, anticholinergic medications, surgical ablations, genetic manipulations, DBS, and electroconvulsive therapy. This means that merely observing improved parkinsonism provided no evidence in support of one mechanism of action compared to any other. This is a consequence of the Inverse Problem, where there are multiple causes for the same phenomenon and thus any changes in the phenomenon cannot be attributed to any particular cause without appealing to information independent of the specific phenomenon. Argument 4.9 likewise is at risk for the Fallacy of Four Terms with two versions of the bridging terms: *DBS in animals* and *DBS in humans*. However, the Epistemic Risk of the Fallacy of Four Terms depends on the Epistemic Distance and Degrees of Freedom between animal and human species and it is an open question as to how great the Epistemic Risk is.

Closer examination of the actual experiments (Montgomery, 2012) used to create the major premise in Argument 4.8 shows these experiments involved in the injudicious use of the Fallacy of Four Terms. At the time, it was not possible for the reporting investigators to record neuronal activity during DBS. Rather, they recorded neuronal activity immediately after discontinuation of DBS and showed reduced neuronal activities. The logical structure of their experiment is shown in Argument 4.9. However, the major premise has to be refined to *DBS in the brain of animals* (bridging term) *results in reduced neuronal activities in the stimulated target recorded immediately after cessation of DBS* (major term). The presumption is that the major term in the major premise would be the same as the major term in the conclusion. Clearly, it is not. Neuronal activity in the major premise was neuronal activity after DBS, while neuronal activity in the conclusion is presumed to be during DBS and they are not equivalent. Thus, there are four terms; in this case, two versions of the major term and the argument are invalid.

Subsequent research demonstrates that what happens immediately after DBS is not equivalent to what occurs during DBS (Montgomery and Gale, 2008).

It is an unfortunate fact of history that fallacious Arguments 4.8 and 4.9 received support from the findings of another set of experiments also involving the injudicious use of the Fallacy of Four Terms. In the early 1980s, intravenous illicit drug users unknowingly injected themselves with the neurotoxin n-methyl-4-phenyl-1,2,3,6-tetrahydropyridine (MPTP), producing irreversible parkinsonism and leading to a powerful animal model of parkinsonism. In the hands of some investigators, the induction of parkinsonism with MPTP produced increased neuronal activity in the globus pallidus and subthalamic nucleus from which the Globus Pallidus Interna Rate Theory of parkinsonism was developed. This theory posited increased neuronal activity in the globus pallidus interna as shutting down neuronal activities in the thalamus, which subsequently reduced neuronal activities in the motor cortex, leading to the slowing of most and absence of other movements seen in patients with Parkinson's disease. These experiments gained additional support by consilience with studies using neurotoxin 6-hydroxydopamine.

ARGUMENT 4.10

Major premise:	*MPTP−parkinsonism* (bridging term) *have increased neuronal activities in the globus pallidus interna* (major term)
Minor premise:	*Humans* (minor term) *have parkinsonism*
Conclusion:	*Humans with parkinsonism* (minor term) *have increased neuronal activities in the globus pallidus interna* (major term)
Inference conclusion:	*Reduction of increased neuronal activities in the globus pallidus should improve parkinsonism*
Inference conclusion and inference:	*Pallidotomy improves parkinsonism by reducing neuronal activity and, via* Argument 4.8, *so does DBS of the globus pallidus interna*

Interestingly, studies published in 1986 (Montgomery et al., 1986) and largely ignored until replicating experiments in 2009 (Muralidharan et al., 2016) demonstrated that administration of MPTP in nonhuman primates did not produce the changes in the globus pallidus interna or subthalamic nucleus postulated by the theory. Indeed, studies comparing neuronal activity in the subthalamic nucleus in patients with Parkinson's disease to those with epilepsy show no difference (Montgomery, 2008) in contradiction to the Globus Pallidus Interna Rate theory.

Facts in the major premise of Argument 4.10 were vindicated by experimental evidence. The minor premise seems reasonable. Then how could Argument 4.10 fail to be reproducible? The answer may lie in the type of parkinsonism produced by the MPTP. Careful titration of MPTP administration just to the point of producing parkinsonism was not associated with changes predicted by the Globus Pallidus Interna Rate theory. Most likely, these changes occurred only when greater doses of MPTP were administered. Thus, the state of neuronal activities associated with "overdosing" animals on MPTP (or 6-hydroxydopamine) is not representative of human parkinsonism. The bridging terms in Argument 4.10 in the major and minor premises are not equivalent and Argument 4.10 is the Fallacy of Four Terms.

The counterargument is that there are just so many neuronal studies in MPTP- or 6-hydroxydopmine-induced parkinsonism in laboratory animals demonstrating neuronal activity changes consistent with the Globus Pallidus Rate theory that surely the theory must be true. Such a position would mean that scientific truth is subject to majority vote, something that most scientists, other than postmodernism-inclined scientists, would disavow. Yet, the actions of many scientists seem to presuppose this to be true by their continued invocation of the Globus Pallidus Interna Rate theory and its variants. It may be because of the concept of correlation implicitly held by these scientists. The fallacies of correlations are discussed later under the topic of Induction and Mill's Method of Induction.

To be sure, there are those who argue that the Globus Pallidus Interna Rate theory and its variants will end up on the "ash heap of history"—eventually. Unproductive irreproducibility is just a symptom of the normal scientific process, the price to be paid for future advances. Thus, the real and only legitimate concern is irreproducibility in the narrowest sense when the exact same experiments are found irreproducible (see Chapter 2: The Problem of Irreproducibility). However, this position smacks of the Solipsism of the Present Moment and a postmodernist notion of what constitutes science. Further, experimentation does not take place in a vacuum free of social, political, economic, and moral costs and lost alternative opportunities. It would only be in the absolute absence of cost that one might afford the notion of "science for science's sake." In the case of the Globus Pallidus Interna Rate theory, it has been nearly 30 years and, despite repeated evidence of its conceptual irreproducibility, the theory continues to be the motivator of research and explainer of the results. How much further would biomedical research in the area of disorders of the basal ganglia—thalamic—cortical system, such as parkinsonism, be had the fallacy of the Globus Pallidus Interna Rate theory not exhausted resources, effort, and, perhaps most importantly, imagination?

In this author's opinion, fallacious Argument 4.8 was taken as proof and the notion that high-frequency DBS inhibits activities was taken as fact. Yet, at the time there was no factual evidence. Nonetheless, the notion had the practical effect as a fact in that it was very difficult for researchers who offered other alternatives with either equivalent or superior explanatory power were prevented from getting grant funding and publication in biomedical research journals. These kinds of fact, generated by such fallacious arguments, are termed here quasi-facts.

The Fallacy of Four Terms is particularly potent in the generation of hypotheses and is analogous to the notion of metaphor. The centrality of hypothesis generation for biomedical research is discussed in Chapter 6: Causation, Process Metaphor, and Reductionism. Introduced here briefly is the role of metaphor in hypothesis generation. The general form of a metaphor contains a target domain and a source domain. The target domain is the hypothesis that derives its degree of certainty from the source domain. In general, the logical form is formulated as *a relates to b as c relates to d*. Note that if *b* and *c* are the same (*b/c*), then the general form becomes *a relates to b/c as b/c relates to d*—this is the form of the syllogistic deduction where *a* is the major term, *d* is the minor term, and *b/c* is the bridging term. If *relates* to were changed to *is* or $=$, a state-of-being linking verb, the metaphor would become the Principle of Identity. As will be discussed later, this form in propositional logic is *if a implies b/c is true and b/c implies d is true, then a implies d is true*, the Principle of Transitivity. If *b* and *c* are not the same, then the general form *a relates to b as c relates to d* is the Fallacy of Four Terms, where *a* is the major term, *d* is the minor term, and *b* and *c* are two versions of the bridging term. This form is also called the Fallacy of Pseudotransitivity.

An example of a very successful use of metaphor, in other words, the Fallacy of Four Terms and Fallacy of Pseudotransitivity, was the work of Mary Broadfoot Walker in 1934. She observed that patients with myasthenia gravis (a condition that causes weakness) were similar to patients treated with curare to treat tetanus, a disease manifest by persistent muscle contractions. The effects of curare could be reversed by administering physostigmine, a cholinesterase inhibitor. The experiment is of the form shown in Argument 4.11.

ARGUMENT 4.11

Major premise:	*Weakness* (bridging term) *is improved with physostigmine* (major term)
Minor premise:	*Patients with myasthenia gravis* (minor term) *have weakness* (bridging term)
Conclusion:	*Patients with myasthenia gravis* (minor term) *improved with physostigmine* (major term)

Critical to the predictive success of the hypothesis generated in Argument 4.11 is the similarity between the weakness of myasthenia gravis and that induced by curare. It is not exactly clear why Walker would select weak patients with myasthenia gravis compared to patients with weakness due to muscular dystrophy or even stroke. There was some discussion that myasthenia gravis was caused by some circulating factor that interfered with neuromuscular transmission—hence a possible similarity to curare that circulates in the blood following an injection. Nevertheless, Argument 4.11 proved to be valid, whereas Argument 4.8 turned out not to be so successful.

An interesting historical aside was Walker's demonstration of the Walker effect in patients with myasthenia gravis. A blood pressure cuff was applied to the forearm and inflated to render the arm ischemic as the muscles of the arm exercised. Once the blood pressure cuff was deflated, muscles elsewhere demonstrated increased weakness. Dr. Walker cited this as evidence that myasthenia gravis was caused by a circulating factor that built up during the exercise that, once released, exacerbated the mechanisms by which myasthenia gravis caused weakness (Johnston, 2005). Perhaps it was Walker's knowledge of curare that became the source domain for the target domain—a circulating factor for myasthenia gravis. From the perspective of understanding the pathophysiology of myasthenia gravis, it was unhelpful, but interestingly, the Walker effect was used as a diagnostic aid for myasthenia gravis. In a sense, the Epistemic Distance and Degrees of Freedom between the target and the source domains were not too large in the metaphor regarding physostigmine treatment, but too far for the notion of a circulating causal factor in myasthenia gravis.

PARTIAL, PROBABILITY, PRACTICAL, AND CAUSAL SYLLOGISMS

The Fallacy of Four Terms is a variant of the valid syllogistic deduction, and the utility of the Fallacy of Four Terms for hypothesis generation has been described previously. Other variants have similarly sacrificed certainty for increased utility. Aristotle recognized the impracticality of the valid syllogistic deduction and offered the practical syllogism, an example, Argument 4.12, of which is shown here:

ARGUMENT 4.12

Major premise:	*Persons who are hungry* (bridging term) *should eat* (major term)
Minor premise:	*Socrates* (minor term) *is hungry* (bridging term)
Conclusion:	*Socrates* (minor term) *should eat* (major term)

Argument 4.12 is invalid, as valid arguments do not result in false conclusions. Socrates may not eat even when hungry. Whence then the invalidity of Argument 4.12? The problem is that the linking verbs are not state of being, rather they are action verbs, signaling intention, such as *should*. From a logical perspective, if *should eat* is replaced by *is eat*, then the syllogism would be valid in form but invalid in conclusion because a state-of-being linking verb is not appropriate. What would be the sense of the premise *persons who are hungry* (bridging term) are eat (major term)?

Using a state-of-being verb, *is* or *are*, enforces the Principle of the Excluded Middle by which deduction obtains its certainty. Linking verbs, such as *should*, cannot compel a state consistent with the Principle of the Excluded Middle. As such, syllogisms with linking verbs other than state-of-being verbs that do not compel the Principle of the Excluded Middle are not valid syllogistic forms but rather represent a variant. As will be seen, such practical syllogisms are very useful, particularly as they inform hypothesis generation (see Chapter 6: Causation, Process Metaphor, and Reductionism). Indeed, in biomedical research, the most critical linking verb used in syllogistic argumentation, the experiment, is *cause* (see Chapter 6: Causation, Process Metaphor, and Reductionism). Variants of syllogistic deduction where the linking verb is *cause* will be referred to as causal syllogisms. As will be seen, the notion of *cause* will become the bridging term in the major and minor premises in arguments that will drive biomedical research. The degree that *cause* as the bridging term is exactly the same in the major and minor premises will determine the Epistemic Risk in entertaining the argument.

This epistemic issue with *cause* as the logical linking verb was well understood by the ancient Greek philosophers such as Aristotle in the Principle of Causational Synonymy. This principle held that whatever mechanisms are in the *cause* must likewise be present in the *effect*. Thus, the notion of *cause* is exactly the same in the major and minor premises and thus no risk for the Fallacy of Four Terms. For example, consider how a hand in a pool of water can move the water. Note that it cannot be just said that the moving hand moves the water as the two are very different except at one level—the level of causational synonymy. It is by the electrons in the outer orbit of the atoms on the surface of the hand that repel the electrons in the outer orbit of the atoms that make up the water molecule that the hand is able to move the water.

Another variant of the syllogistic deduction is the syllogism shown in Argument 4.13:

ARGUMENT 4.13

Major premise:	*All persons who smoke tobacco* (bridging term) *will die of cancer or heart disease* (major term)
Minor premise:	*John Doe* (minor term) *is a smoker of tobacco* (bridging term)
Conclusion:	*John Doe* (minor term) *will die of cancer or heart disease* (major term)

While Argument 4.13 is of a valid form, note that the linking verbs compel the Principle of the Excluded Middle; however, the conclusion is not necessarily true. John Doe may die in an automobile accident. The argument is invalid because valid arguments result in true conclusions. The invalid argument likely is because the major premise is not true. Smokers of tobacco can die by being run over by an automobile. Argument 4.13 can be rehabilitated by reframing as in Argument 4.14, which is called a partial syllogism:

ARGUMENT 4.14

Major premise:	*Some persons who smoke tobacco* (bridging term) *will die of cancer or heart disease* (major term)
Minor premise:	*John Doe* (minor term) *is a smoker of tobacco* (bridging term)
Conclusion:	*John Doe* (minor term) *may die of cancer or heart disease* (major term)

Reframing Argument 4.13 results in Argument 4.14 being less refutable (less irreproducible), for example, by finding a single smoker who died as a consequence of an automobile accident does not help. Considered in Set Theory, the Venn diagrams of Argument 4.14 are shown in Fig. 4.2. In this case, the set of *smokers of tobacco* is a set, but not wholly contained within the set of all those *dying of cancer or heart disease*. While it is clear that the set *John Doe* is wholly contained in the set of *those who smoke tobacco*, it is not clear whether *John Doe* is with that part of the set of *those who smoke* that is also contained within the set of those *who will die of cancer or heart disease* or outside that set.

The partial syllogism can be rehabilitated to provide greater utility by reframing Argument 4.14 as Argument 4.15 described here:

ARGUMENT 4.15

Major premise:	*60% of persons who smoke tobacco* (bridging term) *will die of cancer or heart disease* (major term)
Minor premise:	*John Doe* (minor term) *is a smoker of tobacco* (bridging term)
Conclusion:	*John Doe* (minor term) *has a 60% chance of dying from cancer or heart disease* (major term)

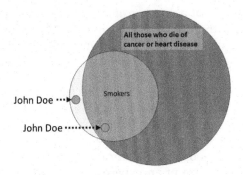

FIGURE 4.2 Venn diagram representation of Argument 4.14. The set of *smokers of tobacco* is not wholly contained within the set of *those who will die of cancer or heart disease*. Some elements in the set of *smokers of tobacco* may be in the regions outside the set of *those who will die of cancer or heart disease*. *John Doe* will be within the set of those who *smoke tobacco* but it is not clear whether this is in the region that also is within the set of *those who will die of cancer or heart disease*. It is possible that *John Doe* lies in the region outside the set of *those who will die of cancer or heart disease* and thus the conclusion may be wrong. A valid argument cannot result in a conclusion that is wrong, hence the argument is invalid.

This form of a partial syllogism provides a probability for the qualifier *some* in the major premise of Argument 4.14. This translates to a quantification of the conclusion of Argument 4.15 being true. While one cannot know whether John Doe actually will die of cancer or heart disease, the probability alone may have sufficient power for John Doe to alter his smoking habits. This quantified version of the partial syllogism is called a probability syllogism.

Since the early 1800s there has been an attempt to quantify intuitions made by physicians and biomedical researchers (Matthews, 1995). These efforts were met with considerable antagonism. Certainly, there were reasonable concerns about inferring from summary statistics of a sample (a subset of the population of interest) to the individual on the part of clinicians. Biomedical researchers objected because purely probabilistic inferences did not rely on or make inferences to underlying mechanistic or causal accounts. Further, anyone with a paper and pencil or computer could render judgments comparable to the most intuitive clinician or researcher by calculating from probabilities alone. The latter concern probably was the most operative because even as probability and statistics became more commonplace, clinicians and biomedical researchers then held themselves as unique interpreters of the quantified data, perhaps to forestall interloping statisticians.

Probability can be established primarily by simple enumeration. For example, a group of observations, n, is obtained and the number of instances of a specific property, s, is determined. The probability of s in the group, p, simply becomes $p = s/n$. But what if this process was repeated? It is likely that a number of different values of p will be demonstrated. Which is the right value of p? The answer to this question is what engendered the discipline of statistics. Going back to the early 1800s, there was great debate about how large an n was required to have any confidence in the value of p. Indeed, the Large Number Theorem holds that with increasing n, the value of p will approach a constant number, but it will not be a single number. In other words, the probability of p given n is not guaranteed to be the same number p given $n + 1$. The role of statistics, in large part, is to give some degree of confidence in the value determined for the probability. For example, the mean (average) effect size of an experimental manipulation is a probability. The standard deviation around the mean (average) is to provide some certainty about the effect size probability. One knows that 95% of all the possible values of the probability, p, will be within plus or minus 1.96 standard deviations, corresponding to a 95% confidence interval.

The use of probability, vouchsafed by statistics, follows from the probability syllogism. As will be seen, a great many errors in statistical analyses of biomedical research experiments follow from logical errors in the variant of the partial syllogism. Methods used to improve the confidence of statistical analyses, and thus reproducibility, will follow from constructing a partial syllogism to be as close to the valid form of syllogistic deduction as possible. Just relying on statistical processing software may not discover the underlying logical failure. Hence, it is important that the biomedical researcher appreciates the logical structure fundamental to the probability and statistical techniques to be employed. These issues are addressed in greater detail in Chapter 5, The Logic of Probability and Statistics.

PROPOSITIONAL LOGIC

In many ways, syllogistic deduction is equivalent to many forms of propositional logic—the latter is of the form *if a implies b is true and a is true, then b is true* (termed modus ponens) or *if a implies b is true and b is false, then a is false* (modus tollens). For example, the Set theoretic version of modus ponens can be construed as *b is a subset of a* or *everything that is true of a also is true of b*. Thus, given that *a* is true, then *b* must be true. Similarly, modus tollens can be construed as *b is a subset of a* or *everything true of a also is true of b, therefore if b is false, then a must be false.* In using modus tollens, one must take care not to commit the Fallacy of Denying the Antecedent, which is of the general form *if a implies b is true and not a is true* (*a* is claimed to be false), *then b is false; b still may be true.*

The great advantage of propositional deductive logic is that it contains logical operators, represented as symbols, that carry out specific transformations to allow very complex and extended arguments. Examples of such logical operators include negation (or *NOT* function), conjunction (such as *a AND b*, represented by *a* ∧ *b*), and disjunction (such as *a OR b*, represented as *a* ∨ *b*). Each operator can be defined in a truth table where every permutation of values for the premises and the conclusion of the argument is defined (see appendix A). For example, in the operator *a AND b*, there are four permutations of possible values of *a* and *b*, *T* (true) and *T*, *T* and *F* (false), *F* and *T*, and *F* and *F*, for *a* and *b*, respectively. The logical statement *a AND b*, (*a* ∧ *b*), has four possible values, one for each permutation of the individual values of *a* and *b*. In this case, the value of *a AND b*, (*a* ∧ *b*), is *T* only when *a* and *b* are both *T*, otherwise *a AND b*, (*a* ∧ *b*), is *F*.

Like syllogistic deduction, propositional logic also conforms to the Principle of the Excluded Middle. Arguments of true premises and valid propositional logic are, in essence, tautologies that do not provide new knowledge except for the recognition that the proposition is a tautology. Like the variants of syllogistic deduction, propositional logic only provides new knowledge with the judicious use of variations on propositional logic that are logical fallacies.

Perhaps one of the most remarkable logical fallacies is the modus ponens form of the Scientific Method, the cornerstone of science since Francis Bacon and institutionalized with the founding of the Royal Society in 1660 and the British Association for the Advancement of Science in 1831, particularly with the influence of William Whewell who championed Bacon's approach to science. One typical description of the Scientific Method is (1) identify a problem or gap in knowledge; (2) review the known knowledge; (3) construct a conceptual hypothesis; (4) derive a testable hypothesis, prediction, from the conceptual hypothesis; (5) test the prediction; (6) if the prediction fails, revise the conceptual hypothesis; and (7) if the prediction is true, take the conceptual hypothesis as true.

The Scientific Method can be framed as a logical chain of events that branch at the point when the truth or falsehood or the prediction is determined. This results in the logical arguments shown:

ARGUMENT 4.16

If a (hypothesis) *implies b* (prediction) *is true and b* (prediction) *is false, then a* (hypothesis) *is false.*

ARGUMENT 4.17

If a (hypothesis) *implies b* (prediction) *is true and b* (prediction) *is true, then a* (hypothesis) *is true.*

Argument 4.16 which is the modus tollens form of a valid propositional logic argument. The conclusion will be true no matter what is substituted for premise *a* or *b*. This is referred to as the modus tollens form of the Scientific Method. Despite the absolute certainty of the modus tollens form of the Scientific Method, it is rarely employed.

Argument 4.17 appears to be quite reasonable and indeed is the form that the vast majority of biomedical experiments are constructed. The argument appears to gain its credibility by its similarity to the modus ponens form of propositional logical certainty. But it is not modus ponens. A modus ponens form would be as in Argument 4.18:

ARGUMENT 4.18

If a (hypothesis) *implies b* (prediction) *is true and a* (hypothesis) *is true, then b* (prediction) *is true.*

But experimentation is rarely, if ever, designed to demonstrate the presence or absence (truth or falsehood) of the hypothesis but rather of the prediction, outside of futility studies. The modus ponens-like form of the Scientific Method is the logical Fallacy of Confirming the Consequence, where the prediction is the consequence of the experiment. The problem with the Fallacy of Confirming the Consequence is that the prediction may be true regardless of the truth or falsehood of the hypothesis. For example, consider Argument 4.19:

ARGUMENT 4.19

If dopamine depletion (hypothesis) *implies that the experimental animal displays parkinsonism* (prediction) *is true and the experimental animal displays parkinsonism* (prediction or consequence) *is true, then dopamine depletion* (hypothesis) *is true.*

Studies have shown that Argument 4.19 may be false. For example, one of the first animal models of parkinsonism was exposure to carbon monoxide or carbon disulfide, which displaces oxygen from hemoglobin. The consequence was anoxia with a predilection to cause damage in the globus pallidus. Thus, the finding of the prediction, that the *animals display parkinsonism*, does not in itself guarantee that *dopamine depletion*, the hypothesis, was the cause.

It may well be that dopamine depletion does cause parkinsonism, but in an experiment where the only outcome is the presence or absence of parkinsonism, the presence of parkinsonism does not guarantee that dopamine depletion was the cause. If the experiments were repeated, it cannot be assured that if dopamine levels were assayed that they would be reduced. Even if they were reduced, the possibility exists that the reduction in dopamine was only coincidental or epiphenomenal, not causal, in which case the premise *a* (dopamine depletion) *implies b* (parkinsonism) cannot be established with certainty. Consequently, there will be risks of unproductive irreproducibility.

The Epistemic Risk of unproductive irreproducibility depends on the Epistemic Distance and Epistemic Degrees of Freedom between the Fallacy of Confirming the Consequence and the valid modus ponens form of the argument. In the case of parkinsonism, the Epistemic Risk is proportional to all potential alternative causes of parkinsonism beyond dopamine depletion. The Fallacy of Confirming the Consequence can be rehabilitated by reframing, as shown in Argument 4.20:

ARGUMENT 4.20

If and only if a (hypothesis) *implies b* (prediction) *is true and b* (prediction) *is true, then a* (hypothesis) *must be true.*

In terms of the production of parkinsonism in Argument 4.19, if all alternatives to dopamine depletion have been excluded, then dopamine depletion must be true in the presence of parkinsonism. Note that it is not sufficient to reframe Argument 4.19, as shown here in the general form of Argument 4.21:

ARGUMENT 4.21

If (a [hypothesis] *OR* c* [alternative hypothesis$_1$] *OR * d* [alternative hypothesis$_2$]) *implies b* (prediction) *is true and b* (prediction) *is true and c is false and d is false, then a* (hypothesis) *must be true.*

Argument 4.21 still is the Fallacy of Confirming the Consequence and *a* still may be false even though *b* is true.

Argument 4.21 just looks like it is true, but to believe that Argument 4.21 is true is to fall victim to the Fallacy of Limited Alternatives. It just seems that eliminating the alternative hypotheses must mean that the remaining hypothesis must be true. Another version of the Fallacy of Limited Alternatives is that if one alternative is found true, then it is falsely inferred that the others must be false (the Fallacy of Affirming a Disjunct). However, the type of *OR* function must be specified. The *OR* * given in Argument 4.21 is the *inclusive OR* function. In this case, *if (a OR * c OR * d) is true,* then at least one among *a, c,* and *d* must be true, although any and all may be true. This means that the truth or falsity of any *a, c,* or *d* does not influence the truth or falsity of the others. The Fallacy of Limited Alternatives, perhaps inappropriately, presupposes an *exclusive OR* function where only one, *a, c,* or *d,* can be true and if one is true, then the others must be false. However, this does not mitigate the Fallacy of Confirming the Consequence, as elimination of *c* or *d* from the argument still leaves the fallacy. As will be seen, translation of the Fallacy of Limited Alternatives, such as Argument 4.21 to probability, becomes the Gambler's Fallacy (see Chapter 5: The Logic of Probability and Statistics).

Because of its symbols representing premises, logical operations, and combinatorics, propositional logic has been extraordinarily useful in constructing valid complex arguments. For example, every digital computer can be represented as vast complex networks of the logical *NAND* operator. The proposition *a NAND b* is true only when both *a* and *b* are false. Thus, anything any digital computer can do is an operation of propositional logic fundamentally (sometimes referred to as Boolean logic after George Boole).

An important combination of premises and logical operators is the Principle of Transitivity, which is of the form *if a implies b is true and b implies c is true, then a implies c* (Fig. 4.3). This is analogous to the mathematical principle where *if a = b and b = c, then a = c*. Again, the operations are a tautology and generate no new knowledge other than the realization that *a* and *c* are the same thing. However, the variant of the valid Principle of Transitivity is the Fallacy of Pseudotransitivity, which is of the form *if a implies b is true and c implies b is true, then a*

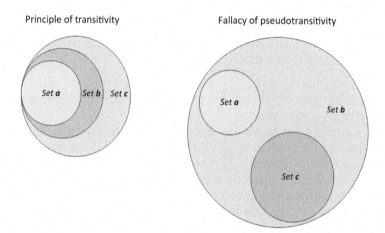

FIGURE 4.3 Venn diagram representation of the Principle of Transitivity and the Fallacy of Pseudotransitivity. The Principle of Transitivity is of the form *if a implies b and b implies c, then a implies c*. Set *a* is contained within set *b*, and set *b* is contained in set *c*. Hence, set *a* is contained within set *c* and thus the proposition is valid. The Fallacy of Pseudotransitivity is of the form *if a implies b and c implies b, then a implies c*. Set *a* is contained within set *b*, and set *c* is contained in set *b*. However, set *a* is not necessarily contained within set *c* and thus the proposition is invalid. It cannot be assured that any element of set *a* will be within set *c*.

implies c is true—this is a false argument. This can be represented in Set theory through the Venn diagrams shown in Fig. 4.3. As can be appreciated, the Principle of Pseudotransitivity is analogous to a metaphor, which is analogous to the Fallacy of Four Terms. Therefore, all the benefits and risks of metaphor and the Fallacy of Four Terms described previously are also true of the Principle of Pseudotransitivity as noted in the discussion of curare weakness and myasthenia gravis given earlier.

INDUCTION

Induction is another form of logic. Whereas in deduction the typical logical flow is from generalities to particulars, induction generally flows from particulars to generalities. It would seem that induction is an important contributor to deduction and to its variants, which is critical to successful and reproducible biomedical experimentation as described earlier. In a sense, induction often generates the hypotheses that are then formulated into experiments. Indeed, many argue, particularly of Baconian science, that induction is central to the scientific enterprise. However, as will be described, it is not so easy and it is in those difficulties that the potential for error and where unproductive irreproducibility may lie.

The most typical form of induction is by enumeration. Multiple observations are gathered based on some set of dimensions and then a different dimension becomes apparent (emerges). A nonbiomedical example is provided for the sake of clarity; however, here and in other chapters, a number of biomedical examples will be offered. A group of ravens is assembled and, lo and behold, they all are black in color. The induction is created as *because all ravens seen are black, then all ravens are black*. The particulars are the set of *ravens observed*, the generalization is that *all ravens are black*. Seems reasonable enough but reframe the induction as a syllogism in Argument 4.22 that follows:

ARGUMENT 4.22

Major premise: *Among the set of ravens observed* (bridging term), *all are black* (major term)

Minor premise: *An unseen bird* (minor term) *is a raven* (bridging term)

Conclusion: *The unseen bird* (minor term) *will be black* (major term)

Interestingly, just positing the minor premise introduces Hume's problem of Induction, from David Hume. How can any unobserved possibility constitute any level of evidence or knowledge or some role in argumentation? Also, Argument 4.22 is at risk for the Fallacy of Four Terms if there are two versions of the bridging term. The first version in the major premise is true by direct (scientific) observation. It is not clear that the bridging term in the minor premise is the same as in the major premise, at least epistemically. Perhaps it is an albino raven that is unseen.

Epistemic Risk can be considered in terms of Epistemic Distance, which in this case relates to how close the number of ravens is observed to the total of all possible ravens. If the set of those observed included every possible raven, then there would be no Epistemic Distance, no Epistemic Risk, and no Fallacy of Four Terms. However, even if it were possible, at least in theory, to observe every possible raven to avoid the Fallacy of Four Terms, how would one know that a sufficient number of ravens were observed? The presumption that one has observed all possible ravens invites the Fallacy of Induction because one cannot know *a priori* whether a nonblack raven exists, unbeknownst to the observer.

Another logical problem with induction is that it may beg the question where the conclusion is already presupposed by the argument, the Fallacy of *Petitio Principii*. For example, if the criterion for assembling the set of observed ravens in Argument 4.22 is that all the observed ravens are black, then the conclusion is trivially true and it is a tautology analogous to saying $a = a$. No new knowledge is obtained. For new knowledge, the criteria, dimensions, used to assemble the set of ravens to be observed must not include their color. The color as a qualifier for being a raven must emerge independently.

But what other observations, independent of color, are used to organize the set of observed ravens? It must be more than having a beak, as other birds have beaks, and similarly for any number of descriptions of birds other than color. Rather, it must be some specific form of beak, feet, or wing that allows the collection of a specific set of observed birds that constitute ravens. Yet, each of these specific forms is a generalization from previous inductions that are subject to the same question. The result is a potential infinite regressive sequence of inductions. This constitutes the *A Priori* Problem of Induction.

A single example here is offered to illustrate the concern. A series of clinical trials on the safety and efficacy of dopamine agonists for the treatment of Parkinson's disease were published in the 2000s. A side arm of those studies involved imaging the dopamine transporter or reuptake systems in neurons of the basal ganglia in an attempt to quantify the number of dopamine neurons in a particular subject. The presumption was that all subjects, all of whom were diagnosed clinically as having idiopathic Parkinson's disease, would have diminished dopamine transporter or reuptake ligand binding detectable on single photon emission computerized tomography or positron emission tomography scans. Surprisingly, approximately 15% of the subjects did not have evidence of dopamine neuron loss.

What to call these subjects? An ad-hoc diagnosis of Symptoms Without Evidence of Dopamine Deficiency (SWEDDs) was made. A group of scientists argued that these patients could not have idiopathic Parkinson's disease, despite the fact that a diagnosis of idiopathic Parkinson's disease was made in the affirmative based on well-established clinical criteria by experts. The argument went from Argument 4.23 to Argument 4.25 as shown here:

ARGUMENT 4.23

Major premise: *Clinically idiopathic Parkinson's disease* (bridging term) *is associated with dopamine depletion detectable on dopamine transporter or reuptake scanning* (major term)

Minor premise: *A set of subjects* (minor term) *has clinically diagnosed idiopathic Parkinson's disease* (bridging term)

Conclusion: *The set of subjects* (minor term) *will be associated with dopamine depletion detectable on dopamine transporter or reuptake scanning* (major term)

Biomedical experimentation demonstrated the conclusion false. The question then becomes the source of falsehood in the conclusion, either an untrue premise or an invalid proposition. Clearly, the minor premise is true as this was a consequence of the structure of the experiments. The falsehood or invalidity must lie in the major premise. But it cannot be in the bridging term such that clinically diagnosable idiopathic Parkinson's disease in the major premise is different from that in the minor premise. This would be an inconsistency unallowable in logic. Rather, the major premise itself must be invalid, and *clinically idiopathic Parkinson's disease* (bridging term) *is **not necessarily** associated with dopamine depletion detectable on dopamine transporter or reuptake scanning* (major term). Argument 4.23 could be reframed as a partial syllogism, as described earlier, which would be of little help. The partial syllogism could be converted to a probability syllogism as shown in Argument 4.24:

ARGUMENT 4.24

Major premise: *Clinically idiopathic Parkinson's disease* (bridging term) *is associated with dopamine depletion detectable on dopamine transporter or reuptake scanning 85% of the time* (major term)

Minor premise: *A set of subjects* (minor term) *has clinically diagnosed idiopathic Parkinson's disease* (bridging term)

Conclusion: *The set of subjects* (minor term) *has an 85% probability of being associated with dopamine depletion detectable on dopamine transporter or reuptake scanning* (major term)

Interestingly, a large number of biomedical scientists were unwilling to accept Argument 4.24 and insisted on Argument 4.25:

ARGUMENT 4.25

Major premise:	*Clinically idiopathic Parkinson's disease and evidence of dopamine depletion* (bridging term) *are associated with dopamine depletion detectable on dopamine transporter or reuptake scanning* (major term)
Minor premise:	*A set of subjects* (minor term) *has clinically diagnosed idiopathic Parkinson's disease and evidence of dopamine depletion* (bridging term)
Conclusion:	*The set of subjects* (minor term) *will be associated with dopamine depletion detectable on dopamine transporter or reuptake scanning* (major term)

Argument 4.25 is circular or tautological. Its validity is solely by the construction (definition) of the major and minor premises. In other words, it is a self-fulfilling prophecy.

What to do with SWEDDs? These scientists insisted that subjects with SWEDDs did not have idiopathic Parkinson's disease, and some of the world's leading Parkinson diagnosticians were incorrect. Alternatively, it is possible that from observations of a group of SWEDDs that an individual property would emerge in retrospect that prospectively justifies excluding SWEDDs from the group of subjects with idiopathic Parkinson's disease, but this rationality cannot be the basis of the dopamine transporter or reuptake scans without inviting circularity.

The only experimental proof of the aforementioned position would be to prospectively assemble a group of subjects with SWEDDs and a group of subjects with idiopathic Parkinson's disease and test for the absence of a dopamine transporter on a reuptake scan. But there does not appear to be any criteria, other than the scan results, by which to differentiate and assemble these two groups; hence the *A Priori* Problem of Induction.

Those scientists who rejected Argument 4.24 likely did so because the argument not only appears to ignore causation but suggests that a notion of causation is unnecessary. This is precisely the objections of medical physiologists in the early 1800s, such as the highly lauded Claude Bernard, to the introduction of probability and statistics in biomedical research (Matthews, 1995). This resistance likely finds parallels in the resistance by many current clinicians to Evidence-Based Medicine as it is held synonymous to randomized controlled trials. If a randomized controlled trial demonstrated a statistically significant difference in favor of administrating a ground-up kitchen sink over a placebo, there would be no rationale within Evidence-Based Medicine to discourage the use of a ground-up kitchen sink; rather, extending the logic of Evidence-Based Medicine, its use should be encouraged, assuming the absence of better alternatives, which does not include doing nothing.

In most instances of classical induction, the generalization induced from a set of particulars must conform to the Principle of the Excluded Middle. What would be the strength of the induction *because all ravens observed have been black then all ravens are black* be if the induction was reframed to *some ravens observed have been black then all ravens are black* or *all ravens observed have been black then some ravens are black*? The latter reframed induction may not be so problematic in a *posteriori manner*. It just is the case that a subsequent set of observed ravens are black. But this is not as helpful to science as would a prediction of or *a priori* knowledge that *some ravens are black*.

Given the Principle of the Excluded Middle, it takes only a single example of a nonblack raven to disprove the induction that *because all ravens observed have been black then all ravens are black*. It does not matter that 1,000,000 observed ravens were black, as the demonstration of a single nonblack raven is sufficient to disprove the induction. The induction *because all ravens observed have been black then all ravens are black* can be salvaged by reframing to *all ravens observed have been black then the vast majority of all ravens are black*. However, to insist that the induction *because all ravens observed have been black then all ravens are black* is to be held correct is to fall victim to a logical fallacy, as there is the possibility of a contradiction (irreproducibility) in finding a single nonblack raven, which is disallowed. This fallacy renders biomedical science subject to a majority vote as described previously.

Scientific inductions can create generalizations that become the axioms of a biomedical logical argument that serves to construct hypotheses and establish experiment approaches. For example, the generalization *all ravens are black* can be used in constructing an experiment following the Scientific Method. It might be of the form *if all ravens are black implies this yet to be observed raven will be black is true, and the bird is found not to be black, then the proposition if all ravens are black implies this yet to be observed raven will be black is true* is false by modus tollens. Note that it may be false not because of the premise that *all ravens are black*, but rather in the implication that *this yet to be observed raven will be black*. The bird actually observed may not be a raven. This would be very helpful to the experimenter because she can go back and examine what it is that caused her to suspect that the *yet to be observed raven* should be a raven and not some other bird. Perhaps the original criteria need revision. The revised criteria become the new axiom of the experimental logic.

In addition to axioms for the logical system in the biomedical experiment, it is necessary to have rules of inference analogous to the logical operators described earlier in propositional logic. John Stuart Mill provided rules of inference in his methods of induction, which are rules for relating induced generalities in his *"A System of Logic"* (Vol. 1, 1843). These methods include the following.

Method of Difference:

If A, B, and C are associated with T, U, and V and if B and C (note *A* absent) *are associated with U and V* (note *T* is absent), *then A is associated with T.* The capital and bold letters represent generalizations for prior inductions. An example would be an experimental manipulation, *A*, in animals with cofounds *B* and *C*, resulting in a predicted effect, *T* and *U* and *V*, whereas the control group had only the confounds *B* and *C* and in whom only *U* and *V* were found. Therefore, manipulation *A* is associated with the predicted effect *T*.

Method of Agreement:

If A, B, and C are associated with T, U, and V and if A, D, and E are associated with T, X, and Y, then A is associated with T. This method can be seen as complementary to the Method of Difference.

Joint Method of Agreement and Difference:

The concomitant use of the Method of Difference and Agreement.

Method of Residual:

If A, B, and C are associated with T, U, and V and if B and C are known to be associated with U and V, then A is associated with T. However, this method assumes exclusive and independent interactions between *A* and *T*, *B* and *U*, and *C* and *V* (in the manner of an *exclusive OR* function described earlier), which may not be the case in a biomedical experiment.

Method of Concomitant Variation:

If A, B, and C are associated with T, U, and V and there is a change in A (ΔA) *and not in B or C and a change is found in T* (ΔT) *but not in U or V, then A is associated with T.* Note that the change in *A* (ΔA) and the change in *T* (ΔT) could be in very different directions. If the Principle of the Excluded Middle is applied to the variables, particularly *A* and *T*, then the only options are *A, NOT A (A), T,* and *NOT T (T)*. The Method of Concomitant Variation would reduce to the Methods of Difference and Agreement. Typically, the Method of Concomitant Variation is in the context of *A* and *T* being a multivalued parameter such as a continuous variable operating over the set of real numbers. In this case, the Method of Concomitant Variation is the statistical method of correlation.

The primary attraction of Mill's method was the potential to impute causality. In the examples given here, *A* stands in some causal relation to *T*. The notion of causality is difficult and is addressed in Chapter 6, Causation, Process Metaphor, and Reductionism. However, conformity to the Principle of the Excluded Middle in the first four methods provides for some certainty and, when combined with the long-standing Principle of Sufficient Reason, that is, things happen for a reason, the implication is that Mill's method provides some support for causal relations. The situation is more difficult for the Method of Concomitant Variation, which does not necessarily invoke the Principle of the Excluded Middle as it is a correlational analysis operating over a range of values.

The Method of Concomitant Variation is prone to the effects of epiphenomena. Consider the situation where A actually is casually related to Z and not T. However, Z is causally related to T. Then it will appear as though A is causally related to T. From the perspective of experiments involving A and T, there is no way to know if A is epiphenomenal to T and not causally related. The risk is when the putative relationship between A and T is then used to hypothesize a relationship between A and Q based on some relationship between T and Q that risks unproductive irreproducibility.

In a set of experiments, electrical signals (local field potentials) are recorded from a Deep Brain Stimulation (DBS) electrode placed in the subthalamic nucleus in patients with Parkinson's disease. The power of an approximately 20-Hz frequency in the electrical signal is measured where power is a measure of the content or amount of the signal at the 20-Hz frequency. A probability density distribution is created with the prevalence (probability) of persons when a specific range of power at the 20-Hz range is constructed. It is known that approximately 15% of patients will have no significant increased power in the 20-Hz signal. The correlation over the probability distribution function likely will result in a non-zero regression coefficient, which, according to the Mill's Method of Concomitant Variation, would argue for a causal role between the 20-Hz signal and Parkinson's disease. But what about the 15% of patients with Parkinson's disease who do not have increased power in the 20-Hz range? Do they have Parkinson's disease? The question is reminiscent of the argument described earlier of whether patients without evidence of dopamine depletion (SWEDDs) on scanning have Parkinson's or some other disease.

A stronger argument can be constructed that utilizes the Principle of the Excluded Middle to provide greater certainty. Based on this principle, one exception is sufficient to invalidate the claim. There is the philosophical notion of a necessary cause. For example, one hypothesizes that the presence of a 20-Hz signal in the electrical recordings is a necessary condition for Parkinson's disease. Thus, it takes only one patient not having increased power in the 20-Hz range to prove that the 20-Hz power is not a necessary condition. Could it be a sufficient condition such that the absence of increased power does not prevent the patient from having Parkinson's disease but the presence of increased power in the 20-Hz range is sufficient to produce Parkinson's disease? It turns out that increased power in the 20-Hz frequency range, such as by DBS, may not be associated with increased parkinsonism and therefore is not a sufficient condition. A condition that is neither necessary nor sufficient is an epiphenomenon, although a correlation will be found. The error would be to hold that the correlation means there is a causal relationship between power in the 20-Hz range in the brain signals and the manifestations of Parkinson's disease.

THE DUHEM–QUINE THESIS

The power of propositional logic is the ability to create long and complex logical arguments that, if valid, the conclusion is true. In that case, one knows that all the premises and propositions in the argument are true and valid. The problem is when the final conclusion is false or the prediction expected is not found. By modus tollens, at least one premise must be false or one proposition invalid. But which? The Duhem–Quine thesis holds that one cannot know which premise or proposition is at fault based only on the failure of the conclusion.

The problem is that virtually even the simplest of biomedical experiments entail a very long and detailed series of interlocking premises and propositions explicitly in the form of demonstrated axioms and rules of inference but also in the assumptions and, most importantly, the presuppositions (Ullian and Quine, 1978). Any one of them could be the cause of the failed conclusion. Consider the failure of a fetal dopamine cell transplant for the treatment of Parkinson's disease. Many investigators focused on the mode of administration, such as the catheters used to deliver the cells, the pattern of implantation, the timing of the cell transplantations relative to the stage of the disease, and the use of adjunctive cell transplants to provide nerve growth factor (Evans et al., 2012). Yet, many of these reasons seem inappropriate, given that imaging demonstrates a robust uptake of dopamine transporter scans suggesting cell survival and proliferation; indeed, many patients developed "run-away" dyskinesia of the kind seen with excessive dopamine from levodopa (and its conversion product dopamine) administration.

Curiously absent from the differential diagnosis of fetal transplant failure is the consideration that it just won't work any better than say lifting the lid of a broken computer and throwing transistors inside. Yet, the Duhem–Quine thesis demonstrates that the possibility that any such transplant will not work in any consistent manner is just as likely as any other reason within the confines of the actual experiment. One sometimes wonders whether a failure to consider the possibility that such transplantations just will not work is an example of willful ignorance. It may be that surgical implantation cannot replicate the precise phylogenetic and developmental structures of normal dopamine neuron interactions, then subsequent experiments likely will fail, a form of irreproducibility.

To be sure, there is a minority of patients with consistent improvement following fetal dopamine transplantation. However, until it is understood why these particular patients had maintained improvement, any attempts at replicating the study invite the Fallacy of Four Terms and unproductive irreproducibility. The critical reason for a better understanding of the particulars in those patients who did respond is to refine the selection criteria and to ensure that the bridging term in the major premise *patients with Parkinson's disease* based on past experiments is modified to *patients with Parkinson's disease and predictor X*. This way, the minor premise bridging term becomes *Parkinson's disease with predictor X* and the minor term, *future patient*, has the same bridging term. One can be reassured then that *future patients* will achieve the same improvement (major term) as those patients in the major premise.

SUMMARY

This chapter proceeded from the premise that knowledge that is absolutely certain will give rise to experiments that are reproducible. Deductive logic was the model for absolute certainty. However, the paradox is that valid deductive logic does not produce new knowledge. Rather, logical fallacies are required. However, the injudicious use of logical fallacies is more likely to give rise to unproductive irreproducibility that is not helpful. A number of variants or fallacies, such as the Fallacy of Four Terms, practical and partial syllogisms, the Fallacy of Confirming the Consequence, and the Fallacy of Pseudotransitivity, can provide new knowledge and hence are of greater utility, but at the expense of certainty and with an increased risk of irreproducibility. Methods to increase or at least assess the probability of reproducibility were reviewed. These included migrating from partial syllogism to probability syllogism. The criticality of sound hypothesis generation and the complicated role of Induction were introduced. The importance of causal syllogism, a derivative of practical syllogism, was introduced.

The Logic of Probability and Statistics

This chapter is not a discussion of probability and statistics per se. There is no discussion of specific statistical tests or methods. Instead, this chapter attempts an explanation of the logic that underlies the use of probability and statistics. Appendix B, Introduction to the Logic of Probability and Statistics provides an introduction to the concepts underlying probability and statistics for those not already familiar with these concepts. Many errors in the use of probability and statistics result from the injudicious use logical fallacies inherent in the use of probability and statistics. Users of statistical software packages in the ideal world are advised to attend to the necessary assumptions, reflecting the inherent logic, that underlie the proper use of a specific statistical method. Too often this does not happen. Even at that, any attention paid to the necessary assumptions often is in the form of a "cookbook" where the user does not appreciate the logical structure of the statistical method in order to understand the caveats as to the necessary assumptions. It is hoped here that by stressing the underlying logic, users of statistics may become more sophisticated in where and when the assumptions apply and where deviation from the assumptions risks unproductive irreproducibility.

Probability and Statistics were initially motivated by gambling and insurance, such as on merchant vessels. The modern origins arose from agronomy (Gigerenzer et al., 1989), but quickly found its way into biomedical research, particularly epidemiology, medicine, and physiology (Matthews, 1995). However, the structure of probability and statistics owes much to logic, particularly syllogistic deduction. Not because early mathematicians and statisticians were logicians, but because all were motivated by the same quest for certainty. Given that there is only one universe in which mathematicians, statisticians, logicians, physicians, and biomedical researchers occupy together, it is not surprising that the structure of the universe should impose its handprint on the methods used to explore the universe by the mathematicians, statisticians, logicians, physicians, and biomedical researchers.

The philosophical school that advocates this position is called Logical Realism, which, as will be discussed, can be applied to probability and statistics. The universe just is the case that logic in its widest connotation would prevail in the descriptions of reality. This is not the same as the school of Logical Positivists. The latter held that all human knowledge would be reduced to and hence drivable from a set of fundamental self-evident axioms and rules of inference. However, the Heisenberg Uncertainty Principle, Gödel Incompleteness, and Alan Turing's work on the Halting Problem argued that the agenda of the Logical Positivists is unobtainable. However, the failure of the Logical Positivists does not undermine the position of Logical Realists in that there is a fundamental, underlying, all-encompassing logic in the structure of the universe that perhaps will be found in the ultimate Theory of Everything in physics, a mathematical logic but logic nonetheless.

THE VALUE OF THE LOGICAL PERSPECTIVE IN PROBABILITY AND STATISTICS

The purpose of this chapter is not to argue any particular perspective on the origin and development of probability or statistics except where such historical notes illustrate the concepts being proposed. Interested readers are referred elsewhere for more information (Gigerenzer et al., 1989; Matthews, 1995). Rather, the position here is that probability and statistics fundamentally can be understood from a logical perspective as was already introduced in Chapter 4, The Logic of Certainty Versus the Logic of Discovery. By understanding the underlying logic, probability and statistics will be less foreign to biomedical researchers, who will be more informed customers of turnkey statistical packages. Indeed, as will be demonstrated, many of the errors in the use of statistics, hence probability, are the result of an injudicious use of logical fallacies. What is needed is a judicious use of logical fallacies that underlie probability and statistics.

Chapter 4, The Logic of Certainty Versus the Logic of Discovery introduced syllogistic deduction as one of the routes to absolute certainty and thus absolute reproducibility in biomedical research. The paradox was that syllogistic deduction does not lead to new knowledge. Thus, what is gained in certainty is lost in utility. This was recognized by the ancient Greeks, such as Aristotle, who extended the utility of syllogistic deduction by use of the partial syllogism. An example of a valid syllogism is seen here in Argument 5.1:

ARGUMENT 5.1

Major premise: *All tobacco smokers* (bridging term) *die of cancer or heart disease* (major term)

Minor premise: *John Doe* (minor term) *is a tobacco smoker* (bridging term)

Conclusion: *John Doe* (minor term) *will die of cancer or heart disease* (major term)

Valid syllogisms are not admittance for a falsity. However, John Doe may die from being hit by an automobile, hence Argument 5.1 is invalid. The falsity does not come from the form of the syllogism but rather the empirical evidence related to the major premise. It is not likely that each and every tobacco smoker necessarily dies from cancer or heart disease.

Argument 5.1 can be rescued by reframing the syllogism as a partial syllogism in Argument 5.2:

ARGUMENT 5.2

Major premise: *Some tobacco smokers* (bridging term) *die of cancer or heart disease* (major term)

Minor premise: *John Doe* (minor term) *is a tobacco smoker* (bridging term)

Conclusion: *John Doe* (minor term) *may die of cancer or heart disease* (major term)

Argument 5.2 is more certain in that should John Doe or any other smoker not dying from cancer or heart disease will not invalidate the argument. But what is gained in utility is lost in certainty. There is no way to know whether John Doe will be among the *some* that die of cancer or heart disease or the other *some* that will not die of cancer or heart disease. Indeed, based on Argument 5.2, there is no way to even guess.

The utility of the partial syllogism in Argument 5.2 can be improved by reframing as the probability syllogism given in Argument 5.3:

ARGUMENT 5.3

Major premise: *80% of tobacco smokers* (bridging term) *die of cancer or heart disease* (major term)

Minor premise: *John Doe* (minor term) *is a tobacco smoker* (bridging term)

Conclusion: *John Doe* (minor term) *has an 80% (0.8) probability of dying of cancer or heart disease* (major term)

Greater utility may be evidenced by the fact that John Doe now considers it prudent to give up tobacco smoking. Alternatively, John Doe, being a skeptic, might ask how does the scientist know there is an 80% probability that tobacco smokers die of cancer or heart disease? The response might be that a number of tobacco smokers who die were studied and 80% of the smokers died of cancer or heart disease. Being a really good skeptic, John goes on to ask how many tobacco smokers were studied? It would be rather embarrassing to the biomedical scientist or clinician to have to admit that only five tobacco smokers were studied. The biomedical scientist or clinician would be much more convincing if 100 tobacco smokers were studied or, better yet, 10,000 tobacco smokers.

Being an excellent skeptic, John goes on to ask how many times was the study replicated and what was the range of results? The biomedical researcher or clinician may respond that 10 similar studies were conducted and the probability of a tobacco smoker ranged from 65% to 95% for a mean of 80%. John then might ask what the *real* probability is, as it cannot be that all probabilities from the studies are *real* probabilities. John then asks what was the distribution of the probabilities? Assume that the biomedical scientist or clinician lists the actual probabilities as 65%, 65%, 65%, 65%, 75%, 85%, 90%, 90%, 95%, and 100%. John might respond by saying as "most of the probabilities were 65% [the mode], I will take my chances and continue to smoke tobacco." Indeed, John may voice skepticism because the median probability is 75%, not 80%, and further, the mode is 65%. Which is to be believed?

METAPHYSICS: ONTOLOGY VERSUS EPISTEMOLOGY AND BIOMEDICAL REPRODUCIBILITY

Clearly, the probability syllogism is not enough in itself. The probability based on enumeration of a sample of observations from a pool of all possible observations, the population, may differ each time a sample is collected and measured. The probability syllogism can be rescued by the application of statistics. However, statistics presupposes a metaphysics, that is, the presuppositions and assumptions that are not directly or solely empiric, that is, based on explicit observations, such as the result of experimentation. An example of a metaphysical presumption is Occam's Razor, which holds that the simplest explanation is most likely true seems reasonable and often guides one favoring one scientific claim over another. However, there is no scientific, logical, or mathematical proof of

Occam's Razor. Occam was reported as saying, "It is vain to do with more that which can be done with fewer" (Russell, 1945). As will be discussed, statistics involves internal validity and is an epistemic issue. In itself, statistics does not provide external validity or knowledge of reality (ontological knowledge).

Metaphysics includes inferences as to the exact nature of reality, and the concerned discipline is called ontology. The other branch of metaphysics includes the discipline of epistemology, which seeks to understand how knowledge is gained. The difference between ontology and epistemology may strike the biomedical researcher as an unnecessary distinction but it is clear from the aforementioned discussion that they are different. The measures of Central Tendency, such as mean (average), median, and mode, are epistemic issues, as these measures are constructed from a sample of observations. For example, the epistemic devices of mean (average), median, and mode of the weight of every living organism could be determined, at least in theory, but what sense would it make in reality (ontology)? It is important to note that a measure of Central Tendency can be constructed from virtually any set of observations without any necessity to appeal to reality, ontology. Which measure, mean (average), median, and mode, truly reflects the nature of the phenomenon? Exactly what the probability of a tobacco smoker dying from cancer or heart disease, for example, is an ontological question.

The discrepancy between ontology and epistemology may be a significant cause of unproductive irreproducibility. For example, epistemological tools often need not necessarily cohere with reality when viewed in their logical form. For example, the modus ponens of propositional logic is *if a implies b is true and a is true, then b is true* is valid and certain if for no other reason than its logical form, regardless of the values of *a* or *b*. But note that the implication *a implies b* is conditional. It must be true that *a implies b* for the conclusion to be true. Now the ontological nature of *a* and *b*, and thus the nature of the implication, *a implies b*, is critical in biology, for example, taking *implies* as meaning *cause*, which results in the proposition that *a causes b*. That *a causes b* is true must be demonstrated by observation or by extrapolation from other observations. In mathematics or formal logic, where *implies* is synonymous with the mathematical equal (=) or the state-of-being linking verb *is* (see Chapter 4: The Logic of Certainty Versus the Logic of Discovery) and typically whether *a implies b* is true is a matter of logic and not contingent on reality. Hence, mathematical logic and formal logic are internal but they have value when they are linked to phenomena in the real external world.

To illustrate the conundrum between logic (epistemology) and empiric fact (ontology), consider the following proposition in Argument 5.4:

ARGUMENT 5.4

If (a or b) implies c is true and (a or b) is true, then c is true, translated to *if ([John believes that abnormality X causes disease Y] or [unbeknownst to John abnormality Z caused disease Y]) implies disease Y, and ([John believes that abnormality X causes disease Y] or [unbeknownst to John abnormality Z caused disease Y]) is true,* then *disease Y is true* [by modus ponens].

The problem of Argument 5.4 is that it only demonstrates that *(a or b)* must be true and not whether *a* or *b* or both is true. Nevertheless, the argument is valid and the conclusion is true. To conclude that *a* must be true is the Fallacy of Affirming a Disjunct. In the case of *disease Y*, John only knows about *a*, *abnormality X causes disease Y*, and not about *b*, *abnormality Z causes disease Y*. The risk is that John may believe *a* to be true without recognizing that it is *b* that is true that ensures a true conclusion, *c*, from the argument. The validity of the argument and the truth of the conclusions are affirmed by the truth of the premises. However, it is premise *(a or b)* that is true and this does not inform whether *a, b* or both are true. It must be a matter of other experiments to determine whether *a* is true.

Further, not realizing that it is actually *abnormality Z* that is the cause of *disease Y*, John constructs the experiment in terms of the Scientific Method to form Argument 5.5:

ARGUMENT 5.5

If (a or b) implies c is true and c is true, then (a or b) is true, translated to *if ([John believes that abnormality X causes disease Y] or [unbeknownst to John abnormality Z caused disease Y]) implies disease Y, and disease Y is true, then ([John believes that abnormality X causes disease Y] or [unbeknownst to John abnormality Z caused disease Y]) is true*—by the modus ponens-like variant of the Scientific Method.

John's experiment was successful—in the presence of *abnormality X*, *disease Y* was found. John may assume, naturally, that *abnormality X* is causal to *disease Y*, which it is not because simultaneously and unknown to John, *abnormality Z* was also present. The next time the experiment is run, *abnormality Z* may not be present because but this would not be recognized. In this case, *disease Y* does not occur, despite the presence of *abnormality X*, and the experiments are held to be irreproducible and likely unproductively so.

The key point is that Argument 5.3 is internally consistent and valid as a formal system, even if it leads to misconceptions that later are demonstrated as ontologically untrue or irreproducible. As seen, the truth of the conclusions depends on the ontological status of the premises. Critically, there is nothing in the logical form—the epistemology—that assures ontological truth, within the confines of the logical argument. Knowledge, presumptions, or assumptions must be imported from elsewhere (see the Duhem–Quine thesis in Chapter 4: The Logic of Certainty Versus the Logic of Discovery).

Similarly, if there is any "truth" in probabilities, it lies in its calculations, that is, its epistemology. For example, one can calculate the probability of a John Doe in all living things, but what would that mean? Also, one can calculate the Central Tendency and variance of any arbitrary collection of observations. But in itself, this is uninteresting; what is interesting is how the probabilities, Central Tendencies, and variances relate to the real world because modern biomedical research presupposes that they do. This presupposition is discussed later.

Generally, in science, there is the presumption of an ontology that is independent of humans; that the ontology, save that which is uniquely human, would still be there even if there were no humans. Such a position is a debated point. Clearly, the notion of a scientific ontology independent of humans is reflected in William Whewell's coining of the term "scientist" in 1834 as someone who does science, suggesting that science is an antecedent to the human scientist. Postmodernists hold that science is what scientists do. Given the frailties of humans, the postmodernist position complicates the notion of reproducibility, tending to make reproducibility whatever scientists want to make it. The problem, even given the presumption that there is an independent ontology or reality, is that human knowledge of ontology is only possible via human implemented epistemic tools.

Given that epistemic tools are, in large part, self-referential or internally consistent, how then are the epistemic tools to map onto reality, that is, ontology? The presumption is that ontology will, at one point or another, make itself known through epistemic tools. Thus, epistemic tools applied incorrectly will produce a version of putative ontology that is not likely to be reproduced upon re-examination, unless the epistemic error is repeated constantly.

Early in the experience of applying probability and statistics to biomedical research, particularly as it relates to humans, was the concept of *l'homme moyen* (average man) described by Adolphe Quetelet in his work of 1835, "*Sur l'homme et le développement de ses facultés, ou Essai de physique sociale*" ("On Man and the Development of his Faculties, or Essays on Social Physics"). The proposition is that true ontology is represented in the average (mean) or other appropriate measures of Central Tendency. What is seen, the epistemology, may vary, but such variety is a consequence of variation around a reality rather than diversity around many realities. In any species, the fundamental components, aspects, and properties are the same and variety is a consequence of variation by which the fundamentals are recombined and instrumental error, which includes perceptual error in humans. Indeed, researchers such as Francis Galton in the late 1800s, using photography, superimposed images of human faces to determine what the average human face was—the composite represented true ontology. Interestingly, in a nod to epistemic uncertainty, Galton stated that "No statistician dreams of combining objects of the same generic group that do not cluster towards a common center; normal or should we attempt to compost generic portraits out of heterogeneous elements, for if we do so the result is monstrous and meaningless" (quoted in Stigler, 2016, p. 39). Essentially, Galton is asking for trust and faith in the statistician, which generally is considered not scientific—note the motto of the Royal Society, *Nullius in verba* (take no one's word for it). Yet, the attempt at reassurance by Galton illustrates the issue of the *A Priori* Problem for Induction (see Chapter 4: The Logic of Certainty Versus the Logic of Discovery).

Many biomedical researchers at the time took great umbrage and, to some degree, ridiculed the notion of the *l'homme moyen*. In the 1800s, the pre-eminent physiologist Claude Bernard suggested taking a sample of urine from every train station in Europe to determine who the average European actually was (Stigler, 2016, p. 35). This is not to say that biomedical researchers did not believe that there were fundamental principles that pertain to all individuals within a species, but only that knowledge of those fundamentals would not be achieved by probability and statistics (Matthew, 1995).

To be sure, at the early days in the development of statistics as its own science or discipline, there were those who thought that the range of variety is the true ontology. Thus, variety was diversity rather than variation. Nevertheless, the notion of variation around a unique ontology remains the central perspective in biomedical research. Indeed, a statistician congratulated the field by being able to disregard the observations so as to see the mean (Stigler, 2016). The consideration of variance is relatively late in biomedical research publications. To be sure, standard deviations often were reported with the means (averages). Yet, for many, the notion of a standard deviation seemed abstract. It is with the publication of confidence intervals that a clear qualification of what the mean (average) means and how it might be interpreted. The use of confidence intervals has only recently taken hold, and it is a tenuous hold at that. As will be seen, biological systems are very complex and operate far from thermodynamic equilibrium, which renders notions of the Central Tendency not only problematic, but not truly reflecting the ontology, thereby increasing the risk of irreproducibility (see Chapter 10: Chaotic and Complex Systems, Statistics, and Far-from-Equilibrium Thermodynamics). Rather, the dynamics of Chaos and Complexity need to be considered.

What is the basis for the position that the mean (average) or some other measure of Central Tendency has ontological legitimacy? Many statistical analyses relate to questions of the Central Tendency, for example, is the Central Tendency (e.g., mean) different between the experimental group and the control group? A critical concept is the Large Number Theorem, which holds that the mean of the sample, $\bar{x}$ for example, of a set of values approaches or becomes asymptotic to a constant value, which is taken as the mean of the population of all possible observations, the symbol for which is μ. But there are at least two accounts of why with increasing n, where n is the number of observations, that $\bar{x}$ should approach a constant value taken as the true mean μ. In other words, as $n \to \infty$, then $\bar{x} \to \mu$. However, the problem is that the additional contribution of x_{n+1} to $\bar{x}$ becomes smaller (assuming x_n operates within a finite range of values), as reflected in the following relation:

$$\Delta \bar{x} = \frac{\sum_1^n x_i + x_{i+1}}{n+1} - \frac{\sum_1^n x_i}{n} = \frac{x_{i+1}}{n+1} + \frac{\sum_1^n x_i}{n+1} - \frac{\sum_1^n x_i}{n}.$$

For n very large

$$\frac{\sum_1^n x_i}{n+1} \cong \frac{\sum_1^n x_i}{n}$$

and then

$$\frac{\sum_1^n x_{i+1}}{n+1} - \frac{\sum_1^n x_i}{n} \cong 0,$$

which means that

$$\Delta \bar{x} \cong \frac{x_{i+1}}{n+1}.$$

If x_i is bounded within a range of values, $\Delta \bar{x}$, the change in mean $\bar{x}$ will become vanishingly small with increasing n.

The mathematics suggests that convergence onto a constant mean value $\bar{x}$ as the number of observations, n, increases need not reflect any unique feature of the set of observations; in other words, not reflect the ontology of the set of observations. Yet, the typical interpretation of the Large Number Theorem is that the reason the mean approaches some constant value typically is that variations in the observations extraneous to the ontology cancel out, leaving the mean as the true ontology. An example is the notion of biological variability. Typically, it is expected that biological variability is Gaussian, that is, a symmetric and diminishing probability of values on each side of the mean. Thus, when the values of the individual observations are summed, the extremes on each side of the Central Tendency cancel each other, leaving only the mean. This view is challenged in Chapter 10, Chaotic and Complex Systems, Statistics, and Far-from-Equilibrium Thermodynamics for Chaotic and Complex systems.

Also contributing to the variety about the mean is instrumental variance, where repeated measures of the same phenomenon result in a Gaussian distribution of actual values. As in the case of biological variability, instrumental variability is thought to "cancel" out, leaving the mean as reflecting the true ontology. This may not be true in complex thermodynamic systems, such as machines, tools, or biological organisms, operating far from a thermodynamic equilibrium, as will be explained in Chapter 10, Chaotic and Complex Systems, Statistics, and Far-from-Equilibrium Thermodynamics. Further, these systems tend to self-organize and, when perturbed, such as in any measurement, may act as a dampened limit cycle oscillator where the amplitudes display a Gaussian distribution. Hence, the Gaussian distribution in biological variability and instrumental variability more truly reflects the ontology, and variance in the observations cannot be dismissed as instrumental variability. The mean or other measures of Central Tendency become abstractions that may be misleading, resulting in irreproducibility.

INDEPENDENCE OF PROBABILITIES AND REGRESSION TOWARD THE MEAN

Probabilities typically are achieved by enumeration over a series of subjects in a sample of a population of concern. For example, it may be the value x_i for some measure X where x_i is the value for each of the n number of subjects. Alternatively, it may be repeated measures of X taken n times for a single subject. The "result" of the experiment is thought to be reflected in the Central Tendency over the range of the values of measurement X, those being x_i where i goes from 1 to n, signifying each subject in the sample. For example, the mean, of $\bar{x}$, is determined by the following equation:

$$\bar{x} = \frac{\sum_1^n x_i}{n}.$$

The variance of the distribution of x_i is given by the following equation:

$$\partial^2 = \frac{(x - \bar{x})^2}{n}.$$

It is critical to note that for most applications of probability calculus and statistics, each value of a measure or observation in a research experiment, such as x_i, is independent of all values or measures of X. For example, considering *measurement* X in a *sample B* of *population C*, the measurement of each subject in the sample must be independent of the measurement in any other sample; similarly, all measurements in an individual subject in a repeated measured experimental design must be independent. In other words, the value or measure of X, x_i, is independent of the value of other measures, x_{i+j}, where j is some integer and x_{i+j} represents some subject or measure other than x_i. This is important if the Central Tendency is to reflect the reality or ontology of the experimental outcomes by being stable, thus helping to ensure reproducibility.

There are a great number of reasons why the value or measure of x_i may not be independent of the value of x_{i+j} such as bias in sampling. Thus, the value of x_i depends on previous values of x_i such as x_{i-j}. Or all values of x_i depend on some factor, y, that biases the selection, for example. However, even unbiased sampling from a randomly distributed population of x_i's can result in a failure of some measure of the system to converge onto a stable value of $\bar{x}$ or Central Tendency. In these situations, the values of x_i relative to x_{i-j} need not converge onto a stable value. This is due to the principle of Regression Toward the Mean. For example, if x_i happens to be far from the $\bar{x}$, then x_{i+1} likely will be closer. Note that the actual values of x_i and x_{i+1} are independent based on random sampling of a population; nevertheless, the relationship between x_i and x_{i+1} is not independent of x_i. If the analysis of interest is constructed in such a way that x_i relative to x_{i-1} is important, then the analysis need not converge on a single stable value and therefore the analysis is extremely difficult to take with certainty as a descriptor of the biomedical system under study.

The effect of a measure that is influenced significantly by the value of x_i relative to x_{i-1} is found in Brownian motion and in Chaotic and Complex Systems (see Chapter 10: Chaotic and Complex Systems, Statistics, and Far-from-Equilibrium Thermodynamics). Brownian motion is the physical phenomenon of a particle, such as a grain of pollen, suspended in a liquid under a microscope that appears to jitter and move through the medium (Fig. 5.1). The movement is related to collisions between the particle and the molecules of the liquid. As the very large number of collisions are small, the particle appears to jitter in a single location; however, at times the particle may make a large jump. As the directions of the collisions are random and small, the movements of the particle will be small and the particle will "hover" in the same location and appear to jitter. However, occasionally a large collision will result in a large movement of the particle.

The movement of the particle depends on the range of collision magnitudes and directions. For example, if the collisions were all of the

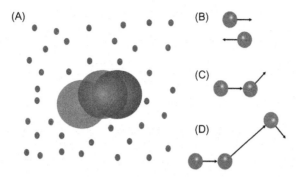

FIGURE 5.1 Brownian motion is a phenomenon exemplified by the movement of a very small particle, under a microscope, that has been placed in a liquid (A). The particle appears to jitter and move through the medium. The direction and magnitude of the particle movement are related to the direction and magnitude of the collisions between the particle and molecules of the liquid. If the collisions are equal and opposite (B), the particle will not appear to move as the collisions cancel out. However, the collisions may not be opposite, in which case the particle will be seen to move (C). As most of the collisions are very small (C), the particle will appear to jitter. Relatively rarely there will be a very big collision that will move the particle through the medium (D). The subsequent collisions are too small to bring the particle back toward the original position.

same magnitude and in a random direction along a one-dimensional line (assuming a Normal or Gaussian Distribution), the particle would remain in relatively the same position as each subsequent collision would move the particle back to the same location by regression toward the mean (Fig. 5.1B). In the case where all the collisions were of the same magnitude and random in all three axes of a three-dimensional space, over time the particle would appear to hover in the same location, analogous to a convergence on the mean according to the Large Number Theorem (Fig. 5.1C). However, if there is sufficient variance in the distribution of collision magnitudes, the particle will make a large jump in space. Because of regression toward the mean, the subsequent collisions will be smaller and the particle will not return to the original location (Fig. 5.1D).

The Brownian motion phenomenon can be modeled as a point process, such as a series of events where the time interval between events is the measure of interest, such as a sequence of neuronal discharges. The interevent interval is analogous to the magnitude of the collision as described earlier for the motion of a particle. The movement of the outcome (a marker along a single dimension) is determined by the interevent intervals (the analogy of collision direction is reduced or projected onto a single dimension). The analysis is shown in Fig. 5.2.

Fig. 5.3 shows results from a series of simulations of varying degrees of irregularity, from very regular to bursting, all drawn from a random

$$Npl = \frac{\sum \frac{|(I_n - I_m)|}{I_m}}{N}$$

Npl = Normalized pointer location

I_n = n^{th} interspike interval

$$I_m = \frac{Number\ of\ spikes}{Time\ of\ sample}$$ = Model interval

FIGURE 5.2 Computations of the Brownian motion indicator. The normalized position pointer is analogous to the spatial location of the particle in the Brownian motion example of a particle suspended in a liquid. The position marker (Npl) is normalized by dividing the absolute value of the difference between each interevent interval, I_n, and the mean interevent interval, I_m.

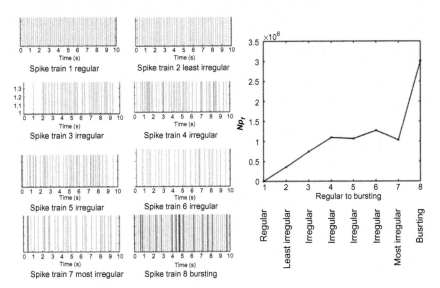

FIGURE 5.3 Results of simulation studies demonstrating the value of the normalized pointer location from Brownian motion analyses for a series of point processes (spike trains) of differing irregularity. As can be appreciated, the Npl correlates well with the degree of irregularity. More importantly, in the more irregular condition, Npl need not converge on a single value and would be counter to the notion of the Large Number Theorem.

distribution with different means and standard deviations. As can be appreciated, the normalized position pointer (Npl) correlates well with the degree of irregularity. The distance traveled by the position marker increases, that is, fails to move closer to 0 (in the case of perfectly equal interevent intervals).

The failure of the pointer to remain at the origin is related to the variance of the interevent intervals, but not just the range of variations in the interevent intervals. Rather, the order in which the interevent intervals are encountered, such as the sequence of collisions for the particle described previously. Thus, distribution of the interevent intervals may be distributed symmetrically around the Central Tendency (the mode interspike interval in the simulation given previously). However, the measure, position marker (Npl), will continue to increase, although at different rates depending on the variance of the interevent intervals.

Situations exist in biomedical research where there are experimental and analytical designs where the measures are not independent in probability or where the sequential relationships between measure are such to cause a drift in the composite measure. Such a drift would render the Large Number Theorem invalid and, consequently, any statistical analyses that depend on the Large Number Theorem of doubtful utility. One potential example is the notion of repeated measures, for example, where the outcomes of repeated experimental measures are considered in aggregate. For example, the incidence of breast cancer in a clinical trial where a Data Safety and Monitoring Committee is periodically examining the incidence of breast cancer (see Chapter 7: Case Studies in Clinical Biomedical Research). It also can be a factor in computational reproducibility (see Chapter 9: Case Studies in Computational Biomedical Research). How often this type of problem occurs in biomedical research is unknown. Nonetheless, it is worth continuous vigilance to prevent.

AVOIDING THE FALLACY OF FOUR TERMS

Despite concerns of how the mean reflects ontology, the mean and other measures of Central Tendency are typical *modus operandi* of biomedical research. Perhaps in the future, tests of Complexity or Chaos will be applied to sets of observations to determine the ontological relevance of traditional statistics, much in the manner that the distribution of the observations helps with deciding between parametric and nonparametric analyses (see Chapter 10: Chaotic and Complex Systems, Statistics, and Far-from-Equilibrium Thermodynamics). In this section, some consideration is given to mitigating the epistemic and ontological risks associated with inferring from Central Tendencies, particularly the mean (average), in order to reduce the risk of unproductive irreproducibility.

In one sense, Central Tendencies, such as the mean (average), are used to "stand in" for the population of interest irrespective of the metaphysical concerns described previously. A mean serum sodium

(Na$^+$) level of 140 milliequivalents per liter (mEq/L) stands in as a reference for the serum Na$^+$ level in normal humans, and significant departures are regarded as indicative of disease. To be sure, some variance around the mean is allowed. Great confidence in the value of 140 mEq/L would be obtained if every "normal" human had their serum Na$^+$ checked and all had a level of exactly 140 mEq/L. In this case, the mean would be for the entire population of interest, that being all humans, and would be represented by the symbol μ, as shown in the equations given earlier. Of course, every new human would have to be tested as well.

The recourse is to examine a sample or subset of the population of interest and determine the mean of the sample, which is represented by the symbol $\bar{x}$ in the aforementioned equations. The strategy then is to infer μ from $\bar{x}$. This creates Argument 5.6 in a general form:

ARGUMENT 5.6

Major premise: *The sample* (bridging term) *had a mean effect of* $\bar{x}$ (major term)

Minor premise: *The population* (bridging term) *had a mean effect of* μ (minor term)

Conclusion: μ (minor term) *is* $\bar{x}$ (major term)

Argument 5.6 is at risk for the Fallacy of Four Terms to the degree that the bridging term, *sample*, in the major premise may not identical to the bridging term, *population*, in the minor premise. However, Argument 5.6 is an idealization, as it is highly unlikely that the population would ever have been examined. Even if conceivable, how would one know because of the Fallacy of Induction?

A more relevant syllogism for the purposes of reproducibility in biomedical research is Argument 5.7:

ARGUMENT 5.7

Major premise: *The sample$_1$ of experiment$_1$* (bridging term) *had a mean effect of* $\bar{x}_1$ (major term)

Minor premise: *The sample$_2$ of experiment$_2$* (bridging term) *had a mean effect of* $\bar{x}_2$ (minor term)

Conclusion: $\bar{x}_1$ (major term) *is* $\bar{x}_2$ (minor term)

If the conclusion of Argument 5.7 is true, then the research is reproducible. However, the risk of the Fallacy of Four Terms increases the likelihood that the conclusion may be wrong and the research held unproductively irreproducible. Note that there are two aspects to the bridging term: *sample* and *experiment*. To the extent that $sample_1$ is different from $sample_2$ and the degree by which $experiment_1$ is different from $experiment_2$ increase the risk of fallacy. However, it may provide greater new knowledge, by avoiding a tautology, in that there may be some mechanism common to both $sample_1$ and $sample_2$ or $experiment_1$ and $experiment_2$ that would not have been known from study of only $sample_1$ or $sample_2$ or when $sample_1$ and $sample_2$ are identical. Similar opportunities hold when $experiment_1$ and $experiment_3$ are different. Such findings suggest that extending the experiments to $sample_3$ or $experiment_3$ likely will be conceptually reproducible, in the broad sense.

There are different uses of the bridging term components *sample* and *experiment*. For example, in a set of experiments, $sample_1$ is made different from $sample_2$, but within the same species, the term species is used in its widest sense (see Chapter 4: The Logic of Certainty Versus the Logic of Discovery), but $experiment_1$ and $experiment_2$ are exactly the same. Such a situation would allow for assessment of the technical reproducibility of the experiment, a valuable exercise. This would be reproducibility in its narrowest sense (see Chapter 1: Introduction). When $sample_1$ is made different from $sample_2$ by a difference in species and $experiment_1$ and $experiment_2$ are exactly the same, this is a test of conceptual reproducibility. When $sample_1$ is the same as $sample_2$ but $experiment_1$ and $experiment_2$ are different, this also may be a test of conceptual reproducibility. When both $sample_1$ and $sample_2$ are different, as well as $experiment_1$ and $experiment_2$ being different, it is not clear what this tests other than confusion and, consequently, has a high risk of unproductive irreproducibility.

Consider the set of experiments where $sample_1$ is made different from $sample_2$, but within the same species, the term species is used in its widest sense, but $experiment_1$ and $experiment_2$ are exactly the same. $Sample_1$ must be different from $sample_2$ in one sense, as otherwise the experiment would consist of reanalyzing the same data and the result would be a tautology. Rather, $sample_2$ must be different from $sample_1$ in actual experience, but the same as $sample_1$ in key properties; in other words, the two samples are from the same population with respect to the key properties—they are conceptually the same. Otherwise an injudicious use of the Fallacy of Four Terms would be the risk. The conundrum is that the population is never examined, only samples of the population.

The strategy is to address those factors that risk making the samples not representative of the population. If the samples are representative of the population, then it is likely the two (or more) samples are representative of each other. Logically, *if $sample_1$ implies* (meaning representative of, a state-of-being relation) *population and $sample_2$ implies* (meaning representative of, a state-of-being relation) *population, then $sample_1$ implies* (meaning representative of, a state-of-being relation) $sample_2$ by the Identity Principle. This argues for the importance of replicating experiments because the goal is to understand the population of interest. However, if the relation of the samples to the population is not a state of being, the logic is the Fallacy of Pseudotransitivity (see Chapter 3: Validity of Biomedical Science, Reproducibility, and Irreproducibility). This means that it cannot be certain, but it might be close enough to be useful. This situation is generic to knowledge gaining and biomedical research is no exception.

Although the population is unknowable, even if only because of the Fallacy of Induction, reasonable strategies can be developed. For example, the Epistemic Distance between the sample and the population diminishes as the sample size increases. As this is true for $sample_1$ and $sample_2$, then as the sample size increases for both $sample_1$ and $sample_2$, the Epistemic Distance between the two samples diminishes, keeping in mind that Complex and Chaotic systems may be different (see Chapter 10: Chaotic and Complex Systems, Statistics, and Far-from-Equilibrium Thermodynamics). This concept can be expressed as a limit, as shown in the following equation, where n approaches infinity (∞):

$$\lim_{n \to \infty} \left(\{sample_1\} - \{sample_2\} \right) \to 0.$$

What the necessary sample size is to sufficiently reduce the Epistemic Distance is difficult to predict. The planned choice of the sample size is determined by the degree of confidence required. The degree of confidence required depends on the consequences of a type I error (finding a difference from a statistical test when a difference does not truly exist) and a type II error (not finding a difference when a difference truly exists). For example, recent discovery of the Higgs boson required a level of confidence of 5 sigmas (where sigma is the standard deviation). In other words, the mean mass of the Higgs boson had to be 5 standard deviations greater than the noise of the measuring device. Why was 5 sigmas chosen? The reason is that in the past when a 3 sigma criterion was used, a type I error was demonstrated. (It is interesting that most statistical tests were more concerned with preventing a type I error than a type II error. From an epistemic perspective, it is not clear why a type I error is dreaded much more than a type II error. Speculating, it may parallel the greater ethical concerns for errors of commission over errors of omission, even though this ethical distinction is without foundation.)

Note that the notion of a limit as applied previously is an example of a Process Metaphor (see Chapter 3: Validity of Biomedical Science, Reproducibility, and Irreproducibility). As with any metaphor, there is a target domain that gets its credibility and meaning by "borrowing" from the source domain. In this case, the target domain, {$sample_1$}−{$sample_2$}, is a concept that is taken as finite and is appreciable but it is understood to become 0 as $n \to \infty$, which never happens. However, one gets a sense of the impossible by taking n as 1, then 2, and so on. The value of n at ∞ is appreciated by just adding one to the largest n thought of at the time. Process metaphors are ubiquitous in biomedical research, for example, in a Reductionist account, and are sources of unproductive irreproducibility.

If, for whatever reason, n cannot be made sufficiently large so that differences in the samples approach 0, then $sample_1$ may be different from $sample_2$ just by chance, but nonetheless the Fallacy of Four Terms still applies. The fallacy applies because the consequences of the samples, that is, the results of the experiment, constitute the major and minor terms in the conclusion. Consider a lake with a large number of fish. The motion, and thus location, of each fish is independent of any other fish (with a notable exception during breeding season). It is highly unlikely that the fish are distributed throughout the lake uniformly. If they were, then one would have to fish in just one small part of the lake in order to estimate the number of fish in the lake. Realizing that it is unlikely that the fish are distributed evenly and, more importantly, being unable to know *a priori* whether the fish are distributed evenly, one would fish in multiple locations. Likely the amount of fish captured at any one location will not be exactly the same as at another location. One then samples multiple locations, takes some measure of the Central Tendency within each sample (note that the total count is the same as the Central Tendency multiplied by the number of samples), and then determines the Central Tendency of the Central Tendencies of the samples (an example of the Central Limit Theorem).

The situation of sampling just described presupposes at least two important metaphysical presuppositions. First, even if one fished in a different location in the lake, how would one know if he is fishing in a different part of the lake based on the dynamics of the fish movements? One approach is estimating how similar the total counts of collected fish are at each location—the greater the similarity means fewer sites would have to be fished. But this presumes that the interactions among the fish are at some equilibrium; in other words, the forces that would cause one fish to move in one direction are offset by forces that would cause another fish to move in the opposite direction. With a sufficient number of fish, the effect would be to disperse the locations of the fish. However, how disperse the fish will be and over what time frame depend very much on the number of fish in the lake (this concept is analogous to the Mean Free Path of a gas particle, Brownian Motion in physical systems, and entropy in thermodynamic systems). Clearly, these dynamics influence how likely the sampling reflects the population (all the fish in the lake). It is critical to note that the aforementioned discussion applies to situations of thermodynamic equilibrium (where the ergodicity theorem applies). However, living biological systems generally do not operate at thermodynamic equilibrium and thus one cannot be sure that the observations will spread "evenly" in the dataset. In this case, selecting samples is very problematic (see Chapter 10: Chaotic and Complex Systems, Statistics, and Far-from-Equilibrium Thermodynamics). To the degree that the sampling does not reflect the population, then the likelihood of the Fallacy of Four Terms and unproductive irreproducibility increases.

The Fallacy of Four Terms may be an issue when, unknowingly, the two samples come from different populations, not just different samples from within the same population as described earlier. Consider Argument 5.8:

ARGUMENT 5.8

Major premise: *Test A detects property B in 90% of subjects* (major term) *in a sample$_1$* (bridging term)

Minor premise: *Sample$_1$* (bridging term) *is the same as sample$_2$* (minor term)

Conclusion: *Test A will detect property B in 90% of subjects* (major term) *in sample$_2$* (minor term)

Consider the situation where the conclusion is false and that the experience with *sample$_1$* is not reproduced in *sample$_2$*. Something is amiss in Argument 5.8. The major premise is an experimentally established fact. This leaves the minor premise as suspect and it may be that *sample$_1$* is not the same as *sample$_2$*. This scenario often appears when the sensitivity and specificity of a test are considered versus the Positive- and Negative Predictive Values.

The fact that *test A* identifies 90% of subjects with *property B* is the sensitivity. Typically, the sensitivity is derived from a sample, 100% of which have *property B*; in this case, *sample$_1$*. While important, the principal value of determining the sensitivity of *test A* is its use in exploring the presence of *property B* in a different circumstance, *sample$_2$*, such that the result of *test A* provides new knowledge of the different circumstance, particularly to provide conceptual reproducibility. However, it cannot be known ahead of time what percentage of subjects in *sample$_2$* actually have *property B* as this may be the purpose of *test A*. This percentage of subjects having *property B* is known as the prior probability as used in Bayes' theorem.

Test A is shown to have 90% sensitivity. But what happens if the prior probability is 50%, meaning that only half the subjects in *sample₂* will have *property B* unlike the 100% of subjects in *sample₁*? Only 45% of subjects in *sample₂* would be detected as having *property B*. This would be the Positive Predictive Value. If the investigator assumes that all subjects in *sample₂* have *property B*, finding only 45% with *property B* would be seen as the experiments being irreproducible as the two experiments involving *sample₁* and *sample₂* should produce the same result under the presumption that, like *sample₁*, 100% of subjects in *sample₂* should have *property B*. But this would be an unproductive irreproducibility. Thus, the reproducibility of the experiments will depend greatly on how similar the prior probabilities are in the two samples on which the same experiment is applied (see Button et al., 2013).

The conundrum is that *test A* cannot be used to determine the prior probability without the experiment becoming a circular argument or self-fulfilling prophecy. There would have to be some other external standard, such as *test C*, but this would have to assess the same mechanisms as would *test A*; further, the same questions that came up regarding *test A* also would affect *test C*. But the same questions can be raised for *test C*, leading to an infinite regress. In the absence of some other external standard, interpretations of the experiments involving *test A* are problematic. This does not mean that the results are unhelpful but just that any inferences must be tentative until future experiments can be performed referencing some future external standard.

Most experimental manipulations are designed to produce an effect. A set of subjects are subjected to a manipulation and the changes of some property of each member of the set are studied. The mean effect on the property is then calculated and compared to the properties studied prior to the manipulation. Alternatively, one set of subjects is subjected to the manipulation and the others are not or subject to a manipulation that is thought ineffective—a placebo. As shown earlier, the question is whether the two sets, observations before and observations after or with and without the experimental manipulation, come from the same population. If the statistical tests demonstrate that they could not have been from the same population, within some allowed tolerance measured in variances, then the population after the experimental manipulation is different than the population before the experimental manipulation and, for example, the null hypothesis is rejected. Further, it is presumed that the subsequent experimental manipulation changed the ontology of the subjects compared to the ontology without the manipulation.

The problem is that the dynamics (changes over time) of the manipulation effect likely are very complex in biological systems. As discussed in Chapter 10, Chaotic and Complex Systems, Statistics, and Far-from-Equilibrium Thermodynamics the complexity and self-organizing dynamics are necessary conditions of life. Thus, the manipulation effects on those Complex systems may vary greatly over time. For some systems, the effects may settle in a steady state, such as equilibrium. For example, a drop of ink in a glass of water may eventually diffuse throughout the water but sampling small volumes of the water too early is likely to give very discrepant results. As a consequence, the variance among the measures may make it statistically difficult to determine whether the addition of a drop of ink had any effect, although the person knows there must be an effect as they put the ink drop in the water—an example of a type II statistical error. Many dynamical systems do not reach a steady state but instead oscillate (meaning changing directions at varying times to maintain a variable, such as blood glucose levels within a specific range in relation to different carbohydrate loads and metabolisms). For such systems, it may be very difficult, if not impossible, to know when to sample.

For some systems, it may be possible to estimate the time course of the dynamics, for example, when a steady state will be achieved and the time suitable for sampling based on causation. However, arguments from causation, derived from the practical syllogism, are problematic, as is discussed in greater detail in Chapter 6, Causation, Process Metaphor, and Reductionism. Consider an experiment where the investigator wants to determine the effect of Parkinson's disease on recordings on neuronal spike activity in the subthalamic nucleus during Deep Brain Stimulation lead implantation surgery. However, prior to the recordings the patient is administered remifentanil, which the investigator knows could confound the neuronal recordings. Consequently, the investigator has to wait a period of time to allow the remifentanil to dissipate. From pharmacology, it is known that it takes approximately five half-lives for the drug to be essentially eliminated. For remifentanil, the plasma half-life is on the order of 10 minutes. This equation is a decaying exponential function, which becomes asymptotic (only negligibly decreasing further) at approximately five half-lives, which would be 50 minutes in the case of remifentanil. Thus, the investigator decides to wait 60 minutes, just to be sure. However, the investigator obtains results that are different from reported previously in subjects not receiving the remifentanil. The experimental experience may be irreproducible and, if so, in a productive manner to give knowledge or in an unproductive manner to gain confusion. The investigator's argument is seen in Argument 5.9:

ARGUMENT 5.9

Major premise: *After 60 minutes* (bridging term) *the effect of remifentanil is gone* (major term)

Minor premise: *My experiments* (minor term) *began after 60 minutes* (bridging term)

Conclusion: *The results of my experiments* (minor term) *were independent of any effect from remifentanil* (major term)

The lack of reproducibility indicates that something in the experiments was wrong. Being the typical modest investigator, she suspected her own research and asked whether her argument (reasoning) was fallacious, particularly whether her reasoning was susceptible to the Fallacy of Four Terms. She asked which of the three terms actually were two versions in disguise? The bridging term and minor term are of experimental or empiric fact and thus not likely to be two versions in disguise. Perhaps it is the major term that is two (or more) in disguise. Could there be more than one effect of remifentanil? She then asked by what measure was the remifentanil effect determined? As it turns out, it is from measurements of remifentanil in the plasma. Is this notion of remifentanil the same as it might affect neuronal activity? In other words, is there a one-to-one correspondence between plasma remifentanil molecules and the excitability of neurons? One merely has to envision how the remifentanil molecules get to and affect the neurons, the latter by binding to the opiate receptors. Thus, there are many steps in between. Further, the process of concern is in the opposite direction. How do the molecules of remifentanil come off the opiate receptors and migrate out of the brain parenchyma back into the bloodstream to be carried away or metabolized? There is considerable Epistemic Distance and numerous Epistemic Degrees of Freedom between remifentanil molecules in the plasma and those molecules binding to the opiate receptors on neurons in the brain.

Framed as a metaphor (the Principle of Pseudotransitivity), the original metaphor is *remifentanil effects are to neurons as remifentanil is to blood plasma*. Again, the Epistemic Risk of asserting this metaphor is significant. The question is whether there is another metaphor with less Epistemic Distance and fewer Epistemic Degrees of Freedom? For example, the metaphor could be constructed as *remifentanil effects are to neurons as remifentanil effects are to an electroencephalogram* (EEG). In turns out that the latter metaphor has been investigated experimentally and that for many older individuals, the time for an EEG to return to near baseline may be considerably longer than five times the plasma half-life

(Minto et al., 1997). The investigator could then use age as a covariate in the analysis of the neuronal activity and then compare her results to those of others to determine whether the studies are reproducible of each other. However, it might have been better if the investigator recognized that virtually every experiment is at risk for the Fallacy of Four Terms, which should have been considered beforehand, perhaps waiting longer after the discontinuation of remifentanil before conducting the neuronal recordings.

The issue of the two metaphors, where one source domain is plasma remifentanil concentrations and the other is remifentanil effects on the EEG, illustrates the fundamental Principle of Causational Synonymy, appreciated since the ancient Greek philosophers but less so by modern biomedical researchers discussed in Chapter 4, The Logic of Certainty Versus the Logic of Discovery. In the situation of the remifentanil, the causal synonymy between neuronal spikes and EEG likely is closer than between neuronal spikes and plasma levels. Examples where violation of the Principle of Causational Synonymy resulted in unproductive irreproducibility may be common.

An extrapolation from the Principle of Causational Synonymy is the Principle of Informational Synonymy. Information defined here is nonrandom state changes, for example, the pattern of alphabetic letters, punctuations, blanks, and special characters that make up a sensible English sentence or the discharge pattern of neurons in the brain driving the specific patterns of muscular activations to produce meaningful behavior. Discussed in greater detail in Chapter 6, Causation, Process Metaphor, and Reductionism a brief example is discussed here. Consider Argument 5.10 made by a number of biomedical researchers and indeed those using pharmacology to infer physiology:

ARGUMENT 5.10

Major premise$_1$: *Administration of levodopa* (major term$_1$) *increases dopamine in the brain* (bridging term$_1$)

Minor premise$_1$: *Increasing dopamine in the brain* (bridging term$_1$) *improves symptoms due to dopamine deficiency by mimicking the effects of endogenous dopamine* (minor term$_1$)

Conclusion$_1$: *Administration of levodopa* (major term$_1$) *improves symptoms due to dopamine deficiency by mimicking the effects of endogenous dopamine* (minor term$_1$)

The effect of this is to establish an identity between *administrating levodopa* and mimicking of the effects of dopamine. This allows *administration of levodopa* to act as bridging term$_2$ (a negation) to allow *Symptom X* among the symptoms of Parkinson's disease to be held as not related to the endogenous actions of dopamine in the continuation of the argument that follows:

Minor premise$_2$:   *Symptom X of Parkinson's disease* (minor term$_2$) *does not improve with administration of levodopa* (bridging term$_2$)

Minor premise$_2$ is now combined with Conclusion$_1$ to result in the conclusion that follows. However, as will be demonstrated, the synonymy between administrating levodopa and the mimicking of endogenous dopamine in Conclusion$_1$ may not be true, thus creating two versions of the bridging term in minor premise$_2$, resulting in invalid Conclusion$_2$:

Conclusion$_2$:   *Symptom X of Parkinson's disease* (minor term$_2$) *is not a symptom amendable to mimicking the effects of endogenous dopamine* (major term$_2$)

Considerable evidence suggests that Conclusion$_1$ is false; consequently, the final Conclusion$_2$ is false. The administration of levodopa results in increased dopamine levels indiscriminately throughout the brain, having a continuous, although varying, effect over a timescale of hours. The normal application of dopamine onto neurons occurs over a space on the order of micrometers with a time course of tens of milliseconds. The information content of these two modes of dopamine application is very different. Consider a value of 0 when dopamine is not acting on the neuron and 1 when dopamine is acting on the neuron. A hypothetical example of the information applied to the neuron normally would be something like 0 0 1 0 0 1 1 1 1 1 1 0 1 0 0 1 1 0 0 1 over 0.2 second, whereas the information applied to the neuron in response to the pharmacological application of levodopa would be 1 1 1 1 1 1 1 1 1 1 1 1 1 1 1 1 1 1 1 1. From an information-theoretic perspective, there is very little information in the dopamine signal from the pharmacologically administered levodopa over the timescale being considered. To equate the two, as was done with Conclusion$_1$, violates the Principle of Informational Synonymy. To be sure, there are at least some aspects of the pathophysiology that do operate over the timescale of pharmacologically administered levodopa, but that is not evidence that other symptoms that perhaps require much greater information over shorter timescales are not dopaminergic in nature. This problem is discussed in greater detail in Chapter 8, Case Studies in Basic Biomedical Research.

It is highly probable that causal mechanisms operate on the basis of the information content. Thus, the Principle of Informational Synonymy could be considered as derivative of the Principle of Causational Synonymy. However, they differ in a significant way. Causal processes may be reversible; however, the loss of information is not unless it is written anew. This is because of the Second Law of Thermodynamic as Applied to Information (see Chapter 10: Chaotic and Complex Systems, Statistics, and Far-from-Equilibrium Thermodynamics). The Second Law holds that in a closed system, entropy can only increase. Information is considered the converse of entropy and thus the Second Law holds that in any closed system, information can only be lost.

Generally, conceptual or broad reproducibility relates to the consistency of general principles or concepts rather than in the specific particulars. When $sample_1$ and $sample_2$ are from the same species, in the most general sense of species, and $experiment_1$ and $experiment_2$ are exactly the same, this is a test of technical reproducibility in the narrow sense. The test of conceptual reproducibility is provided when the samples are of different species and/or if the experiments are different. But it is exactly these types of experiments that have the highest chance of gaining new knowledge but they clearly are at risk from the Fallacy of Four Terms as represented in Argument 5.11:

ARGUMENT 5.11

Major premise: $Sample_1$ (bridging term) *through Experiment 1 supports concept X* (major term)

Minor premise: $Sample_2$ (minor term) *is $sample_1$* (minor term₁)

Conclusion: $Sample_2$ (minor term) *through Experiment 1 supports concept X* (major term)

To the degree that the conclusion is true, the respective tests of *concept X* will be reproducible. However, this requires $sample_1$ to be the same as $sample_2$, but they are chosen to be different.

The resolution comes from the Principle of Causational Synonymy. It is the mechanism(s) in $sample_1$ that matches the mechanism(s) in the effect, the result of the experiment, that is of concern—any other properties of $sample_1$ notwithstanding. Similarly, the mechanism in $sample_2$ must be the same as the mechanism in the experiment that produced the same results. While this is an example of the Principle of Pseudotransitivity of the form *if a implies b is true and c implies b is true, then a implies c is true*, falsely; where *a* is $sample_1$, *b* is the experimental result, and *c* is $sample_2$. However, what is lost in certainty has a chance of gaining new knowledge. The more common the mechanisms between $sample_1$ and $sample_2$ are, the more likely the experiments will be reproducible. It would be most efficient to be as assured as possible that this is the case before the experiments are undertaken. The same concerns apply when the samples are the same but the experiments are different. Maximal reproducibility comes from comport with the Principles of Causational and Informational Synonymy.

THE CONFLATION OF ONTOLOGY AND EPISTEMOLOGY

The purpose of biomedical research is to gain new knowledge of biological reality (ontology) through experimentation (epistemology). But as discussed previously, ontology and epistemology are not synonymous and the danger is that they become conflated, resulting in confusion. The parable of five blind persons describing an elephant (the ontology) by their sense of touch (the epistemology) is apt; a version by John Godfrey Saxe's entitled "Blind Men and the Elephant" (from http://www.allaboutphilosophy.org/blind-men-and-the-elephant.htm) follows:

It was six men of Indostan,
To learning much inclined,
Who went to see the Elephant
(Though all of them were blind),
That each by observation
Might satisfy his mind.

The First approach'd the Elephant,
And happening to fall
Against his broad and sturdy side,
At once began to bawl:
"God bless me! but the Elephant
Is very like a wall!"

The Second, feeling of the tusk,
Cried, —"Ho! what have we here
So very round and smooth and sharp?
To me 'tis mighty clear,
This wonder of an Elephant
Is very like a spear!"

The Third approach'd the animal,
And happening to take
The squirming trunk within his hands,
Thus boldly up and spake:
"I see," —quoth he,— "the Elephant
Is very like a snake!"

The Fourth reached out an eager hand,
And felt about the knee:
"What most this wondrous beast is like
Is mighty plain," —quoth he,—
"Tis clear enough the Elephant
Is very like a tree!"

The Fifth, who chanced to touch the ear,
Said —"E'en the blindest man
Can tell what this resembles most;

Deny the fact who can,
This marvel of an Elephant
Is very like a fan!"
The Sixth no sooner had begun
About the beast to grope,
Then, seizing on the swinging tail
That fell within his scope,
"I see," —quoth he,— "the Elephant
Is very like a rope!"

And so these men of Indostan
Disputed loud and long,
Each in his own opinion
Exceeding stiff and strong,
Though each was partly in the right,
And all were in the wrong!

The last line is worth repeating with comments "Though each was partly in the right [*epistemically*], And all were in the wrong [*ontologically*]!" Also, it is fair to say that it would have been unreasonable to expect the six blind men to be right—ontologically.

In a sense, the blind men conflated what their sense of touch, their epistemic tools, with their predisposition to intuit a reality (ontology). It would be incredible of any person to stop at "this is what I feel with my hands" and not venture an opinion about what it is that the hand is feeling. But how is this any different from a biomedical researcher insisting that the mean (average) effect of some experiment represents some aspect of realty that is ontology? What researcher would stop at saying "the experiment produced a value for the mean given a mathematical operation on the values of these observations?"

As addressed repeatedly, the very large majority of researchers hold that the Central Tendency reflects some reality and, importantly, that the Central Tendency transcends the specifics of the particulars at any point in time, thereby avoiding the Solipsism of the Present Moment. But which measure of the Central Tendency, mean (average), median (the value at which half the observations are greater and half are less), or mode (the most common observation)? The question is resolved when the distribution of values of the observations is Gaussian (or other symmetric unimodal distributions) in which the mean (average), median, and mode are the same. But this is trivially true and does not defeat the question as to which method is the true Central Tendency.

The response is to look at the distribution. For distributions that are skewed, the median may be taken as the Central Tendency. But this is an epistemic response. Would one always insist that the median is the best measure of Central Tendency? What about the case when the numbers of observations are even numbers? The median is then calculated as the average between the two values that bracket the midpoint of the observations ordered in an ascending fashion. Thus, the median value is not even among the observations, but then again, the mean (average) may not be either. The mean (average) is affected greatly by outliers, which are observations far from the mean, because the differences between each observation and the mean are squared. But this feature of the mean may be valuable, as outliers may have the greatest effect on the concept or hypothesis under study. There are a number of different means, $\bar{x}$, such as shown here, where n is the number of observations, x_i, and each type of mean has its own unique value depending on the ontological concerns. Misusing them can result in unproductive irreproducibility.

Arithmetic mean (average)

$$\bar{x} = \frac{1}{n}\sum_{i=1}^{n} x_i = \frac{x_1 + x_2 + \ldots + x_n}{n}$$

Geometric mean

$$\bar{x} = \left(\prod_{i=1}^{n} x_i\right)^{\frac{1}{n}} = n\sqrt{x_1 * x_2 * \ldots * x_n}$$

Quadratic mean

$$\bar{x} = 2\sqrt{\frac{x_1^2 + x_2^2 + \ldots + x_n^2}{n}}$$

Many statistical tests compare the Central Tendency in one sample to the Central Tendency of another sample. However, as discussed previously, the question is what is the certainty of the Central Tendency as the value may change with repeated experiments? What this means is that the Central Tendency depends on the distribution (variance) of the individual values determined by an experiment.

Note that the answer to the question whether *sample₁* and *sample₂* are from the same population is purely based on analysis of the numbers, the epistemology, and not from any sense of reality or specifically some ontological notion of causation invoked by the experimental manipulation. Thus, the answer to the question is a mathematical one, that is, an epistemic answer. Extrapolating from the epistemic answer to an ontological one is fraught with peril. Yet every biomedical researcher makes such an extrapolation and therein lies the risk of unproductive irreproducibility because the ontological extrapolation from the epistemic becomes the testable hypothesis of a subsequent set of experiments.

SUMMARY

The certainty of syllogistic deduction comes at the expense of utility. Valid syllogistic deduction does not provide for new knowledge. In order to gain new knowledge, certainty must be sacrificed to utility. One method used to increase the utility of syllogistic deduction is use of the partial syllogism. However, this is of little help. Also, compared to the valid syllogism, the partial syllogism is a logical fallacy, often the Fallacy of Four Terms. To improve the utility of the partial syllogism, the probability syllogism has been constructed. To improve the certainty of the probability syllogism, statistics are applied. Nevertheless, statistically reinforced probability syllogisms are logical fallacies, particularly the Fallacy of Four Terms. It is only by understanding the logical fallaciousness that unproductive irreproducibility can be avoided.

6

Causation, Process Metaphor, and Reductionism

PRACTICAL SYLLOGISM AND BEYOND

Biomedical research would be completely reproducible or productively irreproducible if it were reduced to deductive logic to which modus ponens or modus tollens applied (see Chapter 4: The Logic of Certainty Versus the Logic of Discovery). Yet, little new knowledge would be gained other than recognition of tautologies. This conundrum led Aristotle to offer the practical syllogism increasing utility but losing certainty. To be explored is the possibility that the necessary use of practical syllogisms in biomedical research may contribute to unproductive irreproducibility.

The certainty of logical deduction is afforded by the use of the Principle of the Excluded Middle and the relationship between premises by state-of-being linking verbs. Loosening of the Principle of the Excluded Middle occurs with the partial syllogism, which is formalized by probability and statistics, the topic of Chapter 5, The Logic of Probability and Statistics. This chapter examines the effects of using action verbs, such as *cause*, in lieu of state-of-being linking verbs in conversion of the syllogistic deduction to the practical syllogism. This transition has a great impact on reproducibility in biomedical research, as the great majority of arguments entailed in experimentation are, in fact, practical syllogisms. Indeed, it is the practical syllogism that gives rise to the modus ponens-like form of the Scientific Method, which in reality is the Fallacy of Confirming the Consequence.

To reacquaint the reader, an example of a valid syllogistic deduction is given in Argument 6.1 and an example of a practical syllogism is given in Argument 6.2.

Reproducibility in Biomedical Research
DOI: https://doi.org/10.1016/B978-0-12-817443-2.00006-9   123

ARGUMENT 6.1

Major premise:	*All humans* (bridging term) *are of all things that are mortal* (major term)
Minor premise:	*Socrates* (minor term) *is a human* (bridging term)
Conclusion:	*Socrates* (minor term) *is mortal* (major term)

Argument 6.1 is valid because the state-of-being linking verbs, *are* and *is*, imply membership in a set. *All humans* is a set, which is completely contained within the set of *all things that are mortal*, and the set *Socrates* is contained within the set of *all humans*. Therefore, *Socrates* is contained within the set of *all things that are mortal*.

ARGUMENT 6.2

Major premise:	*All humans that are hungry* (bridging term) *will eat* (major term)
Minor premise:	*Socrates* (minor term) *is a hungry human* (bridging term)
Conclusion:	*Socrates* (minor term) *will eat* (major term)

Argument 6.2 uses an action (intentional) verb, *will*, instead of state-of-being verbs, *are* and *is*, as the linkage between logical elements. However, it is possible that Socrates, although hungry, may not eat, as the notion of *will* is unconfirmed in the present and only remains a potentiality. As *will* is not actual, it cannot be a state-of-being linking verb. The distinction among verbs, such as *is*, *are*, and *will*, may seem subtle; however, subtlety should not be confused for insignificance— quite the opposite (see Chapter 4: The Logic of Certainty Versus the Logic of Discovery).

The relationship between actuality and potentiality is complicated and has been recognized since the ancient Greeks, such as by Aristotle. The distinction is critical to biomedical research as experimentation starts with potentiality, the hypothesis, that ends in actuality, demonstrated predictions. Errors in the relationships between constructing the potentiality and executing the actuality could be a significant source of unproductive irreproducibility. This issue is discussed more later.

The fundamental axiom of deductive logic is that no argument with valid propositions and true premises can produce a false conclusion. Given that Argument 6.2 can produce a false conclusion, on the occasion that a hungry Socrates does not eat, the argument is invalid and some or all of the premises are false and/or some or all the propositions are invalid. In this case, it is the proposition's dependence on the linking verb *will* that renders the proposition invalid and causes the conclusion to be invalid by argumentation and uncertain.

Paradoxically, invalid Argument 6.2, not valid Argument 6.1, has the greatest scientific potential. If Argument 6.2 is found to be true by demonstration in every case, then the argument says a great deal about the notion of *will*. Also, as the verb *will* is unconfirmed in the present but is proven as a potentiality that will always be realized, *will* becomes synonymous with the notion of a certain cause, perhaps approximating the ontology of *is*. Any biomedical research involving a cause that is certain the linkage verb between its premises will approximate a state of being and will always be reproducible (by modus ponens). Note that one gets just as far by substituting "certainly not the case" for "certain case" and "never demonstrated" for "demonstrated by modus tollens."

Seldom can researchers be sure that the cause in relating their hypotheses to their predictions is a cause that is certainly true or a cause that is certainly false. Thus, the modus ponens and modus tollens forms of certainty generally are unavailable. However, the practical syllogism can be made more rigorous by combining the practical syllogism with the partial syllogism, as shown in Argument 6.3. Further, the

ARGUMENT 6.3

Major premise:	*Some humans that are hungry* (bridging term) *will eat* (major term)
Minor premise:	*Socrates* (minor term) *is a hungry human* (bridging term)
Conclusion:	*Socrates* (minor term) *may eat* (major term)

combination in Argument 6.3 can be made even more robust by converting to the probability syllogism in Argument 6.4. Argument 6.3 is more certain than Argument 6.2 because any instance when a hungry Socrates does not eat will not invalidate Argument 6.3.

ARGUMENT 6.4

Major premise:	*80% of humans that are hungry* (bridging term) *will eat* (major term)
Minor premise:	*Socrates* (minor term) *is a hungry human* (bridging term)
Conclusion:	*Socrates* (minor term) *has an 80% probability that he will eat* (major term)

Argument 6.2 can be reframed in terms of the Scientific Method into the testable hypothesis *if Socrates is hungry* (hypothesis) *he will eat* (prediction), *he eats* (prediction is true), *therefore, Socrates is hungry* (hypothesis is true). Despite the fact that the experiment rests on a logical fallacy (e.g., Socrates may eat for reasons other than being hungry),

nonetheless, the outcome is seen as proving the hypothesis. The point here is to consider what happens if Argument 6.4 is reframed in terms of the Scientific Method. It becomes *if Socrates is hungry* (hypothesis) *there is an 80% probability that he will eat* (prediction), *he eats* (prediction found true), *therefore Socrates is hungry.* But what does this do to the causal notion of hunger—is it or is it not the cause of eating? There are three choices. Either hunger when present: (1) is not causal to eating but rather other factors determine eating and hunger is epiphenomenal; (2) there are cofactors necessary to translate hunger into an effective cause of eating; or (3) hunger is 100% causal to eating, but because of biological or instrumental variability, eating is seen to occur only 80% of the time.

Regarding adjudicating between the alternative explanations why only 80% of those that are hungry eat, there is nothing in the experiment itself that allows one to select among the options. Logically the situation can be represented as *if (a or b or c) implies d is true and d is true, then (a or b or c) is true.* If one selects *a* over *b* or *c* as the explanation, one is engaging in the Fallacy of Affirming a Disjunct. If one is unlucky, an attempt at replication where the structure is *if a implies d is true and d is true, then a is true* when in actuality it was *b* or *c* that was true, then the experiment has a significant risk of being unproductively irreproducible. However, a plausible story, such as hunger causes eating, is intuitively appealing (Philip-Laird, 2013), which is only reinforced with attribution of an 80% probability. Thus, one is likely to conclude that hunger causes eating, but that only biological and instrumental variability causes the observed eating only to occur in 80% of situations.

CENTRALITY OF HYPOTHESIS TO EXPERIMENTATION AND CENTRALITY OF CAUSATION TO HYPOTHESIS GENERATION

It would seem rather obvious that any biomedical experiment would only be as good as the hypothesis. Further, in the situation of productive irreproducibility, then the hypothesis of the irreproducible experiment is clearly wrong (nothing is clear in the situation of unproductive irreproducibility). The concerned scientist not only accepts that the hypothesis was wrong, but wants to know why the hypothesis was entertained to begin with. Yet, there seems to be very little concern expressed by biomedical researchers as to the origins of hypotheses. There are those who suggest psychological causes such as certain predispositions not driven by logic or aesthetic reasons such as beauty or symmetry (Stewart, 2007). Perhaps most often a hypothesis represents a slight variation on a presumed true hypothesis (Kuhn, 1962). Assuming that the prior presumed hypothesis actually is true, then the experiments likely will be reproducible but not contribute much in the way of new knowledge. In other words, there is very little Epistemic Risk and the potential for new knowledge is proportional to the Epistemic Risk.

The very large majority of biomedical hypotheses are posited on a causal relationship between an antecedent and a prediction (consequent): smoking (the antecedent) and death (the consequent and prediction) where demonstration (or refutation) of prediction serves as a test of the hypothesis. There is extensive literature on the notion of causation and the topic will not be exhausted here. Rather, the concern is the risk taken when notions of causation that are too narrow or too broad. Too narrow and the hypothesis becomes a tautology—no Epistemic Risk, yielding little knowledge. Too broad and there may be too great an Epistemic Risk, leading to possible unproductive irreproducibility. Creating a productive sense of causation and recognizing the challenges to doing so are important to future reproducibility and productive irreproducibility in biomedical research. Interestingly, thinkers since the ancient Greeks, and likely before, have analyzed notions of causation and how they framed their questions are germane today, although their conclusions are historically constrained and consequently may be dated.

For the ancient Greeks, causation is derivative of the very definition of a biological or living organism. The key concept that differentiates living from nonliving is the distinction between animate (capable of self-generated movement) and inanimate (moved only when moved by an outside force), although viruses and prions pose challenges to those definitions. Since the time of the ancient Greeks, movement was considered in the widest sense and not just spatially as in locomotion. Change, such as the development of an organism, and the potential of change, such as a seed, were considered variations on the theme of motion. Aristotle grappled with the origin of motion, concluding that everything in motion derives from other things in motion and therefore motion requires a cause. The exception is the Prime or Unmoved Mover, which is the origin of all subsequent motion, but itself is not the consequence of some prior movement—appeals to the Big Bang do not answer the question, at least not intuitively.

The Epistemic Risk associated with positing the causal connection between the antecedent and the consequence in the experiment depends on the nature of causal connection. There are at least two aspects to the Epistemic Risk. The first is epistemic: how is the causal relationship to be proven? An association between the presence of the hypothesis, such as the proposal that disease *mechanism A causes disease B*, and the manifestation of a *disease B*, may result from induction—a type of correlation. But a demonstration that disease *mechanism A causes disease B* depends going beyond induction by enumeration to associations between *mechanism A* and *disease B*. Rather, attempts are made to establish some certainty of causation, for example, through John Stuart Mill's Methods of Induction described in Chapter 4, The Logic of Certainty Versus the Logic of Discovery.

Interestingly, a relationship between the antecedent and the consequence in the hypothesis cannot be exclusive or exhaustive for the purposes of establishing cause. Rather, it requires competing hypotheses with different antecedents and consequences, much in the manner of the importance of the logical construction *if and only if* to rescue the necessary Fallacy of Confirming the Consequence. Consider a set of causal mechanisms, *A*, *B*, and *C*, each thought possibly causal to *X*. If *A* causes *X* and *B* and *C* do not, then *B* and *C* must cause something else, such as *Y* and *Z*. If *B* and *C* did not cause something else but only cause *X*, then either *B* or *C* is *A* and *A* is causing *X* or *B* and *C* would not have been considered. Changes in *A* must be associated with changes in *X* but not in *Y* or *Z*, which is the basis for the Method of Differences, Method of Agreement, and Method of Concomitant Variation. If there was just *A* and *X*, then the association may be happenstance or a tautology, as neither provides a notion of cause.

The principal concern here is cause in the ontological sense. One sense of cause means to bring about, such as an effect, the prediction of which constitutes the biomedical experiment. Explanations of the nature of cause include what is shared between the cause and the effect to be a medium by which the cause creates the effect. Greater dissimilarity between the antecedent and the consequence in the hypothesis increases the Epistemic Risk and thus a greater risk for irreproducibility.

Another notion of the ontological nature of cause derives from the notion of purpose or *telos* (teleology). The notion of purpose implies a value judgment in the context of alternatives, whether actual or potential and whether in practice or in principle. To open the notion of cause based on purpose of *telos* to values is to invite anthropomorphism. *X* causes *Y* because it is in the organism's best interest for *Y* to occur and for *X* to be the cause. Indeed, it is the importance of *Y* that determines *X*.

In one version, *Y* is the actualization of *X*. For example, in motor physiology in complex organisms such as humans, a potentially infinite number of movements of the limb are possible. However, one does not simultaneously execute multiple movements, rather a single movement. The human tendency to anthropomorphize results in a hypothesized process where some agent selects one movement from a large number of movements, just as a human might choose a book from a library shelf of books. That needed role is assigned to the globus pallidus interna of the basal ganglia. It is as though a homunculus (miniature human) resides within the globus pallidus interna, surveys the array of potential movements, and then selects among them.

An overzealous homunculus does not select any movement and bradykinesia (slowed movement) and akinesia (absence of movement) of Parkinson's disease result. A careless homunculus allows unintended movements to escape, resulting in the involuntary movements of disorders such as Huntington's disease. The globus pallidus interna is posited as inhibiting the ventral lateral thalamus, thereby reducing thalamic drive onto the motor cortex, resulting in bradykinesia and akinesia—the Globus Pallidus Rate Theory. Yet, even at the time it was originally proposed, there was reasonable contrary evidence. Nevertheless, the intuitive appeal meant that the Globus Pallidus Interna Rate theory persisted for decades, misinforming hypothesis generation, misinterpreting results, and leading to conceptually irreproducible results, not to mention preventing research by those seeking to explore alternative theories. This story is reviewed in detail in Chapter 8, Case Studies in Basic Biomedical Research.

There may be a tendency to see the irreproducibility of experiments predicated on the agency theory described previously as an isolated case of particulars. However, the particular case is derivative of a specific ontological perspective that is pervasive in biomedical research. The failure of the specific case of the Globus Pallidus Interna Rate theory of an overzealous globus pallidus causing parkinsonism is symptomatic of a deeper flawed general perspective that also is fundamental to other particular cases, thereby risking unproductive irreproducibility in a much wider range of concern. For example, current theories posit an overzealous globus pallidus in terms of excessive beta frequency neural oscillations in the basal ganglia—thalamic—cortical system. As reality would have it, the theory of excessive beta frequency oscillation is also irreproducible because at least 15% of patients with Parkinson's disease do not have increased neuronal beta oscillations. Thus, excessive beta oscillation is not a necessary condition (Chapter 4: The Logic of Certainty Versus the Logic of Discovery). Further, Deep Brain Stimulation (DBS) in the beta frequency does not worsen parkinsonism; hence, increased beta oscillation is not a sufficient condition (Huang et al., 2014). The only conclusion left is that any demonstrated increases in beta frequency oscillations are epiphenomenal.

Historically antecedent to the Globus Pallidus Interna Rate theory was the Cholinergic/Dopaminergic Imbalance theory where a relative excess of acetylcholine (overzealous) and a relative deficiency of dopamine (insufficiently zealous) in the striatum of the basal ganglia were held causal to parkinsonism. This theory also is wrong because the premise of decreased dopamine producing parkinsonism is inconsistent with continued parkinsonism in the face of dopamine replacement, such as by fetal dopamine cell transplantation (Kordower et al., 2017).

All three irreproducible theories just described share the same conceptual notion of causation, which is a one-dimensional push–pull causal dynamics. The pathophysiological mechanism is a relative increase or decrease of some property. One has to wonder whether the common conceptual foundation is the source of the irreproducibility rather than the specifics of each theory (Montgomery, 2012, 2018). Actually, one-dimensional push–pull dynamics are ubiquitous in science, dating back to the ancient Greeks such as Aristotle's notion of the *Contraries*. Galen extended Aristotle's notions to define disease as relative excesses or deficiencies of the humors, and Benjamin Rush in the late 1700s posited that disease was caused by vasodilatation or vasoconstriction. Even today, many genetic disorders are characterized as a gain or loss of gene function (Ségalat, 2007).

As can be seen from this discussion, one mechanism used to make the understanding of causal mechanisms more tractable is to reduce the dimensionality; in other words, reduce the number of ways a cause or phenomenon can vary. Highly complex behaviors such as mood are reduced to a simple one-dimensional push–pull system of a relative deficiency of dopamine or norepinephrine. In some ways, the simplification of complexity to a simpler condition with less dimensions, hence fewer degrees of freedom, also may be seen as an anthropomorphism, that is, reducing the problem to one that is intuitively tractable by humans. Reality is not so obligated, but to insist on such simplification in the generation of hypotheses and predictions for biomedical research is to increase the risk of unproductive irreproducibility.

Teleological reasoning also derives from the realization of a phenomenon from precursors or potentialities. From the example of the role of the basal ganglia to produce movements, the selection of a movement for actual expression presupposes that the eventual movement exists in some latent fashion prior to the actual movement. The existence in a latent form is what Aristotle referred to as potentialities. The realization of the potentiality, a possible movement, is the actuality, the executed movement. Thus, for Aristotle, there were four linked causes that moved from potentiality to actuality. The proximate cause of actualities is the material cause—that which the actuality is comprised. For example, the bricks that comprise a house would be the material cause of the house. Next is the formal cause, that is, organization of the material cause, such as the arrangement of the bricks to form the house. Then there is the efficient cause, that is, the agent that arranges the bricks such as a bricklayer. The final cause is the purpose for which the bricks were arranged—the need for a house. In this schema, it is the final cause that determines the prior links in the chain of causal explanation. Bricks were selected because bricks, as opposed to marshmallows, are better suited to building homes. The problem is assuming some inherent homeness in bricks. If the homeness in bricks is taken too literally, then bricks will not be seen as capable of engaging in anything but homes. It is possible that an experiment is based on the homeness in a brick, but in a circumstance where there is no agent to ensure a home, a bridge may result—an example of irreproducibility.

One only needs to substitute a gene for a brick and the whole edifice of a protein, a cell, a tissue, an organ, and an organism is constructed. A particular gene becomes a particular protein that has "purpose" in a cell that has a purpose in a tissue, then an organ, then a system and so on to the organism. The gene that codes for oxytocin becomes the gene of social bonding. Given a teleological bias, the discussion of genes and social bonding becomes complete and little further attention given—a form of Solipsism. But the gene is not destiny. If it were, the US$3 billion spent on the Human Genome Project surely would have resulted in more than just a few approved gene therapies (discounting genetic tests for diagnosis). In a real sense, the conceptual reproducibility of genetics largely fails, which is not surprising given the lack of one-to-one correspondence between genotype and phenotype, which means the linkage from gene to behavior is very complex and, importantly, the causal mechanism inherent in the gene is insufficient to compel behavior just as in the practical syllogisms discussed earlier where the cause, hunger, is insufficient to compel eating. Now the attention has turned to epigenetics as a means to explicate the chain of causal events from genotype to phenotype.

It seems that humans, since infancy, are prone to think teleologically (Kelemen and Rosset, 2009). Unless trained to avoid default teleological thinking, much in the manner of being trained to give up Aristotelian physics in order to appreciate Newtonian physics, it is not surprising that biomedical researchers would be prone to teleological thinking and thus their research is at risk of unproductive irreproducibility. Indeed, any experimental hypothesis that depends on the notion of a purpose or function, such as *the function or purpose of X is Y*, the latter becomes the prediction and *X* the antecedent. However, many science teachers leverage the child's predisposition to teleology and thus teach from a perspective of teleology (Bernatowicz, 1958). The question is when will the vicious (in the philosophical sense) circularity end?

In some sense, teleology follows from the common conception of the Principle of Sufficient Reason, which holds that nothing happens without a reason, a remarkable claim of the centrality of causation (see Chapter 3: Validity of Biomedical Science, Reproducibility, and Irreproducibility). Perhaps the principle is true if for no other reason than the Conservation of Mass—Energy, meaning that Mass—Energy in total cannot be created (from nothing) or destroyed. Even the creation of subatomic particles in empty space produces both the particle and the antiparticle so in the end it is a zero-sum game. The greater problem is that humans cannot see at the subatomic level, only at the macroscopic level, where what appears to be causation is very different (discussed more fully in Chapter 10, Chaotic and Complex Systems, Statistics, and Far-from-Equilibrium Thermodynamics). Nothing happens without a reason and humans are very clever at inventing reasons. However, evolution requires no *telos*, human organisms form from the same basic laws of physics as a snowflake. While it is hard to impute a purpose or role to the formation of a snowflake, there seems to be relatively little reticence to impute purpose or role to biological processes, humans included.

Biomedical researchers, like all humans, tend to infer causation from temporal associations. For example, if event Y follows event X, X is thought causal to Y; this fallacious thinking is called *post hoc ergo propter hoc*. For example, consider a biomedical researcher who finds an organism behaving differently from expected. On further examination an unexpected protein is found. The natural inclination is to think that the abnormal protein found after the discovery of the unexpected organisms must have existed prior to the discovery of the organism and thus is causally related to the organism being such that it was unexpected. Such reasoning is very appropriate to generate a hypothesis for experimentation to determine whether or not the protein is causal to the expected features of the organism, but it does not constitute proof. Returning to the hypothetical X and Y, to be sure, it would be unlikely that X caused Y if X occurs after Y, but just because X preceded Y does not mean that X caused Y. It could be happenstance or epiphenomenal that Y just happened to follow X. There is an antidote, which is adherence to the Principle of Causational and Informational Synonymy, which is discussed more fully later.

REDUCTIONISM AND THE FALLACIES OF COMPOSITION AND DIVISION

The intuitive appeal, and thus perhaps a psychological need, to impute causation confronts a major epistemic challenge. As is true in the vast majority of circumstances of biological systems, no two phenomena are exactly alike, making attributions to common causes very problematic. At the extreme, it may be necessary to say that there are as many causes as there are phenomena, inviting a type of solipsism. But this approach is the antithesis of biomedical science in particular and science in general. The presupposition of science is that there is an economical set of fundamental causes that combines to produce all phenomena. In physics, the ultimate economical fundamental cause will be the result of the Theory of Everything uniting all the basic forces of physics, such as electromagnetism, strong and weak nuclear forces, and gravitation. To some extent, some of these already have been integrated to a degree, with gravitation appearing to be the holdout. Fortunately, biomedical researchers do not have to wait for a Theory of Everything but operate at a macrolevel resulting from the operations of these fundamental forces. (For the interested reader, the evolution from early thermodynamics in terms of temperature, pressure, and volume to considerations of microstates is a very interesting metaphor. This is addressed briefly in Chapter 10, Chaotic and Complex Systems, Statistics, and Far-from-Equilibrium Thermodynamics.)

The necessity, even if only psychological, of having an economical set of fundamentals by which all biological phenomena can be explicated leads to Reductionism (see Chapter 3: Validity of Biomedical Science, Reproducibility, and Irreproducibility and Chapter 4: The Logic of Certainty Versus the Logic of Discovery). Scientific reductionism requires deconstructing the complex whole, where causal mechanisms are difficult to infer, to simpler parts where causes are more discernible. This technical part of reductionism is relatively easy—one only needs a bigger hammer or sharper knife, metaphorically speaking. The greater problem, even if unrecognized by many (most) biomedical researchers, is how do the subsequent parts relate back to the whole? This question is the center of the philosophical discipline called Mereology attributed to Aristotle—but being a concern from a long-deceased philosopher, who in his own right was a great biologist, does not mean his concerns are not relevant today.

The value of reduction lies in reconstruction. In other words, the whole cannot be explicated whereas the parts can be. The next step is to reconstruct the whole from the parts and impute the explication of the parts, in terms of the economical set of fundamentals, to the whole. As will be seen, this is fraught with difficulty, as evidenced by the irreproducibility of biomedical research that involves inferences to the complex, such as the whole human, from simpler systems where those simpler systems are in vivo, in vitro, or in silico (risking the Fallacy of Four Terms, see Chapter 4: The Logic of Certainty Versus the Logic of Discovery).

Reductionism is at risk for the Fallacy of Division where the properties of the whole are extrapolated to the part and the Fallacy of Composition where the properties of the part are attributed to the whole. The two fallacies are complementary but not reciprocal (analogous to the concept that the specificity and sensitivity of a test are not reciprocal). The problem is addressed in terms of the Principle of Informational Synonymy and the irreversible loss of information in the process of reductionism. Interestingly, these fallacies are inherent in the

Cell Theory developed in the early 1800s and arguably marking the beginning of modern biomedical research. The Cell Theory was based on prior political philosophies such as those of Thomas Hobbes. Indeed, Hobbes used the human body as an analogy to the structure of society where parts of the body stand in relation to the political organization in his "Leviathan, or the Matter, Forme, and Power of a Commonwealth, Ecclesiasticall and Civil" (Hobbes, 1651). The frontispiece of the work conveys the use of the body as a metaphor for the state (Fig. 6.1).

Developers of the Cell Theory invoked the analogy that cells are to citizens as tissues are to villages, as organs are to states, and as the organism is to the kingdom. In a sense, this organization can be considered anatomical. Yet the attribution of purpose, function, or cause was in the reverse direction. The functions and properties of the kingdom were then extrapolated to the constituent parts, ultimately to the individual citizen. In a sense, the physiology of the kingdom was back-propagated onto the citizens. This also occurred in the initial attribution of purpose or function from the organ to the tissue and then the cell.

The situation is fairly clear in organs that are homogeneous histologically. It is fair to interpret the function of the hepatocyte from the function of the liver. It becomes a bit more complex to transfer the function of the heart onto the cells because there are different types of cells, myocytes of the myocardial syncytium and specialized cells for the conduction of electrical signals to excite the myocytes. The two would suggest different purposes, functions, and causes. One would presume that this process is more problematic when applied to an organ as diverse as the human brain, yet the attribution of the properties of the whole to a part has been fundamental to neuroscience ever since the ancient Greeks through the phrenology of Francis Gall, to the localization of function of Pierre Paul Broca, and now finding great currency in neurometabolic imaging, such as functional magnetic resonance imaging (fMRI).

The attribution of functions to specific body structures is not without difficulty. In neuroscience, one only has to consider the number of different functions or properties ascribed to the anterior cingulum. A cursory PubMed search based simply on "function of the cingulate gyrus" and selecting only reviews produces the following functions or properties attributed to the cingulum: (1) paradoxical sleep (Luppi et al., 2017); (2) pain (Coppieters et al., 2016); (3) empathy for pain (Chen et al., 2015); (4) psychosis (Underwood et al., 2016); (5) moral reasoning (Boccia et al., 2017); (6) addiction (Moeller et al., 2016); (7) cognitive flexibility (Bálint et al., 2015); (8) self-representation (Hu et al., 2016); (9) false memory (Kurkela and Dennis, 2016); (10) depression (Northoff, 2016); (11) extraversion−introversion in personality (Lei et al., 2015); and (12) face processing (Gentili et al., 2016). None of this discussion is to doubt the observations obtained or the inferences made. Rather, one

FIGURE 6.1 Frontispiece from Hobbes (Leviathan, or the Matter, Forme, and Power of a Commonwealth, Ecclesiasticall and Civil, 1651). By Unknown, available at: Public Domain, https://commons.wikimedia.org/w/index.php?curid = 226072.

has to wonder about the conceptual method of attributing properties to pieces of brain, particularly if one would hope for a one-to-one correspondence that would affect an economical set of fundamentals.

One sense of reductionism is an ontological one. This ontological reductionism is seen in experiments designed to dissect down to simpler components. The endpoint is a series of observable ontological entities (the result of methodological reductionism) that, when recombined, form an understanding of the phenomena (theoretical reductionism). There is another sense of reductionism that reasonably could be called more an epistemic form as the Process Metaphor discussed in Chapter 4, The Logic of Certainty Versus the Logic of Discovery.

There is an inherent limitation in the Process Metaphor in that it represents an extrapolation as the results of the extrapolation are not confined by boundary conditions as would be the case for interpolation (Fig. 6.2). With interpolation, the hypothesized entity is posited to be between other observable entities and the state of the hypothesized entity predicted is based on the behavior of the observable entities. As seen in Fig. 6.2, a line relating observables that vary over two dimensions is created as the best or most likely relationship. The hypothesized entity is "read" from the line. Unfortunately, a potentially infinite number of lines can contain all the observable entities and produce different values of the predicted entity. There is the presumption that picking the

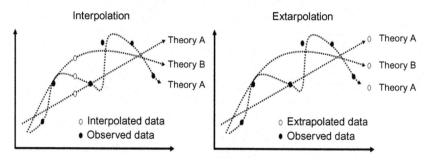

FIGURE 6.2   Hypothetical extrapolation and interpolation between experimental results. With interpolation, hypothesized values (interpolated data) are bracketed by the observed data. It is possible to draw a line through all the observed data points at the hypothesized values read from the line. While a potentially infinite number of lines can be drawn connecting the observed data, some limits or constraints can be applied, such as the degree of complexity of the lines (e.g., limiting the order of polynomial equations used to create the line). One could begin with the lowest order (such as a linear equation) and progress through higher order equations. All the while, one can compute the "goodness" of fit—a Process Metaphor. It is likely one will find progressively better fit with increasing order. It may be that the goodness of fit becomes asymptotic to some order and one can decide what order polynomial is sufficient. The situation is much more difficult in extrapolation where the fitting lines extend beyond the data. *Reproduced from Montgomery Jr., E.B., 2019. Medical Reasoning: The Nature and Use of Medical Knowledge. By permission of Oxford University Press, USA (In Press).*

simplest line is the best, following from Occam's Razor, but this is not certain, particularly when Chaos and Complexity supervene (see Chapter 10: Chaotic and Complex Systems, Statistics, and Far-from-Equilibrium Thermodynamics). The situation is worse when the entity to be predicted is not bound by observed entities, as is the case with extrapolation (Fig. 6.2). In this case, it is even more difficult to select the relationship represented in a fitted curve that would be best.

An example of this conundrum is represented in Fig. 6.3 of the relationship between the severity of Tourette's syndrome and age. Many clinicians have argued that the symptoms dissipate with age and thus invasive treatments such as DBS would be unnecessary as the patient gets older. Actual data published are shown in Fig. 6.3 (Leckman et al., 1998). Note that no patient over the age of 18 was included and even at age 18, patients still were symptomatic. The only way a clinician or scientist could argue that patients would not be symptomatic after age 25 is by extrapolation from the only known data; however, the dangers of

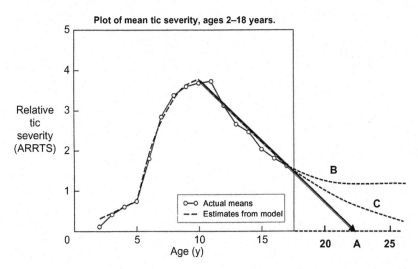

FIGURE 6.3 Plot of tic severity over time for data compiled for the first 18 years of life. One observes that data do not extend to age 25 years, which some experts would establish as the minimum age for DBS consideration. If one extrapolates linearly from the final phase of the model (line A), it appears that no patient aged 23 or older would experience symptoms. Because this clearly is not the case, the extrapolation must deviate from linearity. Yet the appropriate nonlinearity of the extrapolation cannot be known beforehand. Shown here are two hypothetical extrapolations that illustrate the implications. The extrapolation represented by line B suggests a curve asymptotic to a horizontal line, which in turn suggests that tic severity improves no further beyond some age. In such case DBS is warranted because the prospect of a patient improving spontaneously is unlikely. The extrapolation represented by line C, however, suggests continued probability of improvement beyond age 25 years (*Modified from Leckman et al., 1998; Montgomery, E.B., Jr., 2015. Twenty Things to Know about Deep Brain Stimulation. Oxford University Press*).

such extrapolation are demonstrated in Fig. 6.3. Any research based on such an extrapolation would be at risk for unproductive irreproducibility. Jankovic et al. (2010) found 46 adult patients (19 years of age or older) with Tourette's syndrome, of which 8, or 18.6%, experienced no tics prior to age 18 years. Pappert et al. (2003) reported a retrospective study demonstrated on videotapes of 31 subjects (mean age 24.2 ± 3.5 standard deviation years) with Tourette's diagnosed in childhood, 90% continued to experience tics.

OTHER FALLACIES AS APPLIED TO CAUSE

There are a number of other fallacies to which hypotheses of causation are heir to but not necessarily specific to causation, for example, the Fallacy of Limited Alternatives (see Chapter 4: The Logic of Certainty Versus the Logic of Discovery). A variant of the Fallacy of Limited Alternatives is called the Straw Man Argument: *if a or b causes c and c is true and b is blatantly false so as not even to require serious consideration, then a is considered true.* Premise *b* is the Straw Man. The potential for the Fallacy of Confirming the Consequence is compounded by the Fallacy of *Petitio Principii*, or question begging, where the experiment is structured so as to only result in the prediction and thus there is no possibility of rejecting the hypothesis by modus tollens.

Extrapolation of the Fallacy of Limited Alternatives to probability leads to the Gambler's Fallacy, that of confusing an *exclusive OR* for an *inclusive OR* (see Chapter 4: The Logic of Certainty Versus the Logic of Discovery). For example, a card player may think because the last hand produced a valuable set of cards then the next hand is more likely to be a good hand, if the player thinks she is on a streak. In fact, in a fair game, the quality of one hand is not influenced by the quality of the hand before and does not influence the hand later. The hands of cards are independent probabilities. The Fallacy of Limited Alternatives becomes the Fallacy of Affirming the Disjunctive of the form *if a implies (b or c or d) is true and d is found true, then b and c must be false.* The term *(b or c or d)* is the disjunctive. If *b*, *c*, and *d* are independent of each other, then any *b*, *c*, or *d*, but at least one, must be true because *a* is true (modus ponens). The premises *b* and *c* also could be true. Thus, any research that involves the Fallacy of Affirming the Disjunctive risks irreproducibility.

The disjunctive could involve the *exclusive OR* function, in which case only one of the components, *b*, *c*, or *d*, can be true. In this case, demonstrating that *d* is true necessarily means that *b* and *c* must be false. This is analogous to the use of *if and only if* to rescue the Fallacy of Confirming the Consequence. Use of the *exclusive OR* function prevents the Fallacy of Affirming the Disjunctive. Biomedical research confronted where the consequence (prediction) or an implication (the hypothesis) is a disjunctive requires knowledge or demonstration that the function is *exclusive OR* in order to avoid irreproducibility.

DISCIPLINE IN THE PRINCIPLE OF CAUSATIONAL AND INFORMATIONAL SYNONYMY

Humans have remarkable intuitive powers, as evidenced by humans capable of comparing apples and oranges in ways that befuddle the most advanced computers (although the gap appears to be shrinking). Indeed, humans are adept at ignoring logic and supporting logical fallacies when the fallacies are intuitively appealing, for example, by appealing to *teleos* or purpose (Johnson-Laird, 2008).

Research using the notion of cause as the bridging term in the major premise different from the notion of cause in the minor premise risks the Fallacy of Four Terms (see Chapter 4: The Logic of Certainty Versus the Logic of Discovery). Thus, it is important that the notion of cause be exactly the same, which is the Principle of Causational Synonymy. Another statement of the principle is that whatever is in the cause that produces the effect likewise must be in the effect. When causation is dynamic, that is, varying over time, such variation constitutes information and thus the Principle of Causational Synonymy can be extended to include the Principle of Informational Synonymy. Consider neuropharmacology where the argument is that many medications cause normal or restored functions. Many medications often are based on replicating the effects of a neurotransmitter. For example, dopamine derived from levodopa is thought to mimic the effect of dopamine that is lost as a consequence of Parkinson's disease. This rationale can be constructed in the form of a syllogism as shown in Argument 6.5:

ARGUMENT 6.5

Major premise:	*Dopamine deposition* (release from presynaptic terminals) *in the brain causes normal movement*
Minor premise:	*Levodopa causes dopamine deposition in the brain*
Conclusion:	*Levodopa causes normal movement*

Perhaps early in the course of Parkinson's disease, the patient's movements may appear to be normal, but certainly this is not true as the disease progresses. Even when the brain is flooded with dopamine, from either levodopa pills or fetal dopamine neurons, movements are not normal. Therefore, Argument 6.5 must be invalid because a valid argument with true premises cannot result in a falsehood. The question becomes which premise is false or proposition invalid?

One possible cause for the invalidity of Argument 6.5 is in the bridging term, dopamine. It may well be that the notion of dopamine in the major premise is different from that in the minor premise and the Fallacy of Four Terms supervenes. However, there does not seem to be any difference in the molecular form of dopamine as naturally occurring and that obtained by the pharmacological administration of levodopa, so in that sense, the Principle of Causational Synonymy does not appear to be violated. Perhaps the problem lies in the notion of deposition and is discussed in detail in Chapter 8, Case Studies in Basic Biomedical Research. Discipline by asking whether or not a research hypothesis to be subjected to experimentation violates the Principles of Causational and Informational Synonymy can help ensure reproducibility or productive irreproducibility.

SUMMARY

While syllogistic deduction offers maximum certainty, it offers minimum new knowledge. In order to gain scientific utility, complete certainty has to be sacrificed. In biomedical research, this is accomplished by translating the syllogistic deduction state-of-being linking verbs to action verbs, particularly use of the verb "cause." However, this results in an increased risk of irreproducibility.

The risk of irreproducibility can be mitigated to some extent by clear understanding of the notion of cause, particularly its connotations and motivations. Particularly troublesome are notions of purpose, *teleos*, and the temporal order of phenomena, the fallacy of post hoc *ergo propter hoc*. Further, increased reproducibility can be achieved by the disciplined insistence in conformity with the Principles of Causational and Informational Synonymy.

Case Studies in Clinical Biomedical Research

FORBEARANCE OF REPETITION

For those readers who first read Chapters 8 and 9 on Case Studies in Basic Biomedical Research and Case Studies in Computational Simulation, respectively, forbearance is asked for any repetition in the initial part of this chapter. Similar materials are reproduced here as some readers will focus on clinical biomedical research. Those who have read Chapter 8, Case Studies in Basic Biomedical Research and Chapter 9, Case Studies in Computational Biomedical Research may wish to go to the section entitled Clinical Meaningfulness.

This chapter examines a case of unproductive irreproducibility in clinical biomedical research. For purposes here, clinical research is not defined by the materials used, for example, humans versus tissue culture. Nor is clinical research defined in terms of the model used, be it health, disease, or experimentally induced abnormality. Rather, clinical research refers to the immediate purpose directly aimed at the care of living things, such as humans. Clinical research differs from basic biomedical research where the latter refers to the attempt to explicate the fundamental processes of life, with the distant but not insistent prospect of improving the human condition.

The case reviewed is illustrative but not exhaustive of the subject. Also, the primary focus is on the underlying logic, the term used in its widest context. For example, this chapter will not focus on statistical methods except to illustrate how the statistical methods employed illustrate the underlying logical concerns. As discussed in Chapter 5, The Logic of Probability and Statistics, statistics derive from providing a degree of certainty to the probability syllogism, which derives from needing to improve the certainty of the partial syllogism, which was needed to increase the utility of the syllogistic deduction.

Describing specific examples implies no criticism of the investigators, but only criticism of the modes of injudicious logical thinking to which every human is at risk. Rather, the criticism should be seen as an expression of respect and gratitude for the sincere efforts of those scientists and is made in recognition of our common frailties. The following dialogue from Richard Attenborough's movie *Gandhi* (1982) illustrates this point in the context for being a warrior for scientific purity. The original screenplay was addressing Gandhi's impending trip to Pakistan in an attempt to reconcile Hindus and Muslims:

> Margaret Bourke-White: *So what kind of warrior have you been in that warfare?*
> Gandhi: *Not a very good one. That's why I have so much tolerance for the other scoundrels of the world.*

SETTING THE STAGE

Notions of reproducibility were described in the local, narrow, and broad sense (see Chapter 1: Introduction). The local narrow sense relates primarily to replicability—can the exact same experiments using a second set of materials and methods produce exactly the same result, by the same experimenter (local) or a different experimenter (narrow)? In this case, whether different results (as would inevitably occur) constitute "exactly the same" for scientific purposes typically is a statistical consideration. Important are issues related to the sample size and randomization of the subjects, for example. Generally, if the assumption of ergodicity is met, the statistical concerns can be addressed by considerations of sample size (see Chapter 5: The Logic of Probability and Statistics). But, it is important to note that the principle of ergodicity in Complex and Chaotic systems may not hold and, thus, the Large Number Theorem may not hold as well (see Chapter 10: Chaotic and Complex Systems, Statistics, and Far-from-Equilibrium Thermodynamics).

Yet, while the same or different results often are couched in terms of statistical significance, it is far more complicated. Typically, the measure of a statistic used to establish statistical significance is a continuous variable. Yet, the decision, the same or different, generally is a dichotomous variable, which necessitates establishing a threshold or cutoff of the statistic, for example, $P < 0.05$. The concept might be if two studies where the results differ but the P value of the difference is > 0.05, then the two studies can be considered reproducible despite failure to produce the exact same result. However, if the P value associated with the difference is < 0.05, then the difference is significant and the studies are irreproducible with respect to each other.

The critical question becomes how to establish the threshold or cutoff. There is nothing within each study to settle the question. Rather, the threshold or cutoff is established to result in a compromise between the consequence of a type I error and the consequence of a type II error. The critical question is how are the consequences identified—what is the nature of the consequences (Montgomery and Turkstra, 2003)? As clinical biomedical research is directed at affecting the well-being of humans, the consequences must reflect all that is entailed in the well-being of humans. The choice becomes the threshold or cutoff such that a result in a clinical trial of *treatment X* for *disease A*, for example, to one side of the threshold is associated with what a reasonable human would consider a worthwhile benefit—a clinically meaningful difference, which is an ontological question.

The implication for reproducibility in clinical biomedical research is whether the difference between two studies involved in the demonstration of irreproducibility was less than the least clinically meaningful difference. From a clinical perspective, would these studies, although with different results, be irreproducible? Perhaps irreproducible clinical biomedical studies should only be applied when the difference in the results is greater than the least clinically meaningful difference.

It is unlikely that every human with *disease A* who receives *treatment X* will experience a clinically meaningful difference. The critical question becomes what probability of experiencing a clinically meaningful difference with the application of *treatment X* is sufficient to warrant *treatment X*? As can be seen, reproducibility in clinical biomedical research is complicated but attention to these issues is critical. These are the concepts that underlie the recent approaches to clinical noninferiority, nonsuperiority, superiority, inferiority, or equivalence (Walker and Nowacki, 2011).

Generally, with issues of narrow reproducibility, the issue relates only indirectly to the ontological nature of the hypotheses. In other words, the issue is not whether the hypothesis reflects a general principle or universal truth that can be applied to a wider range of instantiations or applications because the testing does not generalize to other instantiations. In a replicability study, the hypotheses are not different and the question predominantly is whether to accept the hypothesis when efforts of reproducibility fail. Thus, the question of reproducibility focuses on method, hence it is an epistemic question. Ontological and epistemological reproducibilities often are quite different. However, some scientists fool themselves by conflating or confusing replicability (narrow reproducibility) with conceptual reproducibility in the broader sense.

To be sure, an experimental question that is not replicable is unlikely to be conceptually reproducible. However, replicability is not a test of conceptual reproducibility or reproducibility in the broad sense. Thus, in most biomedical research, there is a combination of narrow and broad reproducibility that can confuse if not disentangled. The former relates to statistical significance, whereas the latter relates to clinical meaningfulness.

CLINICAL MEANINGFULNESS

An important and very common error is the claim of no difference between two conditions being compared, such as between two alternative therapies, when the experiment fails to reach a level of statistical significance to allow rejection of the null hypothesis. Such studies are not negative studies but rather null studies that are indeterminant. A negative study is one in which the conclusion, that of no difference, is made with high certainty and is an example of productive irreproducibility. Futility studies are an example. Consider an experiment where two groups are studied: one received *treatment A* and the other *treatment B* for *disease X*. The effect and variance, $e_A \pm vA$ and $e_B \pm vB$, of the *treatments A* and *B*, respectively, on a measure, x, of *disease X* is determined. A comparison is made, and the null hypothesis, that being no difference between e_A and e_B, cannot be rejected. It is concluded that no difference exists between therapies *A* and *B* for disease *X*, but such a conclusion is invalid and its truth or falsehood cannot be ascertained. Failure to reject the null hypothesis is not the same as accepting the null hypothesis in a definite sense as would be the case for negative studies. Rejection of the null hypothesis would be a positive study. Note that a null experiment is not a negative or positive experiment, it is indeterminant.

Perhaps a better way to appreciate the null hypothesis rather than framing it as whether the Central Tendency is the same in two or more samples is to ask if the samples came from the same population. The former method, differences in the Central Tendency, often is called the substantive null hypothesis—ontological—while the latter is referred to as the statistical null hypothesis—epistemological. Clearly, the substantive null hypothesis claiming no difference in the Central Tendency is an ontological notion given that the Central Tendency is extrapolated to be the reality of the sample. Arguably, the statistical null hypothesis does not focus on the Central Tendency but the distribution of data in the two samples being compared. The distribution or variance relates to the confidence one can have that the Central Tendency reflects the nature or ontology of the experimental question; hence the statistical null hypothesis is an epistemic concern. As the

Central Tendency represents an abstraction from data thought most clearly representative of the ontology, the substantive null hypothesis likewise is an abstraction.

It may be that the variances, v_A and v_B, are too large relative to the sample size to have a reasonable probability of detecting a difference (the meaning of statistical power) at a predetermined significance level (p value) for the difference in the effect sizes, e_A and e_B, but in truth, the effect sizes are clinically meaningfully different. If the biomedical researcher claimed that *treatments A* and *B* are the same because the null hypothesis could not be rejected, this would be an error. Martin Rees and latter Carl Sagan said, "The absence of evidence is not evidence of absence."

The appropriate approach is to first define the minimum clinically meaningful difference in the effect size. Then, based on the effect sizes and variances found for a given sample size, determine the probability of discovering the preset difference in effect size to a predetermined power and significance level. If sufficient power is found, then a claim of inferiority, superiority, or equivalence can be made (Walker and Nowacki, 2011) with a higher confidence that the results will be reproducible or productively irreproducible.

STATISTICS AND INTERNAL VALIDITY

As discussed in Chapter 5, The Logic of Probability and Statistics, statistics deal with internal validity. Indeed, statisticians maintain that a sharp demarcation is made between substantive and statistical hypotheses, the former related to ontology and the latter related to epistemology. Unfortunately, the two often are conflated, thereby increasing the risk for conceptual or broad unproductive irreproducibility. For example, an experimental argument starts from *hypothesis A* leading to *prediction B*. *A* may be a cause of *B*. The experiment looks for sets of *B*'s in the presence of sets of *A*'s (experimental group) and for sets with *A*'s absent (control group). Typically, some Central Tendency of the measures of *prediction B*'s in both groups is obtained, as well as the variance around the Central Tendency in each group. Differences in the Central Tendencies, such as effect size, are examined in the context of the variances around each Central Tendency. Note that the measure of Central Tendency and variance can be calculated from any distribution of observations, regardless of truth or falsehood of the observations or the validity of the calculations used to produce them. In other words, the Central Tendency and variance are not rooted in any specific way to the ontological nature (reality) of data as they are mathematical abstractions. Any linkage to reality or causal mechanisms, for example, from *prediction B* to *hypothesis A*, requires knowledge external to the statistical results.

What the results of the statistical analyses "mean" is a different question altogether and has to do with the hypothesis and the relationship between the hypothesis and the prediction. Usually, the relationship is causal (see Chapter 6: Causation, Process Metaphor, and Reductionism), and the validity, generally, is not established by the statistics. For example, a hypothesis that is false may still achieve statistical significant results and thought to be true (an example of a type I error). Indeed, that is the point of this entire book and why so much emphasis has been placed on understanding of the inherent logic, particularly the judicious and injudicious uses of logical fallacies. In clinical studies, it is generally appreciated, or should be, that statistical significance does not necessarily mean clinical meaningfulness (Montgomery and Turkstra, 2003). The goal of clinical research is to increase clinically meaningful knowledge, along with the means for assessing clinical meaningfulness.

Because clinical meaning inferred from biomedical experiments is not found in the statistics, the threshold for accepting a statistically significant result must also come from outside the experiment, such as past or anticipated experience. The statistical threshold sets the condition for invoking any sense of meaning—it does not establish meaning.

Decisions about the clinical meaningfulness of any clinical biomedical research typically invoke a dichotomous outcome, either the research result is or is not clinically meaningful in the context of a threshold. Most often, the threshold for statistical significance is taken for the threshold for defining meaningfulness. Typically, a determination of meaningfulness is established by rejection of the null hypothesis. But which null hypothesis, the substantive or the statistical? Typically, it is rejection of the statistical null hypothesis that "stands in" for the substantive null hypotheses. However, the statistical null hypothesis is purely operational—an epistemic assumption—and may not reflect the ontology. Rather, the threshold for statistical significance should be established on the basis of clinical meaningfulness prior to the experiment. The threshold depends on the consequence of a type I and a type II error. The consequences go beyond strictly medical and include social, psychological, economic, political, ethical, and moral concerns (Montgomery and Turkstra, 2003).

Clinical biomedical research was distinguished from basic biomedical research on the basis of intent, not subject material. For example, humans, typically the subjects of clinical research, also can be subjects in basic research where the goal is to elucidate some biological principle or universal claim regardless of any implication for improving the human condition. For example, a trial of *treatment Y* for *disease X* can be conducted solely to see if *treatment Y* has an effect on *disease X* independent from the concern whether *treatment* **Y** would ever be used to treat *disease* **X**. Indeed, establishing the effects of *treatment* **Y** on *disease* **X** may provide insight into questions of physiology and pathophysiology. In addition, the findings may be a proof of concept, perhaps leading to future research to find other treatments that ultimately are used to care for patients. However, the distinction between clinical and basic biomedical research needs to be kept clear.

Clinical research typically involves questions directly related to clinical care. In these cases, the risk of the Fallacy of Four Terms may be quite different in the nature of what "stands in," particularly in a sample of subjects standing in for the population of concern. In the case of biomedical research, great effort is expended to ensure that every subject indeed has *disease* **X** and every subject is compliant with *treatment* **Y**. This is done in order to increase effect size by excluding subjects without *disease* **X** who would not be expected to respond and to reduce variance that would happen should some number of subjects actually not take *treatment* **Y**. However, clinicians understand that patients in clinical practice may or may not have *disease* **X** and may not be compliant with treatment recommendations as subjects in clinical trials are required to be. Nevertheless, the recommendations by clinicians to patients, family members, and caregivers must account for the possibility that the patient may not have the disease and that the patient may be noncompliant as the clinicians cannot know *a priori* that their specific patients will be compliant. A lack of compliance affects the assessment of the ratio of risk to potential benefit, which is fundamental to any treatment decision. For example, the clinician recognizes that *patient* **A** may not be compliant with suggested *treatment* **B**. The potential benefit is reduced by the probability that the patient will not be compliant with the treatment. Similarly, the risk of adverse effects is also reduced by the failure of compliance.

The problem of noncompliance risks the Fallacy of Four Terms, as demonstrated in Argument 7.1 shown here:

ARGUMENT 7.1

Major premise: *X% of patients in sample$_1$ with disease Y entered in a trial and receiving treatment Z (bridging term) had a Q% improvement (major term).*

Minor premise: *Patients in sample$_2$ (or My patient) (minor term) have disease Y and will try treatment Z (bridging term).*

Conclusion: *A second sample (or my patient) (minor term) will have Q% improvement (major term).*

If there are two versions of the bridging term referring to receiving *treatment* Z, then there will be a risk of the Fallacy of Four Terms and the conclusion is invalid, again not true or false but whose truth or falsity cannot be ascertained. The problem is if every patient in the trial received *treatment* Z in the major premise and some subjects in *sample$_2$* (or *my patient*) did not take *treatment* Z, noncompliance, then findings based on *sample$_1$* risk not being reproduced when *sample$_2$* is studied or *my patient* risks not benefiting. Further, findings based on *sample$_1$* may be misleading relative to *my patient*.

To help ensure the reproducibility of the actual care of patients, the sample must reflect the prevalence of noncompliance. If, however, the bridging term in the major premise refers to all patients who entered the study, regardless of whether the patient was compliant, then the conclusion is more likely to be valid. This form of the argument is called an Intent to Treat paradigm versus based on a subgroup of those who were compliant, which is a Per Protocol paradigm. The Intent to Treat paradigm is the most likely to predict clinical use, hence clinical meaningfulness. Indeed, the use of Per Protocol rather than Intent to Treat paradigm may be a major cause of postapproval failures, such as drugs, biologics, or devices recalled by the US Food and Drug Administration (FDA) following pivotal trials that resulted in initial approval. Conflating Intent to Treat with Per Protocol paradigm results in misconceptions, but it is not an error that will be detected by statistical analyses. It is discoverable by critical analyses of the logical arguments entailed in the experiments.

ESTABLISHING CLINICAL MEANINGFULNESS

One approach used to estimate clinical meaningfulness from statistical significance is the Number Necessary to Treat. For example, consider *treatment* X to prevent *disease* Y—the measure is the number of subjects who would be needed to receive *treatment* X in order to prevent one subject from suffering *disease* Y. Translating the Number Necessary to Treat to a clinical decision depends on the costs, in the widest connotation of the term, to treat all those to prevent one case of *disease* Y and the attendant costs for the one case. The decision to dichotomize the clinical decision whether to apply *treatment* X is beyond the clinical trial as it requires the importation of values external to the trial itself. However, dichotomization into the clinical decision requires estimations of clinical meaningfulness, such as the Number Necessary to Treat.

Relating Number Necessary to Treat to reproducibility, two studies may differ in the reduction of risk, for example, and could be considered irreproducible. However, if the difference in risk reduction, while statistically significant, does not materially change the Number Necessary to Treat, are the studies actually irreproducible?

Similar concepts include the Odds Ratio (OR) and the Relative Risk (RR) ratio. The RR ratio is the probability of patients improving on medication divided by the probability of patients improving on a placebo; however, caution about assuming the placebo as inert must be respected. An example of the Odds Ratio is the (proportion with a given trait improved [or worsened] divided by [not improved, or not worsened]) (McGough and Faraone, 2009). Just as in the discussion of the Number Necessary to Treat paradigm, the OR and RR ratio require the importation of values external to the study that generated them. Again, the issue of what threshold in the ratios on which to base a clinical action depends on the cost of setting the threshold too high or too low. Nevertheless, the OR and RR ratio provide means to help in setting a reasonable threshold for clinical significance.

Correlational analyses usually operate on a continuous variable, for example, some *measure A* that correlates with *disease B* such that *measure A* can be used to diagnose the presence of *disease B*. As in all clinical decisions, a dichotomization is necessary, but the relative clinical meaningfulness of the correlation must first be established. One approach is to use logistic regression, where *measure A* is determined for the experimental group, such as a set of subjects with *disease B*, and a control group, such as a set of subjects without *disease B*. In both groups, *measure A* may be a continuous variable. The logistic regression provides a Receiver Operator Characteristics Curve (Fig. 7.1), which plots the true

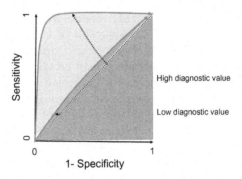

FIGURE 7.1 Hypothetical example of the Receiver Operator Characteristics Curve. The two curved lines connecting the diagonal points represent the plot of the sensitivity versus 1—specificity for a range of thresholds. For example, considered a threshold of some measure, x_t, the resulting proportion of true positives, the sensitivity, is plotted against 1 minus the proportion of false positives (1—specificity). Thus, the consequence of choosing any x_t on true positive and false positive rates can be determined. As can be seen, the plot of a test beneath the top maximally curved line has a high diagnostic value has a great departure from the diagonal. The blue plot of a test with a low diagnostic value has little departure from the diagonal and the area under the plot would be near 0.5. Plots with areas under the curve near 1.0 have the greatest diagnostic utility. In this case, diagnostic utility is directly related to reproducibility.

positives (those diagnosed with *disease B* based on *measure A* who actually have *disease B*, also known as the sensitivity) on the vertical axis versus the false positives (those diagnosed with *disease B* who do not have *disease B*, which equals 1 minus the specificity) for each value of *measure A*. For example, setting value *X* for *measure A* as the threshold likely will have a certain number of true and false positives, but how many false positives relative to true positives that are acceptable depend on the costs, considered in the widest connotation, involved. The diagonal on the Receiver Operator Characteristics Curve (Fig. 7.1) represents if *measure A* was not correlated with *disease B* and the values for the subjects were due to chance. The area under the Receiver Operator Characteristics Curve is a measure of the diagnostic utility, where an area = 1 is a perfect test and an area = 0.5 is just chance. The same situation prevails relative to clinical meaningfulness, as was seen in the Number Necessary to Treat, the Odds Ratio, and the Relative Risk ratio. In the case of the correlation, the threshold on *measure A* can be chosen based on the shape of the curve. In these situations, the threshold is at the inflection point of the curve. But note that this is a statistical decision not rooted in clinical meaningfulness. Alternatively, one optimizes the true and false positive rates to minimize the clinical cost.

Correlational analyses are particularly vulnerable to the Fallacy of Four Terms because the correlation procedure is a mathematically optimizing process. It will find the least sum of squares (a measure of the difference between actual data and the model for the regression) regardless of whether the variables are truly related. This is similar to the fact that the Central Tendency and variance can be calculated from any set of observations even if the observations are duckbill platypuses and humans. The risk is that the correlation may be idiosyncratic or relevant only to the sample on which the correlation was constructed. Consequently, the correlation may not be generalizable, which results in the Fallacy of Four Terms. To avoid the Fallacy of Four Terms, it is necessary to prospectively apply the regression equations to a second set of data, either independent of the initial set or by some randomized resampling of the initial set, to demonstrate the stability of the correlation result. Failure to prospectively test the regression analysis in a second sample increases the likelihood that the experiments that gave rise to the correlation analysis will not be reproducible and indeed risk unproductive irreproducibility.

For exploratory or preliminary studies, it may not be necessary to achieve the statistical threshold that vindicates the clinical meaningfulness. In this case, the primary question is not to extrapolate to a claim of clinical knowledge that will influence clinical decision-making. Rather, the goal may be merely to demonstrate whether further investigation is warranted. In this case, the exploratory or preliminary study is more akin to basic biomedical research. The same issues of experimental design and statistical analysis for studies attempting to demonstrate a clinically meaningful result that must be established prior to the experiments also hold true for exploratory or preliminary studies. Otherwise, these exploratory or preliminary studies will also risk unproductive irreproducibility. In these cases, the prior established cutoff for statistical significance, such as the p value, may be lowered or the sample size calculations based on a lower power $(1-\beta)$. However, data from such exploratory or preliminary studies should not be used in a subsequent study to establish a claim of clinical meaning lest one creates alpha inflation from multiple comparisons, increasing the risk of a type I error.

SPECIFIC FEATURES TO LOOK FOR IN CASE STUDIES

1. Is the experiment structured as the modus tollens-type Scientific Method, which would provide the highest certainty?
2. If the experiment is structured as the modus ponens-type Scientific Method (Fallacy of Confirming the Consequence), how close is the experimental logic to the form *if and only if the hypothesis implies the prediction...*?
   a. Have all reasonable alternative hypotheses been considered? If so, is the experiment constructed as to be able to adjudicate between the alternatives?
3. Will the argument posed by the experiments related to the particulars of the experiment generalize beyond the particulars? For example, the results in a sample must be generalizable to the population of concern. Typically, either the major or the minor premise of the corresponding syllogism relates to the particular and the other relates to the generalization. This difference creates the Fallacy of Four Terms in which there are two notions of the major, minor, or bridging terms. What is the Epistemic Risk due to the two notions of the same term?
   a. What is the Epistemic Distance between the two versions of the same term?
   b. What are the Epistemic Degrees of Freedom between the two versions of the same term?
4. In the syllogistic argument, is the particular a sample of the same species of the population of concern? If so, what is the Epistemic Risk?
   a. What is the Epistemic Distance, for example, how alike are members of the set of the sample to the set of the population of

concern? Do any differences exist in one or multiple domains or dimensions (such as age, sex, gender, or ethnicity)? Are the differences quantitative, thereby Epistemic Distance, or qualitative, thereby Epistemic Degrees of Freedom?

  b. Did the sampling process used to collect the sample mitigate differences in Epistemic Distance and Epistemic Degrees of Freedom between the sample and the population of concern?
    i. Randomization
    ii. Sequential recruitment
    iii. Counterbalanced assignments because of nonstationarity (the measures or effects are not stable over the time of observation)

5. Is the relationship sufficiently Complex or Chaotic such that the Theorem of Large Numbers does not apply?

CASE STUDY—TWO CONFLICTING STUDIES OF HORMONE USE IN POSTMENOPAUSAL WOMEN, WHICH IS IRREPRODUCIBLE?

In studying the issue of irreproducibility, Ioannidis collected a number of clinical research reports of all studies published in the three major or leading clinical journals between 1990 and 2003 that received at least 1000 citations (Ioannidis, 2005). These were the index studies. He then looked for subsequent articles addressing the same hypotheses (issues of concern) and with similar or comparable methodologies. He looked for instances in which the second or later study contradicted the index study categorized as contradicting (the strong case of irreproducibility), showing a different effect sized (and presumably different clinical meaningfulness), replicated, or unchallenged (no comparable study to the index study). Importantly, 45 of the 49 studies collected made a positive claim. Thus, should these studies be refuted, they would represent examples of type I errors. Of the 45 studies, 16% were contradicted, 16% were found to have smaller effect sizes in the subsequent studies, 44% were replicated, and 24% were unchallenged. More interesting is that of those that were challenged (34 studies), and hence presenting an opportunity to test reproducibility, 21% were contradicted, 21% were found to have smaller effect sizes in the subsequent studies, and 58% were replicated.

Ioannidis was careful to point out that the results of his analysis of the cases of contradiction or change in effect size did not indicate whether the index or subsequent studies were correct (Ioannidis, 2005). However, this represents a serious problem. Does one then conduct a third tie-breaking study? How about going for three out of five? Note that resorting to a meta-analysis may not be sufficient to adjudicate the discrepancies, particularly in the case of contradictions (Walker et al., 2008). Where does it end and at what cost? Clearly, some other analysis, calculation, or logic is necessary. But that analysis cannot be in terms of the results, that is, choosing from the different results as this would beg the question. Rather, one looks internally in each study to see if there is something in the rationale that allows a preference of one study over the other. However, this is dangerous, as discussed previously with respect to the Duhem—Quine thesis (see Chapter 4: The Logic of Certainty Versus the Logic of Discovery).

An interesting case relates to the use of hormones for the prevention of cardiovascular disease in postmenopausal women (note that the author is not an expert in cardiovascular disease or women's health). The index case selected by Ioannidis was the Nurses' Health Study (Stampfer et al., 1991). The subsequent study was the Women's Health Initiative study (Rossouw et al., 2002). The Nurses' Health Study found a 44% relative risk reduction in coronary heart disease events. However, in the Women's Health Initiative study, estrogen and progestin significantly increased the relative risk of coronary events by 29%. Ioannidis also pointed to a third study that also demonstrated an increase (Hulley et al., 1998). This third study will not be considered here, as the point is not to decide whether it is an ontological fact that postmenopausal women taking estrogens are at an increased or decreased risk of a cardiovascular heart disease event. Rather, the purpose here is to contrast the logic or rationale in the studies and how these differences influenced the interpretation of the results.

As will be discussed, the issue of postmenopausal hormone replacement is a critical question that centers not only on the ontology, whether there is benefit or harm, or only on epistemology, how does one know whether the reported differences are true or false, that they are real? To be sure, the purposes were to address the ontological issues in the context of the epistemic issues. However, as these were clinical biomedical research studies, there was a purpose, which was to affect treatment of postmenopausal women. Thus, the studies have to be understood in that clinical context. The ethics must be considered in the context of type I and type II errors. What is the medical, sociological, psychological, moral, economic, and political costs of not providing hormone replacement to postmenopausal women if there is a benefit versus those costs of providing replacement if there is no benefit (for a more general discussion, see Montgomery and Turkstra, 2003)? These considerations will be returned to later.

The Nurses' Health Study involved a cohort of 121,700 female registered nurses who completed questionnaires every 2 years and who were then followed for 10 years. The questionnaires assessed the postmenopausal use of hormones. Study endpoints included nonfatal myocardial infraction, fatal coronary heart disease, coronary-artery bypass grafting or angioplasty, fatal and nonfatal stroke, and total cardiovascular mortality. Medical record examinations and follow-up interviews or letters were done. The results reported were "Overall, the age-adjusted risk of major coronary disease among current estrogen users was about half that of women who had never used estrogen, with a relative risk of 0.51 (95% confidence interval, 0.37–0.70; $p < 0.0001$). For former users, the age-adjusted relative risk was 0.91 (95% confidence interval, 0.73–1.14; $p = 0.42$)." Interpretation of the finding regarding former users of hormones was problematic, as calculating a p value in the absence of any effect or departure from randomness is difficult. Currently, methods would use the estimated minimum clinical significant effect and then use the variance and sample size to determine the power of avoiding a type II error.

Assuming that the age-adjusted risk for former users is no different than nonusers, then the logic of the conclusion follows from Mill's Methods of Induction, specifically the Method of Differences (see Chapter 4: The Logic of Certainty Versus the Logic of Discovery). In this case, in condition 1, A and B are associated with X and Y, while in condition 2, C and B are associated with Z and Y. A can be women with a postmenopausal use of hormone, B is all other factors, C is former hormone users, X is less risk of cardiovascular disease, Y is all other diseases, and Z is unchanged risk of cardiovascular disease. From Mill's Method of Differences, A is associated with X, that is, postmenopausal hormone users had less risk of cardiovascular disease.

In the Women's Health Initiative, 373,092 women were screened, from which 8506 were randomized to treatment with estrogen and progestin and 8102 received placebo (Rossouw et al., 2002). The primary outcome was the incidence of coronary heart disease and the incidence of breast cancer was the secondary outcome, interestingly described as the primary adverse outcome. How the latter term differs from a second primary outcome or from a secondary outcome and its impact on statistical analysis is unexplained. In other words, the study was designed to have sufficient statistical power to detect the primary outcome measure, but it was unclear if it was sufficient to detect a secondary measure.

Typically, items described as secondary outcomes are not considered when determining pre-hoc sample sizes in order to assure sufficient power, presumably to make a study directed at the primary measures less demanding and reducing alpha inflation due to multiple comparisons (roll the dice enough times and you will get 7's), thereby reducing statistical power. However, after the fact, most articles discuss findings in the secondary measures as though they were primary measures without caveats or qualifications. It would seem to be a dodge. Indeed, purported negative findings in randomized controlled trials that are underpowered and found irreproducible in subsequent clinical use are the major reason why the FDA withdraws drugs from the market. In the Women's Health Initiative, subjects were followed biannually by completing a questionnaire. Any clinical events were evaluated in the clinic. The primary outcome was a hazard ratio of 1.29 for coronary heart disease (95% confidence interval, 1.02−1.63) for women randomized to estrogen and progestin relative to those randomized to placebo. The inference is an increased risk from hormones in contradistinction to the results of the Nurses' Health Study.

It is likely that the "take away" inference from the Women's Health Initiative would be that postmenopausal women should not take hormones from the standpoint of cardiovascular disease risk [considered as proposition *not Hormones* (¬*H*)], while the Nurses' Health Study suggests they should [considered as proposition *Hormones* (*H*)]. On the face of it, the inferences of both studies cannot be true and both cannot be false based on the Principle of the Excluded Middle (also called the Principle of Bivalence). Also, each study cannot be both true and false. As the physician cannot both prescribe and not prescribe hormones to the postmenopausal woman seeking advice and treatment, how is the physician to decide? Note that one cannot attempt to not decide, as the failure to provide a prescription is tantamount to making a decision not to prescribe. Furthermore, from the perspective of informed consent, the physician is obligated to discuss all reasonable options for the patient and all sides of the issues.

Some physicians will look to the salience of the respective studies but, unfortunately, salience may have more to do with newness or the most recent, thus favoring the take away inferences of the Women's Health Initiative study and not prescribe hormones. Similarly, studies that comport with the current ethos of the primacy of randomized placebo-controlled trials also may influence which study is taken for guidance. A PubMed review of citations of the two studies showed that the Nurses' Health Study was cited 149 times, while the Women's Health Initiative was cited over 2000 times. Remarkably, of the 140 of the most recent publications that cited the Women's Health Initiative, only three also cited the Nurses' Health Study as well. The wisdom of such preference will be discussed later.

Rather, there must be some other means with which to adjudicate between the two studies. The means to adjudicate are not just the conclusion, as this would be the logical fallacy of Confirmation Bias. Rather, there must be something in the logical chain of reasoning within each study and one of the studies must bear a fatal flaw. The problem is that the study alone cannot indicate which link in the chain of reasoning is false, thereby causing the failed conclusion. This follows from the Duhem−Quine thesis. For example, consider the

Principle of Transitivity *where a implies b and b implies c and c implies d,* where *d* is taken as the conclusion. The conclusion *d* is found false, which means that any of the premises, *a, b,* or *c,* must be false and/or any of the propositions, *a implies b* or *b implies c* or *c implies d,* must be invalid. The falsehood of *d* does not point to any individual premise or proposition.

Perhaps the most cited concern for irreproducibility is laxed statistical analyses. Thus, one approach to adjudicating between the contradictory studies is to contrast the statistical methods. In both cases, the primary outcomes were variations on relative risk, that is, the risk for coronary heart disease in postmenopausal women who took estrogen and those who did not. The ratio is derived by simple enumeration of those with and without coronary heart disease during the period of observation. This appears relatively straight forward, and assuming the methods for definition of the conditions—what is meant by postmenopausal and what constitutes "taking hormones"—there appears to be relatively little grounds to favor one study versus the other.

Perhaps one might look at the confidence intervals around the estimated relative risk. Wider confidence intervals indicate a larger variance, which would undermine confidence in the calculated relative risk. The sample size is a critical factor in estimating the variance and hence the confidence intervals. The question becomes: how does one compare the samples in the two experiments? Note that because the primary issue is hormone use, the duration of hormone use in each subject is the critical variable, which turns out is quite complicated (Grodstein et al., 2001). One approximation may be person-years, which is the duration of taking the hormone for each subject summed over all the subjects in the sample. For the Nurses' Health Study, the sample contained 337,854 person-years and the Women's Health Initiative contained 84,991 person-years. The Women's Health Initiative did not report the total person-years but rather the incidence of various outcomes was reported as the rate per 10,000 person-years; consequently, the total person-years had to be calculated based on the incidences and ratios reported (Rossouw et al., 2002). From this perspective of sample size, greater confidence would seem to be in the estimate of relative risk reported by the Nurses' Health Study.

Could the adjudicating flaw lie in the respective methods, such as in the definition of menopause? For example, surgically induced menopause may tend to occur among younger women who may have different confounding risk factors for cardiovascular disease compared to older women with natural menopause. The Nurses' Health Study included both natural and surgical postmenopausal women. The Women's Health Initiative study excluded subjects who had hysterectomies, which may have confounded comparisons of the two studies.

The Nurses' Health study accounted for risk factors related to cardiovascular disease, such as the prevalence of diabetes and hypertension in the study groups, and did not find obvious differences. The Women's Health Initiative study reported more extensively on confounding cardiovascular disease risk factors. There were no risk factors whose difference between the group treated with estrogen/progestin and the group treated with placebo was at the level of $p < 0.05$. However, there were a number of factors whose p value of the percent differences was < 0.10 but > 0.05, suggesting a between probability of 5 and 10% that the difference was due to chance. Further, a note of caution is needed as very large sample sizes tend to underestimate the variance and thus the p values. These risk factors include aspirin use, history of cardiac vascular procedures, and history of stroke, all of which were greater in those subjects randomized to placebo. Note that the difference between having a history of myocardial infarction was 0.7% in the treated group versus 1.0% in the placebo group with an associated p value of 0.14. A reasonable concern is that the placebo group had a greater risk for cardiovascular disease compared to the treated group but perhaps a lower risk based on aspirin use, 20.1% versus 19.1%, in the placebo and treated groups, respectively. Unfortunately, there is no way within the study to determine whether these differences are clinically meaningful and, in whose favor, placebo or treated group, they would fall.

Based on an analysis of confounding risk factors, there seems to be little reason to disqualify either of the studies. A similar analysis would suggest that ascertainment of data was not a significant factor in either study. Ninety-three percent of the treated group and ninety-four percent of the placebo-treated subjects completed the Women's Health Initiative Study. The follow-up rate in the Nurses' Health Study was 88.4% for nonfatal outcomes and 98% for fatal outcomes. Also, both studies were prospective, although one was placebo controlled and blinded. Nonetheless, being placebo controlled and blinded does not necessarily convey superiority of one particular study over another if both studies achieve the same "protections" from bias (see Chapter 5: The Logic of Probability and Statistics).

Other methodological issues used to account for the disparities between the Nurses' Health Study and the Women's Health Initiative have been considered. For example, the duration of hormone use and the timing of its use relative to menopause have been suggested as important factors (Schierbeck, 2015). The discussion here is not to defend any particular position or offer any clinical recommendations, but rather to explore the structure of the studies to illustrate issues of reproducibility.

As can be seen from the aforementioned discussion, there are no "smoking gun" statistical flaws that would provide an explanation for the irreproducibility. The implication is that measures proposed by journal editors and grant administrators to minimize unproductive irreproducibility would be of little avail in this particular case of irreproducibility. Where then to find an explanation?

Perhaps the problem lies in the syllogistic deduction that compares the two studies. The deduction can be framed as in Argument 7.2:

ARGUMENT 7.2

Major premise: *Patients in sample A* (bridging term) *had less cardiovascular disease events* (major term).

Minor premise: *My patients (sample B) are like patients in sample A* (bridging term).

Conclusion: *My patients (sample B) have less cardiovascular disease events.*

In this argument, sample *A* is from the Nurses' Health Study, whereas sample *B* is from the Women's Health Initiative. The conclusion is wrong in that the sample from the Women's Health Initiative did not have less cardiovascular events. The wrong conclusion suggests that one of the premises may be invalid. Suspicion falls on the minor premise, as the major premise was empirically validated in the Nurses' Health Study and there were little methodological or statistical flaws to invalidate the major premise. If sample *A* is different than sample *B*, the minor premise cannot be valid.

Both studies attempted to carefully characterize their samples, particularly with respect to obvious risk factors for cardiovascular events. The authors of the Nurses' Health Study attempted to account for other factors, including access to medical care. Interestingly, ethnicity was not reported. However, there are other possible confounds that differ among the samples of both studies. Interestingly, a demographic study of registered nurses responding to a mailed questionnaire in 1976 demonstrated that respondents typically were relatively young, bicoastal in residence, and likely to have postsecondary education (Barton et al., 1980). A similar analysis of the sample in the Women's Health Initiative was not provided, although the study did reference an article that discussed the purpose and general structure of the Women's Health Initiative and did not provide specific information relative to the postmenopausal study (Design of the Women's Health Initiative clinical trial and observational study, 1998). Nevertheless, it is reasonable to suspect that the two samples likely are different in important ways. Thus, the syllogistic argument that attempts to relate the two studies, given in Argument 7.2, is invalid to the degree that the two samples differ. It is possible to estimate the degree of difference based on Epistemic Risk or in probability terms; however, the latter will require common and quantifiable metrics.

While both studies cannot be true relative to each other, that is, expected to have the same results and conclusions, it is possible that both studies are true internally relative to the premises (data) and propositions (methods of analysis) within each study. Nonetheless, when confronted by the individual patients before them, the clinician must come to some decision that necessarily involves consideration of both studies. The argument for the clinician then becomes Argument 7.3:

ARGUMENT 7.3

Major premise 1: *Sample A is associated with risk reduction with hormone treatment.*

Major premise 2: *Sample B is associated with risk increase with hormone treatment.*

Minor premise: *My patient is more like sample A (exclusive) or is more like sample B.*

Conclusion: *My patient will be treated with hormones if my patient is more like sample A and not be treated if my patient is more like sample B.*

The decision becomes which study sample is relevant to the patient before the clinician? Will the clinician's subsequent experience for the individual patient likely to be reproducible of the study, that is, predict the response to hormones? This estimation is very dependent on the variability of the subject criteria for each study. For example, if the sample is very homogeneous, then it is relatively easier to estimate the degree the sample is representative of the patient. Unfortunately, neither study (and no others known to this author) provides the clinician with assistance. However, a follow-up study of the Nurses' Health Study suggests that a short-term use of hormones may increase the risk, while long-term use may lessen the risk (Grodstein et al., 2001). The Women's Health Initiative was stopped after an average duration of 5.2 years, while the Nurses' Health Study was conducted over a 10-year period of observation. Perhaps the irreproducibility is resolved when the duration of hormone treatment is factored in. However, such post-hoc reanalysis of a study always is problematic because of the issue of alpha inflation secondary to multiple comparisons and the risk of Confirmation Bias related to the Fallacy of Confirming the Consequence.

There is an interesting curiosity in the Women's Health Initiative that did not seem to garner much attention. The Women's Health Initiative was stopped earlier than planned. It was stopped by the data and safety monitoring board (DSMB), which noted a significant increase in the rate of breast cancer in the group randomized to hormones. Prior to the actual conductance of the study, the monitoring board established thresholds for safety that, if exceeded, would necessitate discontinuation of the study. There were 10 interim analyses with the final analysis demonstrating that the safety threshold for breast cancer had been exceeded. Note that the threshold for cardiovascular hear disease did not cross the threshold. The study authors wrote "On the basis of these data, the DSMB concluded that the evidence for breast cancer harm, along with evidence for *some* [italics added] increase in CHD [cardiovascular heart disease], stroke, and PE [pulmonary emboli], outweighed the evidence of benefit for fractures and possible benefit for colon cancer over the average 5.2-year follow-up period" (Rossouw et al., 2002). When addressing the cardiovascular heart disease risk, the qualification of *some* is disconcerting. If the qualifier *some* is an indication of some reservation regarding the risk of cardiovascular heart disease, then the main conclusion of the Women's Health Initiative is suspect and consequently cannot be used as evidence of irreproducibility of the Nurses' Health Study.

The point here is not to debate the ethics of the decision to stop the study. However, there are two interesting considerations. The study did demonstrate that should a subject develop breast cancer, she is more likely to have been treated with estrogen/progestin. However, as others have pointed out, serious cardiovascular heart disease is greater than the risk of breast cancer. As Schierbeck noted, "In Europe, 22% of all women die from coronary heart disease (CHD) alone, and 52% of mortality in European women is due to cardiovascular disease (including CHD, stroke and other CVD). Cancer is the cause of death in 18% of all European women, and breast cancer accounts for 3% of the total number of deaths" (Schierbeck, 2015). Assuming a 44% risk reduction following from the Nurses' Health Study and an increase in breast cancer of 26% based on a Hazard Ratio of 1.26, then there could be nearly 10 fewer deaths per 100 women from cardiovascular heart disease compared to an increase of 1 death per 100 women from breast cancer, as rough estimates. The caveat necessary to express is that this author is not an expert in breast cancer. However, one wonders whether the Women's Health Initiative study would have been stopped if the DSMB used clinical meaningfulness (relative risk for death) rather than statistical significance as the threshold. To be sure, the monitoring board only used data from the study as a basis for their decision, perhaps rightfully so. In that case, the monitoring board did not see any potential net gain in lives in terms of cardiovascular heart disease. Consequently, there was little choice, perhaps, but to end the study. Given the aforementioned discussion, it is not clear that continuing the study would have fully resolved the question of hormone use by postmenopausal women. However, clinicians cannot use such "technicalities" in deciding what to do for a patient. As discussed previously, the ultimate decision is an ethical one and it does not appear that these issues were widely considered.

More interestingly, it took 10 multiple interim analyses for the monitoring board of the Women's Health Initiative to discover when the risks exceeded the pre-established thresholds and reach a decision to end the study. Repeated multiple interim analyses constitute what is called multiple comparisons (Ludbrook, 1994). This raises the risk of a type I error of detecting a statistically significant difference when a difference does not truly exist. The analogy is rolling dice enough times that you will win, but likely only after a long string of losses. Thus, it is possible that the incidence of breast cancer or cardiovascular heart disease events could rise above the threshold based solely on chance.

The authors of the Women's Health Initiative argued that several of the interim analyses demonstrated either exceeding the thresholds or a trend in the direction of exceeding the thresholds. The implication is that even if one interim analysis demonstrated a type I error due to chance, surely multiple other interim analyses would not demonstrate such an error and the experience over the series of interim analyses would be a true indication of the state of the study. For example, a series of 10 tosses of a fair coin may result in a statistically significant increase in the number of "heads" that could be interpreted as a biased coin (a greater risk for cardiovascular heart disease or breast cancer). A second series of 10 tosses is more likely to result in a statistical equivalent to a 50% probability of "heads" consistent with a fair coin (no increased risk for cardiovascular heart disease or breast cancer), as would a third, fourth... series of 10 coin tosses, based on regression toward the mean. This would be expected if the probability of tossing a "heads" with one flip is independent of the probability of tossing a "heads" on a subsequent flip. However, if the series of coin tosses included data from the series that produced a statistically significant departure from chance, comparison of the series of coin tosses is not independent and the assumption that multiple comparisons (series of coin tosses) reflect the true nature of the coin is not warranted. In other words, the series containing the outlier (an unexpectedly greater number of "heads") contaminates the total experience if it is retained in the experience. Unfortunately, the methods used in the interim analysis were not described in sufficient detail to determine whether the findings had or did not have a high risk of a type I error. There are methods for dealing with multiple comparisons, particularly when they involve repeated measures, a discussion of which is beyond the scope of this effort and the reader is referred to publications elsewhere.

The use of hormones in otherwise healthy postmenopausal women dropped dramatically with publication of the Women's Health Initiative Study, but conclusions continue to be discussed and questions raised (Mikkola et al., 2017). It is not the intent of this text to offer any resolution of the controversies and certainly not to suggest any clinical course or management. Rather, this exercise is intended to explore issues of reproducibility in clinical trials. However, the analyses given here raise an interesting question going back to the study by Ioannidis that prompted the analysis (Ioannidis, 2005). Ioannidis' inference is that the Nurses' Health Study was irreproducible in light of the Women's Health Initiative study; however, a reasonable alternative interpretation is that the Women's Health Initiative was irreproducible in light of the Nurses' Health Study. But based on the analysis given previously, neither of these may not be the case as the studies may be too dissimilar. Thus, it is intriguing that the study claiming irreproducibility itself may be irreproducible. Then again, the analysis offered previously may be flawed and yet another analysis may demonstrate that the one presented here is irreproducible. Where does it end?

It just may be that no final and complete analysis is possible, particularly given the complexity of biological organisms and the fact that no theories or methods of analyses escape necessary simplifying assumptions and presumptions (see Chapter 3: Validity of Biomedical Science, Reproducibility, and Irreproducibility). Yet, at the same time, the necessary assumptions and presumptions that allow "progress" in coming to grasp the biological phenomena at the same time distort the very thing one is trying to grasp (the concept of Instrumental Uncertainty).

In the end, weighing the outcomes from competing findings may be less about the findings and more about the manner by which the findings were obtained. However, this puts the nonexpert reader at a significant disadvantage, being inexperienced in the intricacies of experimental design and statistical inference in a specific instance. Yet, the nonexpert is expected to act on the studies when the postmenopausal woman sits before the clinician and asks for help. The nonexpert can only depend on the article for insights to assess the soundness of the experimental design and statistical analyses; hence, depend on the authors. But how are the authors to know how much detail to put into the paper? Many journals now have authors putting more information into supplemental materials to reduce the size of journals, with the supplemental information being off-loaded to the Internet. It is an open question whether this facilitates understanding by the nonexpert.

Even with off-loading greater information to the Internet, the question becomes how much information should be put into the Internet-supplemental material? This does not answer the original question, it only changes the question. There is at least one clear answer, that is, for authors to compare and contrast their study relative to other studies, which certainly would highlight relevant issues of experimental design and statistical inference. It is interesting that this author could not find any mention of the original Nurses' Health Study in the report by the Women's Health Initiative (Rossouw et al., 2002). What was mentioned was the follow-up study that re-examined the cardiovascular heart disease risk over time (Grodstein et al., 2001).

Both studies are described and cited only rarely in any subsequent relevant study. It is as though the only aspect of each that received attention was the inferences in the conclusions. This raises serious questions about the place of scholarship in clinical biomedical research. Scholarship is far more than just running experiments and reporting results. One concerned about biomedical research has to be concerned. There may be many reasons why scholarship does not appear to be pursued. There is a tendency since the scientific revolution marked by the founding of the Royal Society in 1660 where the experiments spoke for themselves and the results taken as obvious. It may be held that the experiments need no interpreters from the past. Indeed, one of the major roles played by the Royal Society was public demonstrations of experiments, although the public was highly selective (Schaffer and Shapin, 1985).

It is not as though the practice of mentioning past and particularly contrasting studies had not been done before. Indeed, Aristotle would exhaustively discuss prior philosophers and their positions in great detail before offering his own. Rather, this unfortunate tendency of modern scholarship is relatively recent in modern science. Charles Bazerman analyzed scientific reports in the *Philosophical Transactions of the Royal Society* from 1665 to 1800 (Bazerman, 1997, pp. 169–86) describing four stages. During the period of 1665–1700, scientific papers were uncontested reports of events. From 1700 to 1760, discussions centered over the results were added. More theoretical aspects were addressed during the third period, 1760–1780, where papers "explored the meaning of unusual events through discovery accounts" (Bazerman, 1997, p. 184). From approximately 1790 to 1800, experiments were reported as claims for which the experiments were to constitute evidence, thus risking the Fallacy of Confirming the Consequence. Importantly, reports centered on claims derived solely from the experimenter's efforts and not "recognizing the communal project of constructing a world of claims.... Although the individual scientist has an interest in convincing readers of a particular set of claims, he does not yet explicitly acknowledge the exact placement of the claims in the larger framework of claims representing the shared knowledge of the discipline" (Bazerman, 1997, p. 184). Perhaps taking Aristotle's example could help resolve or at least put in perspective the current crisis of irre-producibility in biomedical research.

SUMMARY

As clinical biomedical research seeks to affect decisions that translate into better conditions for humans, the purpose is an important consideration. The tradeoffs between utility and certainty operate in clinical research but the certainty is interpreted in the context of clinical meaningfulness. However, experiments in themselves only attest to statistical significance and cannot comment on clinical meaningfulness. Statistical significance is internal to the experiment, whereas clinical meaningfulness requires the importation of knowledge that inescapably entails ethical and moral considerations. This dilemma is addressed by incorporating thresholds for minimal clinical meaningfulness in establishing statistical thresholds for significance.

As statistics was a means to "rescue" the probability syllogism, which was intended to rescue the partial syllogism, many statistical errors reflect the fundamental necessity of having to use logical fallacies. Consequently, comparisons between clinical studies that give contradictory results would appear to be clear indications of irreproducibility. However, rarely is it that simple, as the case of postmenopausal use of estrogen can attest.

Case Studies in Basic Biomedical Research

FORBEARANCE OF REPETITION

For those who come to this chapter having read Chapter 7, Case Studies in Clinical Biomedical Research or Chapter 9, Case Studies in Computational Biomedical Research on clinical research and mathematical and computational modeling and simulations, respectively, a bit of repetition of material is necessary for those reading this chapter without the benefit of having read Chapter 7, Case Studies in Clinical Biomedical Research or Chapter 9, Case Studies in Computational Biomedical Research. Those who have read Chapter 6, Causation, Process Metaphor, and Reductionism or Chapter 8, Case Studies in Basic Biomedical Research may wish to jump forward to the section entitled "What Is Basic Biomedical Research?"

PURPOSE

This chapter includes a case study of productive and unproductive irreproducibility in basic biomedical research. For purposes here, basic research is not defined by the materials used, for example, humans versus tissue culture. Nor is basic research defined in terms of the model, be it health, disease, or an experimentally induced abnormality. Rather, basic research refers to an immediate purpose that is not directly aimed at the care of things, such as humans. Instead, basic research refers to the attempt to explicate the fundamental processes of life, with the distant but not insistent prospect of improving the human condition.

Reproducibility in Biomedical Research
DOI: https://doi.org/10.1016/B978-0-12-817443-2.00008-2   165

The primary focus is on the underlying logic, the term used in its widest context. For example, this chapter will not focus on statistical methods except to illustrate how the statistical methods employed illustrate the underlying logical concerns. As discussed in Chapter 5, The Logic of Probability and Statistics, statistics derive from attempts to provide a degree of certainty to the probability syllogism, which derives from needing to improve the certainty of the partial syllogism, which was to increase the utility of the syllogistic deduction.

Describing specific examples implies no criticism of the investigators but only criticism of the modes of injudicious logical thinking to which every human is at risk for. Rather, the criticism should be seen as an expression of respect and gratitude for the sincere efforts and is made in recognition of our common frailties. The following dialogue from Richard Attenborough's movie *Gandhi* (1982) illustrates this point in the context for being a warrior for scientific purity. The original screenplay was addressing Gandhi's impending trip to Pakistan in an attempt to reconcile Hindus and Muslims.

> Margaret Bourke-White: *So what kind of warrior have you been in that warfare?*
> Gandhi: *Not a very good one. That's why I have so much tolerance for the other scoundrels of the world.*

SETTING THE STAGE

As discussed in Chapter 5, The Logic of Probability and Statistics, statistics deal with internal validity. For example, an experimental argument starts from hypothesis *A*, leading to prediction *B*. *A* may be a cause of *B*. The experiment looks for sets of *B*'s in the presence of sets of *A*'s (experimental group) and sets in which the *A*'s are absent (control group). Typically, some Central Tendency of the measures of prediction *B*'s in both groups is obtained, as well as the variance around the Central Tendency in each group. Differences in the Central Tendencies, such as effect size, are examined in the context of the variances around each Central Tendency. Note that the measure of Central Tendency and variance can be calculated from any distribution of observations regardless of truth or falsehood of the observations or the validity of the calculations used to produce them. In other words, the Central Tendency and variance are not rooted in any specific way to the ontological nature of the data—they are mathematical abstractions. Any linkage to reality or causal mechanisms, for example, from *prediction B* to *hypothesis A*, requires knowledge external to the statistical results.

What the results "mean" is a different question altogether and has to do with the hypothesis and the relationship between the hypothesis and the prediction. Usually, the relationship is causal (see Chapter 6: Causation, Process Metaphor, and Reductionism), and the meaning, generally, is not established by the statistics. For example, a hypothesis that is false still may achieve statistical significance in the results from the actual experiments based on the hypothesis; type 1 errors are examples. Indeed, that is the point of this entire book and why so much emphasis has been placed on inherent logic, particularly the judicious and injudicious use of logical fallacies. In clinical studies (see Chapter 7: Case Studies in Clinical Biomedical Research), it is generally appreciated, or should be, that statistical significance does not necessarily mean clinical meaningfulness (Montgomery and Turkstra, 2003). The goal of clinical research is to increase clinically meaningful knowledge, and clinical studies have the means for assessing clinical meaningfulness. Likewise, statistical significance in basic research does not necessarily translate into scientific meaningfulness, but such a statement begs the question that there is something like scientific meaningfulness analogous to clinical meaningfulness.

THE VALUE OF A TOOL FROM ITS INTENDED USE

The value of any tool can only be measured against its intended use. A sledge hammer may not be of much value for placing a nail in a wall by which to hang a picture, and a jeweler's hammer may not be much help in removing a concrete walkway. In the case of basic biomedical research, at least two purposes can be described, and each has different implications as to what defines reproducibility, typically through the choice of statistical measures. One purpose is ontological, making a claim as to some reality. The other purpose can be construed as epistemic where the intent is to validate a method, approach, or the reasonableness of posing a hypothesis for subsequent research. These experiments exist to determine the feasibility of a technology, method, or experimental design. Some experiments are preliminary because the costs to pursue experiments definitive to claim new scientific knowledge are too costly. This author often has heard researchers argue that costs, in the widest connotation, force them to start with lower species, such as rodents, when the ultimate knowledge claim to be demonstrated or refuted relates to humans. But the question is how many resources are wasted on experiments pursued in lower species, such as rodents, that result in unproductive irreproducibility when applied to humans because of the Fallacy of Four Terms (see Chapter 4: The Logic of Certainty Versus the Logic of Discovery)?

How then to judge the reproducibility of experiments with different intended purposes, hence meaningfulness? As discussed frequently, the meaningfulness is not derived solely from statistical significance, if only because it is meaningfulness that sets the threshold for what is considered statistically significant. Meaningfulness is derived from the consequences of an experiment that extend beyond the actual experiment as otherwise solipsism is risked. Thus, the consequences of an experiment to claim new scientific knowledge are different than the consequences of an experiment to determine the feasibility of an experimental design, technology, or methodology. Does this not mean that reproducibility also must be considered in the context of meaningfulness? Does this not mean in the broad sense of reproducibility that criteria for what is reproducible or not depend on the notion of meaningfulness, either ontological or epistemic?

These issues of meaningfulness are appreciated more readily in clinical biomedical research (see Chapter 7: Case Studies in Clinical Biomedical Research) where meaningfulness can be gauged on the impact on human lives. Even this is difficult, for example, what extension of life in the context of a terminal illness is meaningful or what are the consequences of a false positive or false negative, and with these concerns then establishing the p value for statistical significance (Montgomery and Turkstra, 2003)? For example, one could reasonably argue that there should be a more lenient or larger p value for statistical significance in experiments on rodents of a potential therapy for a human disease as the rodent research is considered more as an epistemic exercise. This might be done to avoid abandoning the potential effective therapy due to a type II error. Thus, the p value becomes the threshold for continued research, ultimately leading to human trials. Even at that, the choice of the p value cannot be too lax because there are consequences of a false positive.

What is clear is that it seems unreasonable to paint all biomedical research with the same brush, such as a default $p < 0.05$. Thus, applying the same criterion with respect to establishing reproducibility to every experiment will be problematic and potentially unproductive. To be sure, these issues complicate the efforts of biomedical researchers, but it is unlikely that Nature feels compelled to accommodate scientists. As Albert Einstein was reported to have said, "everything should be made as simple as possible, but not simpler!"

WHAT IS BASIC BIOMEDICAL RESEARCH?

In reality, the distinctions among basic, clinical, and computational biomedical research are blurred at best. With respect to any distinction between basic and clinical biomedical research, studying pathological conditions, such as a disease or vivisection, has been used since antiquity to understand basic biological processes. This is particularly true in neuroscience, as will be demonstrated in the following case study. To this day, most inferences to normal physiology have been by extrapolation from pathological conditions. Perhaps a reason for the reliance on pathological conditions comes from the complexity of the biology, such as neurological and psychological behaviors, and the Inverse Problem. As a result, the use of Mill's Methods of Induction, specifically the Method of Differences (see Chapter 3: Validity of Biomedical Science, Reproducibility, and Irreproducibility), are required to make mechanistic inferences, as demonstrated in Argument 8.1:

ARGUMENT 8.1

Observation 1:   *Conditions A, B, C, and D are associated with phenomena W, X, Y, and Z.*

Observation 2:   *Conditions B, C, and D are associated with phenomena X, Y, and Z.*

Conclusion: *Condition A is associated with phenomenon W.*

In this hypothetical example, Observation 1 could be considered the normal or intact state. Observation 2 could be in the pathological state where condition A is damaged or diseased, whether intentionally or as a result of natural causes such as disease or injury. In the normal or intact state, there is no way any of the conditions, A, B, C, or D, could be related to any of the phenomena, W, X, Y, or Z. It only is when *condition A* is removed, such as in the pathological state, that *condition A* could be associated with *phenomenon W*, which is absent in the pathological state. The hypothetical example described in Argument 8.1 could be changed such that Observation 2 is the normal state by which Observation 1 is the pathological state. In the normal state, there is no *phenomenon W*. It only is with the addition of *condition A* in the pathological state that *phenomenon W* is manifest.

This approach was formalized in the reasoning of the neurologist John Hughlings Jackson. He held that the symptoms and signs of neurological disorders, in this case phenomenon W, were the expression of either excessive function (positive symptoms and signs) or loss of function (negative symptoms and signs)—in this case *condition A*. In the case of positive symptoms and signs, there was a relatively straightforward inference from the symptom or sign of a neurological disease to the function of that part of the nervous system affected by the pathology. In the case of negative symptoms and signs, the function of the part of the nervous system affected by the pathology could be inferred from the inverse of the symptoms and signs. For example, jerking of the limbs associated with an epileptic seizure originating from the precentral (motor) cortex reflected the function of the precentral (motor) cortex in the generation of movement. Similarly, paralysis associated with damage to the precentral (motor) cortex reflected a loss of function, the inverse of which is the function to move the limb. As will be demonstrated in the case study that follows, biomedical research of pathological states can be considered basic research when the experiments are designed to elucidate basic biological mechanisms even if those mechanisms are inferred indirectly or inferred as the converse of the symptoms and signs of the pathological process such as disease. For example, biological mechanisms related to the functions of genes can be inferred from knock-in and knock-out genetic manipulations (Nilles and London, 2007; Picciotto, 1999).

SCIENTIFIC MEANING VERSUS STATISTICAL SIGNIFICANCE

Because the scientific meaning inferred from biomedical experiments is not found in the statistics, the threshold for accepting a statistically significant result must also come from outside the experiment. However, it is that statistical threshold that sets the condition for invoking any sense of meaning. The choice of the threshold derives from what the effects of a type I error and a type II error have on the anticipated scientific meaningfulness of the experiment. From the example of the Higgs boson, the choice was driven out of concern for a type I error (finding a difference when a difference does not truly exist, such as the difference of the weight of the Higgs boson from that due to chance). The past use of three sigma led to a claim that was not reproducible, a type I error.

It is interesting that type II errors (not finding a difference when one truly exists) do not appear to elicit the same level of concern as a type I error. Indeed, in the large majority of cases (probably nearly all), a failure of reproducibility begins with an initial positive claim to some new knowledge, and thus, the vast majority of concerns raised have been with respect to type I errors. In part, this may parallel the notion that errors of commission (causing a problem by taking an action) are somehow worse than errors of omission (allowing a problem by not acting) in ethics. Such a distinction generally is not valid. Perhaps the notion in biomedical science is that if a potential discovery is missed because of a type II error, then eventually the discovery will be discovered—as long as someone is willing to pay the bill. At the very least, concerns about waste would be reasonable.

The importance of disambiguating statistical significance from clinical meaningfulness is seen in recent developments in clinical biomedical research (see Chapter 7: Case Studies in Clinical Biomedical Research). Increasingly, clinical studies are reporting results based on recent developments in concepts of inferiority, superiority, noninferiority, nonsuperiority, and a combination of the latter two creating equivalence (Walker and Nowacki, 2011). No longer is equivalence defined as failure to reject the null hypothesis based on any effect size and given variance, a null study rather than a negative study. Rather, this approach requires first the establishment of an effect size that is clinically meaningful. Then the purpose is to demonstrate, based on power analyses, whether data should have been able to show a clinically meaningful difference in light of the variance in the outcome of the study. Any study with insufficient power is a null study from which no conclusions can be drawn (see Chapter 1: Introduction). It is not an instance of irreproducibility, rather it is one of indeterminacy. Similar approaches would be helpful in basic biomedical research.

The failure to disambiguate statistical significance from scientific meaningfulness obscures the issue of reproducibility in the local, narrow and broad senses. The narrow sense relates to replicability, which critically hinges on statistical significance. However, multiple repetitions of an experimental manipulation, whether in single or multiple subjects, are not likely to produce the exact result every time. Thus, some degree of irreproducibility is inescapable even in experiments that are considered reproducible when retested by replication or generalization. Yet, how much irreproducibility is allowable to be considered reproducible? The wise biomedical scientist looks to the degree the inevitable irreproducibility impacts the scientific inferences to be drawn. Intolerable irreproducibility is when the difference between studies is as great or greater than the least scientifically meaningful difference, whereas reproducible irreproducibility is when the difference between the studies is less than the least scientifically meaningful difference.

Biomedical researchers appear reluctant to make a distinction between statistical significance and scientific meaningfulness. Indeed, the default position of most biomedical researchers seems to be that any finding that is inherently different from what would happen by chance is meaningful, typically to a default level of $p < 0.05$. Some researchers may be of the opinion that data "speak for themselves" and that considerations of meaning are an unnecessary throwback to the metaphysical speculations of the medieval scholastic natural philosophers—the position of the early modern scientists who founded the Royal Society of London in 1660. However, this position is untenable (Montgomery, 2018).

Perhaps the difficulty is in understanding what is meant by scientific meaningfulness. Yet, there is scientific meaningfulness, as recognized by international awards for scientific contributions, such as the Nobel and Lasker prizes and the Fields Medal. Perhaps these recognitions are possible only because of the extreme nature of the contributions that received the recognition and therefore are not generalizable to the work-a-day requirements of the biomedical research enterprise. Perhaps the situation is analogous to the statement by US Supreme Court Justice Potter Stewart about obscenity: "I know it when I see it" (*Jacobellis* v. *Ohio*, 378 U.S. 184, 1964). The difficulty of the definition does not mean that it does not merit consideration. Indeed, the future of biomedical research depends on an allocation of resources that require some form of prioritization, which at least implicitly addresses the issue of scientific meaningfulness.

Some biomedical researchers may oppose the inclusion of scientific meaningfulness lest the notion of value judgments should sneak into the scientific enterprise. This is a controversial area that has been addressed by a number of authors (Laudan, 1984; Machamer and Wolters, 2004). Interestingly, the issue of scientific irreproducibility has been welded, perhaps as a cudgel, to raise doubts about the unique nature of science (Saltelli and Funtowicz, 2017). However, given the finite nature of biomedical research funding and the potentially infinite number of experiments that can be concocted, some prioritization must be done, which, of necessity, requires value judgments. However, the history of science demonstrates that value, as measured at the time experiments are considered, often does not predict the value in the future (described in the very entertaining book by Byson, 2003). There was great skepticism about the value of transistors when they were first developed. But it cannot be just "science for the sake of science."

Unfortunately, the notion of value in biomedical research conflates many different types of values. There are moral and ethical values—should a human be cloned, or should the stockpile of smallpox viruses be destroyed to ensure extinction of the species? The sense of value to be pursued here is the human value associated with scientific meaningfulness, but this is distinct from the sense of scientific meaningfulness intended here. The sense here is interpreted as the possible contribution to scientific knowledge and theory of demonstrating the hypothesis to be true or false, in the modus tollens sense, with experimental demonstration of the prediction divorced from ethical, moral, economic, or political sense. How much do the anticipated results of the experiment support and endanger all the reasonable relevant theories? Perhaps the greatest value is in those experiments whose prediction, when they fail to be demonstrated, do great violence to the currently favored theories—in the manner of modus tollens. Also of great value are hypotheses whose predictions resolve conundrums and discrepancies in current knowledge and theory when demonstrated.

REPRODUCIBILITY AND THE WILLINGNESS TO IGNORE IRREPRODUCIBILITY

As no hypothesis or experiment is perfect, a certain degree of tolerance (forgiveness) is necessary lest the accusation of irreproducibility is applied in a counterproductive manner. Yet, humans, including biomedical researchers, have an uncanny ability to not even see irreproducibility, particularly when what actually is irreproducible is intuitive (Johnson-Laird, 2008). The Copernican notion of planets in circular orbits around the sun still did not eliminate the need for epicycles (corrections or fudge factors depending on one's perspective) and, in many ways, was no better than the Ptolemaic earth-centered system. Nonetheless, the Copernican system came to dominate (epicycles were not eliminated until Johannes Kepler introduced elliptical orbits).

In his historical analysis of science, Thomas Kuhn (1962) raised the issue of when does a faulty theory finally reach a point with contrary evidence that the theory is thrown over and new theories are considered? Clearly, failure to throw out a paradigm that produced irreproducibility only increases irreproducibility in the future. Kuhn, as well as Paul Feyerabend (2010), demonstrated by historical analysis that many times the answer was political, sociological, or psychological and not "science" driven. Some have argued that a wider degree of tolerance for an emerging science may be necessary to ensure that it survives long enough to prove or disprove itself, as suggested by Thomas Kuhn (1962). Perhaps a greater tolerance for cases of irreproducibility would be reasonable, as discussed later for exploratory studies. Yet, precisely opposite should be the approach to challenging what is considered established knowledge. This author presented data to demonstrate that the Henneman's Size Principle regarding the progressive engagement of muscle unit activations with increasing force was not true (Montgomery, 2016). The response from a reviewer was that the Size Principle was considered a law of nearly 60 years and only the most extreme contrary data could be accepted to disprove the principle. But, longevity of a theory is not an indicator of truth, as otherwise everyone would still believe the earth is the center of the solar system. Nor does scientific truth majority rule.

For exploratory or preliminary studies, it may not be necessary to achieve the statistical threshold that vindicates the scientific meaningfulness. In this case, the primary question is not to extrapolate to a claim of scientific knowledge. Rather, the goal may be merely to demonstrate whether further investigation is warranted and whether it is a potential source of scientific meaningfulness. The same issues of experimental design and statistical analysis for studies attempting to demonstrate a scientifically meaningful result that must be established prior to the experiments also hold true for exploratory or preliminary studies. Otherwise, these exploratory or preliminary studies will also risk unproductive irreproducibility. However, in these cases, the prior established cutoff for statistical significance, such as the p value, may be lowered or the sample size calculations based on a lower power $(1-\beta)$. In a sense, such changes reduce the risk of irreproducibility, giving the new hypotheses some breathing room. However, data from such exploratory or preliminary studies should not be used in a subsequent study to establish a claim of scientific meaning lest one creates alpha inflation from multiple comparisons, increasing the risk of a type I error (see issue of repeated measures and the Brownian motion effect in Chapter 5, The Logic of Probability and Statistics).

Regardless of whether scientific meaningfulness can be explicitly defined and quantitated, even a vague notion of scientific meaningfulness with all the caveats and disclaimers should be considered in the design of the experiment, just as it should be in the design of the statistical analyses. However, while clinical research has measures to explicate and quantitate clinical meaningfulness (see Chapter 7: Case Studies in Clinical Biomedical Research), it is not so clear in basic research. Nevertheless, no experiment that is poorly designed from a logical perspective, in the broad sense of the term logic as discussed in this text, can be scientifically meaningful.

SPECIFIC FEATURES TO LOOK FOR IN CASE STUDIES

1. Is the experiment structured as the modus tollens-type Scientific method, which would provide the highest certainty?
2. If the experiment is structured as the modus ponens-type Scientific method (Fallacy of Confirming the Consequence), how close is the experimental logic to the form *if and only if the hypothesis implies the prediction...*?
   a. Have all reasonable alternative hypotheses been considered? If so, is the experiment constructed as to be able to adjudicate between the alternatives?
3. If the argument posed by the experiments relates to the particulars of the experiment, it must be able to generalize beyond the particulars. For example, the results in a sample, such as a group of rodents, must be generalizable to another larger group within the species and to any other species for which the set of rodents "stands in," such as humans. In this sense, the particular may be a sample of the

population or "stand in" for the population of interest. Typically, either the major or the minor premise relates to the particular and the other relates to the generalization. This difference creates the risk for the Fallacy of Four Terms in which there are two notions of the major, minor, or bridging terms. What is the Epistemic Risk due to the two notions of the same term?

    a. What is the Epistemic Distance between the two versions of the same term?

    b. What are the Epistemic Degrees of Freedom between the two versions of the same term?

4. In the syllogistic argument, is the particular a sample of the same species as the population? If so, what is the Epistemic Risk?

    a. What is the Epistemic Distance, for example, how alike are members of the set of the sample to the set of the population of interest? Do any differences exist in one or multiple domains or dimensions (such as age, sex, gender, or ethnicity)? Are the differences quantitative, thereby Epistemic Distance, or qualitative, thereby Epistemic Degrees of Freedom?

    b. Did the sampling process used to collect the sample mitigate differences in Epistemic Distance and Epistemic Degrees of Freedom between the sample and the population of interest?

        i. Randomization

        ii. Sequential recruitment

        iii. Counterbalanced assignments because of nonstationary (the measures or effects are not stable over the time of observation)

5. In the syllogistic argument, is the particular a sample that "stands in" of a different species as the population of interest? If so, what is the Epistemic Risk?

    a. Are there species differences, reduced preparations, or computational modeling and simulations?

    b. What is the Epistemic Distance, for example, how alike are members of the set of the "stand in's" to the set of the population of interest? Do any differences exist in one or multiple domains or dimensions? Are there unique circumstances that may increase Epistemic Risk, for example, special species of inbred animals where presumed genetic differences may underestimate the extent of the differences? Are the differences quantitative, thereby Epistemic Distance, or qualitative, thereby Epistemic Degrees of Freedom?

6. What is the relationship between the hypothesis and the prediction? If causal, does the relationship conform to the Principles of Causational and Informational Synonymy?

7. Is the relationship sufficiently Complex or Chaotic such that the Theorem of Large Numbers does not apply?

CASE STUDY—PATHOPHYSIOLOGY OF PARKINSONISM AND PHYSIOLOGY OF THE BASAL GANGLIA

The clinical phenomenology of disorders of the basal ganglia has been central to understanding the physiology of the basal ganglia specifically and the nervous system more generally. In this sense, research in the clinical phenomenology can be considered basic biomedical research (Brown and Marsden, 1998; Marsden, 1982; Marsden and Obeso, 1994; Thompson et al., 1988). This is particularly true of Parkinson's disease. Indeed, there are few neurological syndromes where there has been an apparent consilience of clinical phenomenology, histopathology, neurochemistry, and, to some extent, electroneurophysiology. Yet, despite the remarkable consilience, the field has been rife with irreproducibility in the broad or conceptual sense related to issues of basic science, as will be demonstrated. This case study was chosen because of the author's training and experience in this area.

Biomedical research into the pathophysiology of Parkinson's disease would appear to be an archetypical example of the success of the biomedical research paradigm. The milestones of that success include increased basic scientific knowledge about the: (1) anatomy of the basal ganglia and the basal ganglia—thalamic—cortical system; (2) the neurotransmitters mediating the functional connections; (3) demonstration that nigra-striatal neurons projecting from area substantia nigra pars compacta to the striatum utilize dopamine as the neurotransmitter; (4) degeneration of the substantia nigra pars compacta as a hallmark of Parkinson's disease; and (5) replacement of dopamine lost consequent to degeneration of the substantia nigra pars compacta reverses the symptoms, signs, and disabilities of Parkinson's disease. The intuitive "connections" between each of the bits of knowledge just described make a compelling story showing the triumph of biomedical research, but only when told in this manner. But that is not how it happened.

The history suggested previously is a Whiggish history written late in the course of time by the "victors." To be sure, this was the impression left by Yahr and colleagues writing "The demonstration within the past decade of the presence of dopamine in the brain localized chiefly in the striatum and substantia nigra and the subsequent discovery of its depletion have provided the rationale for new therapeutic approaches to parkinsonism" (Yahr et al., 1969). As will be demonstrated, this is not the case. Note that Yahr and colleagues wrote their article in 1969 after Cotzias and Papavasiliou published the first reports of effective levodopa therapy for Parkinson's disease in 1967 where previous attempts since 1961 had failed. History can be a great teacher but only if told truthfully.

The actual course of events (reviewed in Carlsson, 1971; Fahn, 2015) was quite different and likely owed its starting point to the discovery that reserpine produced akinesia in rodents and that the administration of racemic (D-L)DOPA reversed the akinesia. In 1958, this led to the suggestion of the metaphor (Fallacy of Pseudotransitivity) *if DOPA implies* (improved) *akinesia* (reserpine induced in rodents) *and if human Parkinsonism implies akinesia, then DOPA implies* (improves) *human parkinsonism* (Carlsson, 1959). Whether or not the demonstration of improvement in rodent catatonia consequent to the administration of DOPA meant the mechanisms were related to the conversion of DOPA to dopamine is unclear. Note that the argument to be advanced by experimentation could have been *if (DOPA increases dopamine) implies improved catatonia is true and improved catatonia is true, then (DOPA increases dopamine) is true.* Carlsson recognized the uncertainty caused by the appeal to the Fallacy of Confirming the Consequence, writing "the ability of DOPA to counteract the hypokinetic action of reserpine. Whether this action of DOPA is entirely due to formation of dopamine, or whether formation of noradrenaline contributes to the effect, remains an open question" (Carlsson, 1959).

In 1959, Carlsson wrote, "It is interesting to note that reserpine, which depletes dopamine from the corpus striatum, may produce a syndrome very similar to parkinsonism" (Carlsson, 1959). In 1961, Birkmayer and Hornykiewicz reported on the effect of DOPA on the akinesia of Parkinson's disease. In 1960, Ehringer and Hornykiewicz reported that the brains of patients with Parkinson's disease had reduced dopamine, along with reduced noradrenaline and 5-hydroxytryptopan. It was not until 1961 that Bertler reported that dopamine was found in the substantia nigra pars compacta. To be sure, pathological changes in the substantia nigra in patients with Parkinson's disease were noted in 1953 (Greenfield and Bosanquet, 1953).

It is difficult to reconstruct the intellectual timeline of the thinking about the pathoetiology of Parkinson's disease just on published reports. One does not know about informal conversations among the various actors. However, it is important to appreciate that there were alternative and competitive conceptions. There were differing opinions about the underlying pathology associated with Parkinson's disease as noted by Greenfield and Bosanquet (1953). The conundrum of equating dopamine depletion with Parkinson's disease was discussed in the condition known as Symptoms without Evidence of Dopamine Depletion in Chapter 4, The Logic of Certainty Versus the Logic of Discovery. Further, there were a number of different animal models for the hypokinesia and akinesia of Parkinson's disease (reviewed by Poirier, 1971). Nevertheless, it appears that the initial observations of the effect of DOPA on reserpine-induced catatonia in rodents was the precipitating event in the evolution of levodopa for the treatment of Parkinson's disease.

Interestingly, the use of levodopa for the treatment of Parkinson's disease in humans was an example of conceptual or broad irreproducibility, at least circa 1966. Initial studies of levodopa for humans with Parkinson's disease were a failure, clearly not replicating the beneficial effects in rodents, thus implicating the Fallacy of Four Terms (Fahn, 2015). It was not until 1967 that Cotzias and colleagues demonstrated a benefit from levodopa, but only after what was at that time administration of very high doses of levodopa. When asked, Cotzias said the reason for their success was that they could follow patients in the hospital for months, which allowed them to avoid side effects, which had limited so many previous studies. But why that explains how Cotzias and colleagues were able to achieve such high doses of levodopa, it does not explain why Cotzias wanted to achieve those doses.

Interestingly, Cotzias was not operating under the assumption that levodopa was replacing the lost dopamine. Rather, Cotzias endeavored to replace the neuromelanin pigment known to be lost from the substantia nigra pars compacta—perhaps influenced by his interest in melanin, manganese, and manganese-induced extrapyramidal syndromes (Cotzias, 1966; Cotzias et al., 1964a,b).

In retrospect, Cotzias and colleagues were right for the wrong reason. However, the curiosity here is the conceptual irreproducibility between the animal studies and the initial human studies. Further, it is difficult to understand this irreproducibility as productive or unproductive. It is clear there was some injudicious use of a logical fallacy. One could argue that the animal and initial human studies represented the Fallacy of Four Terms as in Argument 8.2:

ARGUMENT 8.2

Major premise:	*Parkinsonism* (in reserpine-treated animals) (bridging term) *improves with levodopa* (major term).
Minor premise:	*Human patients* (minor term) *have parkinsonism* (bridging term).
Conclusion:	*Human patients* (minor term) *should improve with levodopa* (major term).

The likely problem is that there are two versions of the bridging term, that is, *parkinsonism* in rodents is not the same as *parkinsonism* in humans. Hence the conclusion is invalid and, in the case of the initial clinical trials of levodopa, the conclusion was shown to be false. Alternatively, the source of the fourth term in the Fallacy of Four Terms may be the major term *levodopa*. It just may be that the mechanisms of action of levodopa in the rodent are not the same mechanisms in the human. There is ample precedent for such differences in species, as was the case in the discussion of thalidomide in Chapter 1, Introduction.

Consider, however, the situations with later clinical trials with high-dose levodopa. The syllogism becomes that in Argument 8.3:

ARGUMENT 8.3

Major premise:	*Parkinsonism* (in reserpine-treated animals) (bridging term) *improves with levodopa* (major term).
Minor premise:	*Human patients* (minor term) *have parkinsonism* (bridging term).
Conclusion:	*Human patients* (minor term) *should improve with high-dose levodopa* (major term).

The fact that the conclusion is true necessarily means that the premises must be true and the argument valid. There cannot be the Fallacy of Four Terms. This means that the form of parkinsonism in rodents must be the same, relative to the effect of levodopa, as the form of parkinsonism in humans treated with high doses of levodopa. Further, the true conclusions of the syllogism mean that the levodopa use in rodents is the same as the high dose of levodopa in humans.

This discussion demonstrates the irreproducibility of the claim of a conceptual irreproducible levodopa effect in reserpine-treated rodents and the initial human trials. This is not unlike the scenario in the postmenopausal studies of estrogens and cardiovascular heart disease discussed in Chapter 7, Case Studies in Clinical Biomedical Research. Had the irreproducibility of the initial human treatment trials of levodopa held sway, it is difficult to know if levodopa would have been available to patients with Parkinson's disease today. From a practical perspective, it is great news that Cotzias was not cowed by the failure of the early trials of levodopa. However, it is important to know that his relative disregard for the previous levodopa trials had nothing to do with the early trials but rather with the presumed mechanisms of action, dopamine versus neuromelanin replacement. In some way, the situation is similar to Galileo's praise of Copernicus for ignoring the prevailing arguments against the heliocentric position, particularly when predictions from the Copernican model were less reproductive of the actual movements of the planets than the geocentric system of Ptolemy and the movement of objects on earth (Feyerabend, 2010). Sometimes it is better to be lucky than just smart (or logical).

The experiences described here did much to enshrine the notion of Parkinson's disease as a dopamine deficiency where increasing the amount of dopamine in the brain was the means of relieving the symptoms, signs, and disabilities of Parkinson's disease. Indeed, this was the rationale for the trials of fetal dopamine neuron transplantation in humans with Parkinson's disease. The syllogism underlying the trial is shown in Argument 8.4:

ARGUMENT 8.4

Major premise:
Parkinsonism (major term) *improves with increased amounts of dopamine* (bridging term).

Minor premise:
Fetal dopamine neuron replacement (minor term) *increases dopamine* (bridging term).

Conclusion:
Fetal dopamine neuron replacement (minor term) *should improve parkinsonism* (major term).

Prospective controlled studies demonstrated the conclusion false—patients with Parkinson's disease did not improve following a fetal dopamine neuron transplant (Olanow et al., 2003). It was not the case that the fetal-derived dopamine neurons did not take root or failed to produce dopamine in the basal ganglia. Indeed, many of the transplanted patients developed involuntary movements (runaway dyskinesia) analogous to what is seen in patients given excessive doses of oral levodopa.

As the conclusion of syllogistic Argument 8.4 was false, either some premise was false or the structure invalid, such as the Fallacy of Four Terms. The question is which term—major, minor, or bridging—was in actuality two terms leading to a total of four terms? It is unlikely that there are two versions of the major term, parkinsonism. The major term, *parkinsonism*, in the major premise are those patients with Parkinson's disease who improved with the oral administration of levodopa with the inference that this led to increased dopamine in the basal ganglia. There are little external reasons to suspect that these patients are fundamentally different from the patients with parkinsonism (major term in the conclusion). It is not likely that there are two versions of the minor term, *fetal dopamine neuron replacement*, that are different in the minor premise than that in the conclusion. Both arise from the same experiment.

Suspicion falls to the bridging term—*increased dopamine*. Perhaps it is the manner by which the dopamine was increased. Indeed, discussions subsequent to the demonstrated failure of fetal dopamine neuron transplantation in the controlled trials focused heavily on the manner by which the fetal cells were transplanted, as well as on patient selection (Barker et al., 2013). Note that the issue of patient selection is not to suggest that the failure of syllogistic Argument 8.4 was due to the major term, parkinsonism, as being different in the major premise and conclusion. Rather, the issue is that a certain type of patient, the exact type to be defined, is more suited for fetal cell dopamine neurons and, in the future, stem cell transplant.

According to the Duhem—Quine thesis (see Chapter 4: The Logic of Certainty Versus the Logic of Discovery), the failure of fetal dopamine neuron transplants in the prospective control trials does not provide information as to which among the various factors is at fault. Information external to the trials must be brought to bear, and indeed, has been, as will be demonstrated. Curiously absent, among all the discussions of all the factors and elements of the logic underlying the use of dopamine cell transplants, is the possible fact that such transplants just will not work. There is nothing in the research experience that says dopamine cell transplants will work once the technicalities have been worked out. To be sure, there have been small-scale open label studies with positive outcomes, but the uncontrolled nature and the placebo effect render such studies suspect. Further, examination of actual data increases suspicion as demonstrated in Fig. 8.1 from the review by

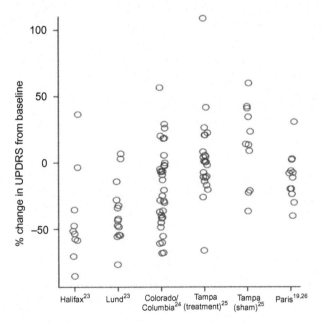

FIGURE 8.1 Change in Unified Parkinson Disease Rating Scale (UPDRS) scores for patients enrolled in ventral mesencephalic transplant trials. Data are taken from studies undertaken at five North American and European centers. Circles represent the percentage change from baseline in the UPDRS score for every patient at every center. Source: *Reprinted with permission from Barker, R. A., Barrett, J., Mason, S. L., & Björklund, A. (2013). Fetal dopaminergic transplantation trials and the future of neural grafting in Parkinsons disease. The Lancet Neurology, 12(1), 84—91. doi:10.1016/s1474-4422(12)70295-8.*

Barker et al. (2013). Yet, the tone of the review by Barker and colleagues was optimistic about dopamine cell transplants and never mentioned the possibility that it just may be that such transplants will not work.

In a sense, the conceptual irreproducibility seen in the fetal dopamine transplant studies represents unproductive irreproducibility in that the results could not point with confidence to a specific direction for future research. Consider the situation where there are five factors thought relevant to the success of a dopamine cell transplant. There are 31 different combinations of those five factors, which means that, potentially, 31 different clinical trials would be necessary to explore each factor and combination of factors. While not impossible in principle, it is unclear whether this would be possible in practice. Thus, some prioritization of the factors is necessary but the information on which to base such prioritization cannot come from the previous controlled trials.

It would seem the motivation propelling future studies is the notion that dopamine replacement, in this case by cell transplant, just *should* work. To be sure, the success of orally administered levodopa would be supportive of this faith, but perhaps it is not that simple. A clinical trial was conducted of three different doses of levodopa and a placebo in patients with Parkinson's disease not already taking medications for the disorder and who were early in the course of their disease (Fahn et al., 2004). Fig. 8.2 shows the group changes over the subsequent 42 weeks. The purpose of the study was to determine whether levodopa worsens

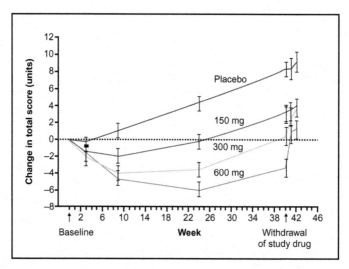

FIGURE 8.2   Changes in total scores on the Unified Parkinson's Disease Rating Scale from baseline through evaluation at week 42. Source: *From Fahn, S., Oakes, D., Shoulson, I., Kieburtz, K., Rudolph, A., Lang, A., et al., 2004. Levodopa and the progression of Parkinson's disease. N. Engl. J. Med. 351 (24), 2498–2508.*

parkinsonism over time, and the primary outcome measure was the measure of the Unified Parkinson Disease Rating Scale (UPDRS) following discontinuation of the levodopa. The point here is that following an initial improvement, the disease severity as measured by the UPDRS worsened and for the group treated with 150 and 300 mg of levodopa as the total daily dose, they worsened to the same disease severity as when they started. To be sure, the placebo group worsened and the gap in the UPDRS between those on the placebo and those treated with levodopa was maintained throughout the study. Thus, it is reasonable to infer that levodopa does improve the symptoms of Parkinson's disease, but it also suggests that the manner of improvement is not simple.

One could argue that the differences within the three doses reflect just a dose—response curve. However, there was no difference in the change in UPDRS at two weeks among the groups and no differences between the 300 and the 600 mg doses up to eight weeks. The question then becomes why the departure in the UPDRS scores after eight weeks? It may just be that with the progression of the underlying disease, patients just required more dopamine that first outstripped that provided by 150 mg of levodopa then that provided by 300 mg and finally that provided by 600 mg.

Research on the physiology of dopaminergic neurons inferred from pharmacological administration of dopamine, via levodopa, or by dopamine neuron transport necessarily involves the Fallacy of Four Terms. The Epistemic Risk is determined by how closely the normal dynamics of dopamine action match the dynamics of pharmacological or neural transplant dynamics. The Epistemic Risk can be estimated by applying the Principles of Causational and Informational Synonymy. Permit a crude analogy, one does not fix a broken computer by lifting the lid and throwing transistors into the computer, whether by methods of low spatial and temporal specificity—pharmacological administration of dopamine—or dopamine cell transplants of unknown spatial and temporal specificity. Clearly, the organization and dynamics of the computer are far too complex for just a bunch of transistors thrown in. It is reasonable to suggest, based on the Principles of Causational and Information Synonymy (see Chapter 6: Causation, Process Metaphor, and Reductionism), that the dynamics (changes in state over time) that parallel the normal dynamics of dopamine physiology would be more likely to improve parkinsonism. Just such a notion has been implicit in much past and current thinking into the physiology of the basal ganglia. It is just that the dynamics are nearly static or, at the very least, operate over very slow (minutes to hours) timescales. Thus, there would appear to be no conceptual discrepancy between the dynamics of pharmacological administration of dopaminergic agents and normal dynamics.

The perception of slowly changing or static normal dopamine dynamics led clinicians/scientists to argue for continuous levodopa therapy (Olanow and Stocchi, 2017). The authors wrote, "The concept is based on the hypothesis that intermittent doses of standard oral L-dopa/carbidopa do not restore brain dopamine in a physiological manner and contribute to the development of motor complications" with the implication that continuous dopamine administration was more physiological. They wrote, "These observations have led to the hypothesis that continuous delivery of L-dopa with stable plasma levels might restore brain dopamine in a *more physiological* [italics added] manner than intermittent oral levodopa doses and avoid motor complications." They cited a study employing microdialysis sampling from the caudoputamen of rodents (Miller and Abercrombie, 1999). However, the investigators in that study sampled interstitial dopamine levels over 15 minutes. Based on the Nyquist theorem, any dynamics (changes) faster than 30 minutes would not have been detected reliably. As an analogy, picture vigorous and rapid dancing illuminated by a flashing stroboscopic light. As the flashes of light are slower than the dance movements, one sees strange, although interesting, dance movements, certainly not what would be seen if the light was flashing much faster so as to have a continuous effect. Hence, the aforementioned statement by Olanow and Stocchi (2017) that the normal dynamics of dopamine action are static is unfounded.

The claim that an unchanging dopamine release in the brain is a direct contrast from actual recordings of dopamine neurons in the substantia nigra pars compacta in nonhuman primates is shown in Fig. 8.3 (Schultz, 1986). In these experiments, a nonhuman primate was trained to rest its hand on a key. In response to a signal, for example, the door to a box containing a food reward opening, the animal would reach for the food. The time the key was released indicated the animal starting an arm movement to reach the reward. Simultaneously, dopamine neurons were recorded and the time of each action potential neuronal discharge, presumably reflecting the time course of dopamine release into the striatum from the dopamine neurons in the substantia nigra pars compacta, could be plotted over time for each trial, producing a raster display. Multiple trials could be collapsed into a histogram to demonstrate average trends over time.

As shown in Fig. 8.3, some neurons demonstrated an increase in the number of action potentials at the start of the movement that persisted over many seconds. However, other increases lasted on the order of 0.1–0.3 second (Fig. 8.4). Clearly, the activity of the dopamine neurons and presumably the time course of dopamine release from the dopamine neurons were not static and changing over timescales on the order of seconds and not more than the 15 minutes required to have been

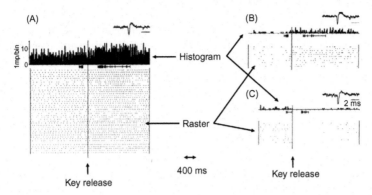

FIGURE 8.3 Analysis of recording of action potential discharges from dopamine neurons in the substantia nigra pars compacta in a nonhuman primate while the animal performs a task. The animal is trained to hold down a key. In response to a cue, the opening of a door to a box that contained a food reward, the animal moved its limb so as to release the key and reached up to retrieve the food reward. There are two displays for each of the three neurons (A, B, and C) represented. In the raster, each dot represents the discharge of the neuron. Each row is a trial of the task. Thus, one can appreciate the timing of neuronal discharges relative to the key release. As can be seen for neurons A and B, there is an increase in neuronal activities and presumably a release of dopamine in the striatum on the order of several seconds. For neuron C, there is a reduction of neuronal activity over a few seconds (Schultz, 1986).

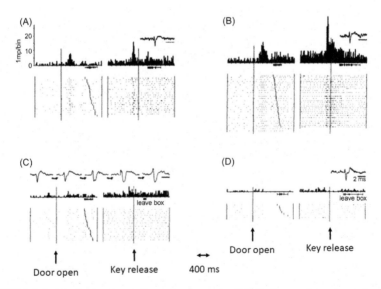

FIGURE 8.4 Analysis of recording of action potential discharges from dopamine neurons in the substantia nigra pars compacta in a nonhuman primate while the animal performs a task as described in Fig. 8.3. In this case, the responses of the neurons to the cue to move, door open, and to the onset of movement, key release, can be noted. As can be seen for neurons A, B, and C, there is an increase in neuronal activities and presumably a release of dopamine in the striatum for less than a second. For neuron D, there is a reduction of neuronal activity over a few seconds (Schultz, 1986).

detected by the microdialysis experiments described previously. Thus, there was no foundation for the claim that the normal dynamics of dopamine action were static or operating over long timescales. It is not clear whether the authors advocating continuous dopaminergic administration were aware of the findings of neuronal recordings and thus it is not clear whether this represents Confirmation Bias in the review of the scientific literature. However, these issues were raised in prior publications (Montgomery, 1992).

It was not only the dynamics of dopamine neurons of the substantia nigra pars compacta that were thought to be static or relative, but also the dynamics of neurons in the globus pallidus interna. In both cases, the inferences to static physiology were derived from the clinical phenomenology of disorders of the basal ganglia, particularly Parkinson's disease with its paucity of movement and Huntington's disease with its excessive movement. The presumed output station, the globus pallidus interna, was thought to act like a gate. With the gate open, intended movements were initiated. With the gate closed, unintended movements were prevented. In Parkinson's disease, the gate was considered to be too tightly shut, resulting in a lack of movements. With Huntington's disease, the gate was held open, allowing unintended or involuntary movements.

The concept of the globus pallidus interna acting as a gate was translated to a concept where the globus pallidus interna neurons inhibit the neurons of the thalamus, which in turn inhibit the neurons of the motor cortex, thereby reducing drive to the lower motor neurons and muscles—the Globus Pallidus Interna Rate theory. In Parkinson's disease, the neurons of the globus pallidus interna were thought to become overactive as a consequence of degeneration of the dopamine neurons of the substantia nigra pars compacta. Further increased neuronal activity within the globus pallidus interna was, in part, attributed to increased neuronal activity within the subthalamic nucleus, also consequent to degeneration of the substantia nigra pars compacta. With the development of animal models of parkinsonism, particularly utilizing the neurotoxin N-methyl-4-phenyl-1,2,3,6-tetrahydropyridine (MPTP), it was possible to directly test the hypothesis of increased neuronal activity of the globus pallidus interna (Montgomery, 2007).

Most early studies utilizing MPTP demonstrated increased neuronal activity in the globus pallidus interna (DeLong, 1990; for review, see Montgomery, 2007). However, one study did not show predicted changes in neuronal activity in other structures within the basal ganglia and thalamus compared to recordings prior to the administration of MPTP (Montgomery et al., 1986). Note that it is not that the study by Montgomery and colleagues was "negative," that is, failed to demonstrate any abnormalities. The animals' behaviors were clearly abnormal, and neurometabolic imaging using C^{14}-2-deoxyglucose autoradiography in these same nonhuman primates demonstrated changes as others have noted with MPTP-induced parkinsonism. Studies nearly 30 years later demonstrated no changes in baseline neuronal activities in the globus pallidus interna in nonhuman primates treated with MPTP (Muralidharan et al., 2016), as well as the expected changes in other structures within the basal ganglia–thalamic–cortical system (see Montgomery, 2007). In both the studies by Montgomery et al. (1986) and Muralidharan et al. (2016), the animals were dosed with MPTP to produce mild parkinsonism. Muralidharan and colleagues demonstrated a more severe parkinsonism where the changes in baseline neuronal discharge frequencies were consistent with the Globus Pallidus Interna Rate theory. At least in the narrow sense of irreproducibility, it is likely that the source was the excessive use of MPTP in those that demonstrated increased neuronal activity in the globus pallidus interna and the subthalamic nucleus and attributing the changes to the usual parkinsonian state, thus the Fallacy of Four Terms.

Irreproducibility in the narrow sense can be constructed as the syllogism in Argument 8.5:

ARGUMENT 8.5

Major premise: *Administration of MPTP* (bridging term) *increases neuronal activity in the globus pallidus interna in experiment A* (major term).

Minor premise: *Experiment B* (minor term) *administers MPTP* (bridging term).

Conclusion: *Experiment B* (minor term) *should increase neuronal activity in the globus pallidus interna* (major term).

In this case, the conduct of *experiment B* did not produce increased neuronal activity in the globus pallidus interna, thus the conclusion is false, demonstrating a logical failure, due to either the premises or the construction of the argument. It is not hard to see that this represents the Fallacy of Four Terms in that *administration of MPTP* in the major premise was likely *administration of excessive MPTP*, while *administration of MPTP* in the minor premise was *administration of not excessive MPTP*.

Findings of a lack of changes in baseline discharge neuronal activities would constitute irreproducibility in the narrow sense. Importantly, does the irreproducibility also suggest a broader conceptual irreproducibility that could impact other research, thereby placing those studies at a risk of irreproducibility? Indeed, other forms of producing parkinsonism in a nonhuman primate that allow recordings of neuronal activities in the globus pallidus include administration of neuroleptics, and electrolytic lesions of the medial forebrain bundle interrupting the nigrostriatal pathway did not produce increased neuronal activities in the globus pallidus interna (Filion, 1979). Further, the expected increases in neuronal activity in the subthalamic nucleus of humans with Parkinson's disease compared to those in patients without parkinsonism, in this case patients with epilepsy, were not found (Montgomery, 2007), which were similar for dystonia (Hutchison et al., 2003) and Huntington's disease (Tang et al., 2005).

Despite demonstrated narrow and broad irreproducibility of increased neuronal activity of the globus pallidus interna as an explanation of parkinsonism, the Globus Pallidus Interna Rate theory continues to be used to influence experimental design and interpret findings. The reasons are multiple and complex, including psychological, such as Confirmation Bias and the seductiveness of intuitively appealing notions (Johnson-Laird, 2006). Whatever the reason, the failure to appreciate the demonstrated irreproducibility risks future unproductive irreproducibility as the same experiments (excessive use of neurotoxins) are used continually. However, there is a potential cause for not recognizing irreproducibility stemming from the misunderstanding of statistical inference and logic as will be discussed later.

Even as the Globus Pallidus Interna Rate theory is falling out of favor, new theories are arising that in reality are also being found irreproducible because they share fundamentally the same sense of dynamics. The dichotomization of pathophysiology of the Globus Pallidus Interna Rate theory reflect and the notion of Parkinson's disease as a relative deficiency of dopamine display one-dimensional push–pull dynamics. As this dynamic has been demonstrated conceptually irreproducible, it is likely that other theories, such as one that posits excessive neural oscillatory activity in the 20–30 Hz range or increased synchronization of neuronal activity in the globus pallidus interna, have and will continue to be found irreproducible (Montgomery, 2016). Further, therapies designed based on such one-dimensional push–pull dynamics would be at risk for irreproducibility. Note that this is not to say that such future studies are going to be irreproducible; rather, such studies are at a higher risk. Such studies should be constructed so that if the studies fail to demonstrate narrow reproducibility, the irreproducibility will be a productive form of conceptual irreproducibility, and hypotheses based on one dimensional push-pull dynamics in the case of the physiology of the basal ganglia–thalamic–cortical system can be abandoned.

An example of this concern is Deep Brain Stimulation (DBS) when initial studies on the mechanisms of action were found irreproducible. Initially, high-frequency DBS was thought to inhibit the structures stimulated based on a Fallacy of Pseudotransitivity that lead to the Fallacy of Confirming the Consequence. Early recordings of neuronal activities affected by DBS demonstrated the expected reduction of neuronal activities; however, in a circumstance that invited the Fallacy of Four Terms. Initially, there were no means to record neuronal activities during DBS because of the artifact caused by DBS. Consequently, the investigators used the neuronal activities recorded just after discontinuation of the DBS as reflecting what happens during DBS—the fallacy of *post hoc ergo propter hoc* (because it is seen after means that it has to have been there before). As luck would have it, many studies demonstrated reduced neuronal activities immediately following discontinuation of DBS. As methods were developed to remove DBS-induced artifacts, subsequent studies demonstrated increased neuronal activities during DBS (Montgomery, 2012; Montgomery and Gale, 2008).

The involved Fallacy of Four Terms is seen in the syllogism based on the Fallacy of Pseudotransitivity in Argument 8.6:

ARGUMENT 8.6

Major premise:	*Parkinsonism is improved* (bridging term) *by pallidotomy that reduces neuronal activity in the globus pallidus interna* (major term).
Minor premise:	*DBS* (minor term) *improves parkinsonism* (bridging term).
Conclusion:	*DBS* (minor term) *is equivalent to pallidotomy that reduces neuronal activity in the globus pallidus interna* (major term).

The hypothesis for subsequent experimentation is of the form *if DBS reduces neuronal activity implies improved parkinsonism and parkinsonism is improved with DBS, then DBS reduces neuronal activity*. The conclusion is false as it has been demonstrated that DBS increases neuronal activity in structures such as the globs pallidus interna and the subthalamic nucleus. There must be a premise or proposition (statement) that is false or invalid, respectively. It is not the premise that *DBS improves parkinsonism* as this is a demonstrated fact. Rather, the false premise is that *DBS reduces neuronal activity*.

The power of the Fallacy of Pseudotransitivity as illustrated should not be underestimated in that it still informs experimental design and data interpretation. Its power is supplemented with consilience with the one-dimensional push—pull dynamics of the intuitively appealing Globus Pallidus Interna Rate theory. Further, this conceptualization is also the basis for other studies related to DBS, such as the possible role of DBS in depression based on the notion that DBS would reduce overactivity of the subgenu cingulum inferred from neurometabolic imaging (Hamani et al., 2011). Initial open label studies were found irreproducible when subsequent controlled studies failed to demonstrate a benefit (Holtzheimer et al., 2017). It is interesting in that Holtzheimer and colleagues, acknowledging the failure of the study, stated in the abstract that "Future studies are needed to investigate factors such as clinical features or electrode placement that might improve efficacy." Conspicuously absent is any suggestion that DBS for depression just may not work. Indeed, the authors wrote, "Given the strength of the preliminary data leading to this study, and the emerging data suggesting ways to optimize targeting for this intervention, the negative outcome of this trial should not be interpreted simply as a failure of subcallosal cingulate DBS for treatment resistant depression." Note that there is nothing in the controlled trial that allows such a statement. Rather, justification comes only from the positive effects of the open label study. The authors suggest that the duration of the depression or targeting in the controlled trial may be factors that explain the irreproducibility, but based on the Duhem–Quine thesis, these factors are not privileged over any other consideration, such as DBS just won't work. These discussions are reminiscent of the discussion of dopamine cell transplantation for Parkinson's disease.

With regards to the Globus Pallidus Interna Rate theory, the argument can be made that an incredible number of studies in a large variety of animals using multiple means of producing parkinsonism, most notably the neuron toxins MPTP and 6-hydroxydopamine, have demonstrated increased neuronal activity in the globus pallidus interna or its homologue. If truth was a majority vote, then the Globus Pallidus Interna Rate theory would be unassailable. Yet there may be more than just "herd psychology" at work. Certainly, if one were to correlate the severity of parkinsonism with the increase in globus pallidus neuronal activity, there would be a correlation beyond that associated with chance, as demonstrated in the results of the study by Muralidharan and colleagues (2016). Thus, even if a few studies demonstrate no change in baseline neuronal activity in the globus pallidus interna or subthalamic nucleus in parkinsonian subjects, clearly the significant correlation could argue for some cause and effect.

The correlation between severity of parkinsonism and increased neuronal activities in the globus pallidus interna and subthalamic nucleus could be considered evidence of cause and effect only in a very simplified notion of cause and effect. Philosophers make the distinction between necessary and sufficient causes (see Chapter 6, Causation, Process Metaphor, and Reductionism). A sufficient cause is one that alone produces the effect. A necessary cause is one that must be present for there to be an effect. Even if there is only a single case of parkinsonism without increased neuronal activity in the globus pallidus interna or subthalamic nucleus, increased neuronal activity in the globus pallidus interna or subthalamic nucleus cannot not be a necessary cause. If there is a single case of increased neuronal activity in the globus pallidus interna of subthalamic nucleus without parkinsonism, then such increased activity cannot be a sufficient cause (for which there is considerable evidence but the review of which is beyond the scope of this effort). Therefore, a phenomenon that is neither necessary nor a sufficient cause must be epiphenomenal, even if it is correlated. Understanding cause, in terms of necessary and sufficient, could help in recognizing irreproducibility, which would convert unproductive irreproducibility into that which may become productive.

SUMMARY

Basic biomedical research may have many purposes and each purpose may require different experimental designs and data analyses. Consequently, what is considered to be reproducible or not is also driven by the intended purpose. If the outcomes of one experiment are to make a claim to new scientific knowledge, it may be irreproducible in both the narrow and the broad sense. However, if the same experiment is intended to demonstrate feasibility of design, technology, and method, that experiment may not be irreproducible as greater tolerance should be afforded.

Reproducibility in the narrow sense relies heavily on notions of statistical significance, but scientific significance derives from the consequence of false positives and false negatives. The consequences shape the scientific meaning of the studies. The notion of meaning is even more central to the notion of broad reproducibility, that is, conceptual reproducibility.

To be sure, a great deal of concern in the scientific community arises from frequent demonstrations of unproductive irreproducibility. At least as much a concern is the failure to recognize both productive and unproductive irreproducibility, even when evidence of such is clear in the published literature. The causes are complex and often are difficult to recognize as many times they are not found in obvious statistical errors or lack of transparency. Indeed, many of the causes are necessitated by having to import other theories, hypotheses, and predispositions outside of the experiment in question, yet these outside factors can be powerful, as demonstrated in the case study.

Case Studies in Computational Biomedical Research

SCOPE OF COMPUTATION IN BIOMEDICAL RESEARCH

Interestingly, the term "computer" was actually applied to humans (www.nasa.gov/feature/jpl/when-computers-were-human). For example, in the computational modeling of the neuronal action potential by Hodgkin and Huxley described in this chapter, the calculations were conducted by a large staff of humans, particularly women. Now it is difficult to overestimate the importance of machine computation in biomedical research. First, virtually any quantitative research depends on mathematical analyses, which today virtually necessitates the use of computers. Given the ubiquity of machine (digital) computers in the conduct of biomedical research, it would not be surprising that the role of digital computing in the reproducibility of biomedical research would come under scrutiny. In the everyday experience, the expectations of highly precise digital computing are absolute replicability (precision), at least by appearances. Concerns regarding computer reproducibility would seem strange. Yet, as will be seen, there is reason for concern and caution as discussed here.

To date, most of the concerns related to computational reproducibility that have centered on the role of computation are quantitative analyses of biological data. Concerns of irreproducibility center on the sufficient transparency of the computational methods so as to allow tests of reproducibility (Garijo et al., 2013; Lewis et al., 2016; Sandve et al., 2013). In a sense, this issue of computational reproducibility is one where the hypothesis to be tested is relatively independent of the mathematics. Rather, the mathematics serve as a check on the

Reproducibility in Biomedical Research
DOI: https://doi.org/10.1016/B978-0-12-817443-2.00009-4

hypotheses. There is another sense of computational biology in which the mathematics serve as a metaphor for the biology where the mathematical relationships serve as the source domain of the metaphor and the biological inference is the target domain. In other words, biological hypotheses are generated from mathematical or computational models. In such metaphors, the source domain, the mathematical or computational model, is that which provides insight into the target domain, the biological hypothesis (Montgomery, 2018). As will be shown here, the nature of ions acting across the neuronal cell membrane is just as described by the set of differential equations. The set of equations was not derived from the biology but rather from a different metaphor, that is, transatlantic cable conveys telephone and telegraph information. Thus, the mathematics of the source domain in the metaphor for the biology of the target domain can be a source of unproductive irreproducibility in biomedical research.

There are a number of different aspects of computational biomedical research. In addition to the operations of quantitative analyses of biological data, other notions include bioinformatics, computational simulations, and computational modeling. The argument here is one dimension on which these notions differ is the source of the hypotheses. Bioinformatics typically is applied to large datasets of biological data such as genomes. Thus, bioinformatics focuses on data mining and using innovated mathematical tools for exacting new insights from biological data. Computational simulations allow studies of vast numbers of biological scenarios constructed from mathematical or computational algorithms leveraging the potentially vast memory and computer speeds of modern digital computers. An example is genetic algorithms modeled on evolution. Typically, genetic algorithms involve relatively simple interacting mathematical representations of biological agents, a means of altering the mathematics of the interactions typically biologically inspired (such as mutations) and a biologically inspired measure of success. The more successful variants of the mathematical agents then are taken as a test of the biological hypothesis represented in the creation of the mathematical representations. A similar scheme is found in machine learning of neural network computing. The key here is that the nature of the computational simulations in these cases is derived biologically. The mathematics primarily are a means to represent the biological agents.

Computer simulations need not first be derived from biological inferences. A form of computational modeling that can lead to computational simulations can proceed from "first principles" such as mathematics and physics. The biological insights become derivative from the first principles as the mechanisms in the mathematics are translated to the biological mechanisms, as demonstrated in the work of Hodgkin and Huxley described later. As also will be seen, biological knowledge provides the critical constraints to shape the mathematical derivations into biological models. The problem is that the mathematical models are so powerful that nearly any biological problem cast as a mathematical problem can be solved, but in very nonbiological manners. It is the nonbiological methods in successful computational modeling that become a critical source of irreproducibility in computational biomedical research. This creates a major problem of the Fallacy of Confirming the Consequence (see Chapter 4: The Logic of Certainty Versus the Logic of Discovery). The fallacy is *if computational mechanism A implies biological inference B is true and biological inference B is true, then computational mechanism A is true.* This is an invalid argument. It is not that *computational mechanism A* is not true but rather one cannot have certainty that it is true and, therefore, it is at risk for unproductive irreproducibility.

Central to computational modeling is the construct of the implication *computational mechanism A implies biological inference B.* The risk is that the implication can become a tautology (true by definition) if *computational mechanism A* is fundamentally the same as *biological inference B.* For example, one theory of the physiology of basal ganglia of the brain is that the activity of neurons in the globus pallidus interna serves to facilitate intended movements and prevent unintended movements. Following from this notion of physiology, the pathophysiology of Parkinson's disease is that the neurons of the globus pallidus interna are overactive, thereby preventing intended movements in addition to unintended movements (called the Globus Pallidus Interna Rate theory, see Chapter 8: Case Studies in Basic Biomedical Research). Computational models can easily be constructed of reciprocal (inhibitory) actions of the globus pallidus interna neurons onto the thalamic neurons, which drive the motor cortex to produce movement, such that an increase in globus pallidus interna neuronal activity is considered to reduce the activity of thalamic neurons. Thus, any simulation involving an increased activity of globus pallidus interna neurons will be associated with a reduction in thalamic neuronal activity. In a sense, thalamic neuronal activity is defined by the globus pallidus interna neuronal activity. Thus, it is a tautology that increased neuronal activity in the globus pallidus interna will be associated with decreased neuronal activity in the thalamus.

The problem with the tautological nature of the computational modeling thus far is that reduced thalamic neuronal activity is taken as the proxy for the biological phenomena of reduced movements in Parkinson's disease. The argument is perceived as *a implies b and b implies c and thus a implies c* where *a* is overactivity of globus pallidus interna neurons, *b* is underactivity of thalamic neurons, and *c* is reduced movements of Parkinson's disease. But, note that *b* is defined (inversely) as *a*, and *b* is taken as the proxy for *c*. The argument becomes *if a = b and b = c, then a = c*. But if *a = b* and *b = c*, then *b* can be substituted in the argument, which becomes *if b = b and b = b, then b = b*. This is tautological but importantly is internally consistent (true). Consequently, the argument will not be found irreproducible as long as internal consistency is the only or most important criterion. Further, the model will be thought true and, consequently, seen as evidence for the biology that the model was seeking to explicate. It may be reproducible in the narrow sense but likely fail reproducibility in the broad or conceptual sense. The Globus Pallidus Interna Rate theory has been proven wrong (Montgomery, 2007, 2012) and clearly then the concept in the computational modeling of the pathophysiology of Parkinson's disease is wrong, despite being internally consistent (Montgomery, 2016).

The power of computational modeling and simulation risks producing intuitively appealing solutions to biological problems. They are intuitively appealing because they represent Confirmation Bias as a result of the Fallacy of Confirming the Consequence. Perhaps for this reason the reproducibility of computational biomedical research where the mathematics is the target domain to explicate the biology has received relatively little attention, unlike the role of computations in the quantification of biological phenomena. Nevertheless, the solutions to the biological problems instantiated in the mathematical mechanisms can be highly nonbiological. Such solutions likely are going to be conceptually irreproducible.

IMPORTANCE OF MATHEMATICAL AND COMPUTATIONAL MODELING AND SIMULATIONS

This chapter reviews the situation where computational biomedical research uses mathematical and computational procedures to elucidate biomedical concerns. This notion goes well beyond the use of mathematics, particularly statistics as used to quantify data and relations between data in terms of statistical inferences. Rather, mathematical relations become the source domain in the metaphor in which biology, the target domain, is to be explicated (for a discussion of metaphors, see Chapter 4: The Logic of Certainty Versus the Logic of Discovery and Chapter 6: Causation, Process Metaphor, and Reductionism). While the importance of mathematics and computational science in the development of concepts is readily appreciated in physics and increasingly in chemistry, there also is an expanding role in the biomedical sciences

Different from clinical and basic biomedical research, computational biomedical research proceeds or derives from mathematical or algorithmic fundamentals rather than induction from empiric observations in order to recreate or explicate biological processes. Note that empirically derived knowledge is central to the process but that the empirically derived knowledge is recast in mathematical terms. Thus, the dynamics of fundamental biological processes are extrapolated to a different level of biological explanation driven by the properties of the mathematical and algorithmic fundamentals. The history of sciences using mathematics has been opportunist, taking prior mathematical conceptualizations that seem to fit the physical problem rather than generating those mathematical conceptualizations to resolve the physical problem (Bochner, 1966). The primacy of mathematics demonstrates its independent capabilities rather than being derivative of empiricism.

It is important to note that this definition of mathematical and computational modeling and simulation is a very different concept from the mathematical and computational description of the biological phenomena. The distinction is made clear by a referral to physics where Johannes Kepler was able to proceed from detailed astronomical observations by Tycho Brahe in the early 1600s to describe the planetary orbits as ellipses, thus describing a number of more precise mathematical properties of the orbits. However, it was Isaac Newton who answered Edmund Hailey's question about how Newton knew the orbit of the moon was elliptical—Newton replied that he had derived it (by applying his newly developed calculus to the two-body problem in physics). The following is Abraham de Moivre's account:

> In 1684 Dr Halley came to visit him at Cambridge. After they had been some time together, the doctor asked him what he thought the curve would be that would be described by the planets supposing the force of attraction towards the sun to be reciprocal to the square of their distance from it. Sir Isaac replied immediately that it would be an ellipse. The doctor, struck with joy and amazement, asked him how he knew it. Why, saith he, I have calculated it. Whereupon Dr Halley asked him for his calculation without any farther delay. Sir Isaac looked among his papers but could not find it, but he promised him to renew it and then to send it him (www.mathpages.com/home/kmath658/kmath658.htm).

Mathematical and computational modeling and simulation have long been greatly appreciated in disciplines such as physics. In the past 40 years, mathematical and computational modeling and simulations have played a much greater role in understanding biochemistry. Honest examination would suggest that mathematical and computational modeling and simulations have not yet become mainstays in biomedical research. Yet, as biological systems become increasingly recognized as Complex or Chaotic systems, traditional methods of mathematical descriptions, such as Central Tendencies and variance, become increasingly problematic (see Chapter 10: Chaotic and Complex Systems, Statistics, and Far-from-Equilibrium Thermodynamics). Ultimately, precise explication will necessitate advanced mathematical and computational algorithmic analyses.

A distinction here is made between mathematical and computational algorithmic modeling and simulations. Mathematical modeling generally is reflected in an attempt to define a set of equations, typically partial differential equations, which model the dynamics (changes in the states or situations over time). The set of equations may be organized in the manner of a mathematical proof, which typically depends on the Principle of Transitivity linking the premises, which is of the form *if a implies b is true and b implies c is true, then a implies c is true.* This is the basis for the substitution procedure in mathematical proofs. In this case, a series of premises, beginning with a mathematical description of the biology to be explicated, is then linked to a series of more fundamental mathematical operations, each of which carries the explication to a more fundamental level, in the sense of analytic geometry. An alternative is to start at the most fundamental level and build to the more complex description of the biological phenomena, in the synthetic geometry notion. Again, in each the transition has more to do with the mathematical properties than the biology.

Computational modeling and simulations do not necessarily derive from the mathematics per se, but rather use mathematics as cogs in a machine with the cogs organized in a manner so as to be a metaphor for the biomedical phenomena. The cogs in the machine are allowed to turn and where they are at the completion of the process is hoped to provide the biological insight—analogous to synthetic geometry. Thus, in the computational modeling and simulations in this sense, the outcome is not preordained, as would be the case in a mathematical proof. Yet, despite the distinctions between mathematical and computational modeling and simulations, both share many of the same issues of irreproducibility if only because modern mathematical modeling and simulations are run on computers, generally electronic devices. The nature of the electronic devices will be shown to play a role in the reproducibility of both mathematical and computational modeling and simulations.

THE NOTION OF IRREPRODUCIBILITY IN MATHEMATICAL AND COMPUTATIONAL MODELING AND SIMULATIONS

At first, it would seem that mathematical and computational modeling and simulations would be at the least risk for irreproducibility. After all, is not $2 + 2 = 4$ true no matter how many times one adds $2 + 2$, is it not always 4? If one runs a computer program that adds $2 + 2$, does it not always result in 4 no matter how many times the computer program is run? This concern relates to reproducibility in the narrow sense (replicability). For sufficiently complex mathematical and computational problems, the answer is such modeling and simulations need not produce the exact same result, thus even in the narrow sense of replicability. There may be situations where a very complex version of $2 + 2$ may not equal 4 reliably. Mathematical and computational modeling and simulations may not be reproducible in the exact sense. How then to regard reproducibility in the narrow sense (Bailey et al., 2016)? As will be demonstrated, mathematical and computational modeling and simulations are also at risk for irreproducibility in the broadest sense, particularly conceptual reproducibility.

Consider the irrational number π (pi), which can be expressed as the ratio of a finite length of the circumference of a circle divided by twice the finite length of the radius. Yet, π is not a finite number, as the decimal numerical representation of π goes on forever without any digits repeating themselves. One could use the concept of π in a geometrical sense to calculate the circumference of any circle, x, given a radius y from a reference circle whose circumference w and radius z are known. The circumference would be $x = 2 y \times (w/2z)$. In special cases, where y is an integer multiple, n, of z, the circumference would be $n \times w$. This would be exact (within the limits of Instrumental Uncertainty in the computer sense discussed later) and reproducible if y and w are finite (what would it mean to say that y and w are infinite?). In other circumstances, circumference x would be related to the ratio of y to z, multiplied by w, which may or may not be a rational number.

The case just given proceeds from a given finite circumference w and radii y and z and the stipulation that y is an integer multiple of z from which the circumference x is calculated. Alternatively, one could determine circumference x by multiplying 2 times y times π, but what is the value of π? One expression of π may be 3.14159265358979323846, but this would result in a different value of x than if π is taken at 3.14159265358979323846264338327950288441971. Thus, the reproducibility of the measurement of circumference x depends on the choice of π but what difference in calculated circumferences is allowed (tolerated) so as to assert reproducibility? One could argue that if the consequences of using either value of π are inconsequential, then any difference in the calculated circumference x is still within the range of reproducible— irreproducibility considered still reproducible. One could always round off the resulting values and force them to be the same, but that raises metaphysical issues. For example, does 2.5 get rounded to 2 or 3? Certainly, by most conventions, 2.5 is rounded to 3 but there is no rationale other than one's math teacher made them do so. Note that the consequence of the mathematical simulation of circumference x depends on the context. For example, the use of either value of π when the radius is

1 mm may be less likely to be consequential compared to the radius of the earth. Just as in clinical and basic research, meaning (the effects of the consequences) is critical. Thus, reproducibility is considered within the context of meaningfulness, not the consequence of the numbers or statistics generated.

The issue of π and the choice of its value reflect the rounding or truncation (limiting the decimal places in the value of π) problem of numeric calculations. For example, calculating the circumference of the earth will be quite different if the value of π is rounded or truncated to 3.14159265358979323846 from 3.14159265358979323846264338327950288841971. As will be seen, the problem of rounding or truncation is virtually inescapable with digital computers, for example, when the representation of a number exceeds the number of bits that can represent the number. This invariability is true of irrational and transcendental numbers. For example, every number in a typical digital computer is represented as a binary code containing bits of information containing a 0 or 1, much like a decimal number of bits containing values of 0, 1, 2, 3, 4, 5, 6, 7, 8, or 9. The position of the bit in a string of bits determines the multiple of power of the base of that bit. In the decimal system, the number 123 contains three "1s", two "10s", and one "100" where 1 is 10^0, 10 is 10^1, and 100 is 10^2. The binary number (as used in digital computers) 10101 contains one "1", zero "2s", one "4", zero "8s", and one "16" where 1 is 2^0, 2 is 2^1, 4 is 2^2, 8 is 2^3, and 16 is 2^4. Truncation or rounding is necessary as the number of bits used to represent a piece of data is limited. There are various ways of attempting to compensate for these limitations in the manner by which data are represented, such as integer or floating-point representations [analogous to integer versus scientific notion to represent large (or very small) values].

SOURCES OF IRREPRODUCIBILITY IN THE NARROW SENSE

Digital computers manipulate data and instructions based on a binary code that consists of bits of information. The value of each bit in the series is held in an electronic switch. When the switch is in one position, it has a value of "1" and when in the other position it has a value of "0". The switch cannot have any other state. Electronically, the value of the switch is determined by the voltage on the switch. For example, a voltage on a switch greater than 2 V is considered "high" or a value of 1. Below 0.8 V, the switch is considered "low" or a value of 1. But for intermediate voltages the value is not clear and often is not stable. Consequently, the valve may alternate between 1 and 0. The analogy is when one screws in an old-fashioned lightbulb with the switch on.

Many times, the lightbulb will flicker until it is tightened in the socket. A similar phenomenon can occur in electrical switches that contain data in the digital computer, which can change the value of data and consequently the results of the computational analysis.

The flickering of the switch state can be considered noise. There are many sources of noise (for the interested reader, these include thermal, shot, flicker, burst, and transit-time noise). There are a number of techniques to minimize such noise involving the use of hysteresis in transitions between states such as between "0" and "1". Nonetheless, any digital computer is prone to noise that can affect reproducibility. The problem is that the effects of noise are compounded with repeated iterations of changing states of the binary bits. Thus, a simple computer program will be less affected by these sources of noise whereas a very long reiterative computation may contain considerable noise, perhaps to the point that the output is irreproducible in the narrow sense.

In an interesting way, noise in computer bits can be considered analogous to biological variation, where no two individuals of the same species and sex are exactly the same. However, while biological variability can be managed, in some degree, by statistical analysis, it is not clear that such approaches are effective in the face of computational variability (irreproducibility). Management of biological variability can appeal to the Large Number theorem (see Chpater 5 The Logic of Probability and Statistics), which states that any description of some sample (such as some measure) converges onto the mean. However, this presupposes that each observation is independent of any other observation, that is, the outcome of some measure in one subject in the same species is independent of the measure of some other subject in the same species. However, if the observations are not independent, then the Large Number theorem may not hold and one cannot have confidence that with sufficient sample size there will be a convergence of the mean onto a certain value of the Central Tendency (see discussion of multiple comparisons, repeated measures, and Brownian motion in Chapter 5: The Logic of Probability and Statistics and Chapter 7: Case Studies in Clinical Biomedical Research). In a computation, an error in one data representation due to noise in a bit can be perpetuated and indeed enlarged by repeated reiteration in a computational routine. Thus, the error is propagated and cannot be considered independent. The result would be unproductive computational irreproducibility.

An analogy can be made to Brownian motion as discussed in Chapter 5, The Logic of Probability and Statistics. Applying the analogy to computations, while the noise effects on the status of bits in the representation of data may be random, the effects are propagated and additive. The net effect (representing the sum of the iterations) will not be a convergence onto some measure of the Central Tendency but rather a drift in the manner of the pollen particle in Brownian motion. As the "path" of the Brownian effect cannot be predicted, the computational outcome cannot be predicted. Presumably, the effect is small and negligible. Nevertheless, it is a potential source of irreproducibility, and the criteria for reproducibility must include some tolerance (such as a confidence interval in traditional statistical analysis).

COMPILATION VERSUS RUNTIME ERRORS

Two general types of errors in computing are compilation and runtime errors. Compilation errors typically are due to the syntax of the computer program and are often detected as the computer program written by the programmer is converted to machine language that can be run on the computer. However, runtime errors generally are not due to syntactical errors and thus are not flagged by most computer program compilers or interpreters that create the machine language code. Many of the sources of error noted previously represent runtime errors and consequently may escape notice by the computer scientist. Indeed, the first suspicion of a runtime error may be when the results do not correspond to the computer scientist's expectations. There may be a tendency to think that the lack of correspondence reflects a runtime error referable to the computer programming, which then tempts the computer scientist to "tweak" the computer program so that a more palatable outcome is achieved. Clearly, this would be an exercise in Confirmation Bias reflecting the Fallacy of Confirming the Consequence. As a logical fallacy, it puts the results of the computational modeling and simulations at risk for subsequent unproductive irreproducibility.

Alternatively, the computer scientist can analyze the program for flaws that are not syntactical in the computer language. Rather, there may be logical flaws or misrepresentations of biological data or processes. Interestingly, a demonstration of logical rigor is thought to suggest ontological validity of the outcomes. However, this is suspect. The logical rigor is internal to the project just as statistical validity is internal to the data and the experimental design. Just as statistical rigor does not necessarily mean clinically or biologically true or meaningful conclusions, logical rigor in the computational model or simulation does not attest to the biological validity or meaningfulness. Thus, even a logically rigorous computational model or simulation is at risk for broad irreproducibility in the context of biological or clinical meaningfulness. Rather, logical and statistical rigor relate more to precision than to accuracy and one can be precisely inaccurate (wrong).

COMPLEXITY AND CHAOS AND UNDERDETERMINATION IN COMPUTATIONAL MODELING AND SIMULATIONS

It is likely that the mechanisms underlying computational noise are nonlinear and that the vast number of computations of mathematical or computational models and simulations translate the nonlinearity to a Complex System. Thus, the concerns for reproducibility in Complex systems in Chapter 10, Chaotic and Complex Systems, Statistics, and Far-from-Equilibrium Thermodynamics likely are applicable to mathematical and computational modeling and simulations. Indeed, the first inklings of Complexity and Chaos were noted by Jules Henri Poincaré (1854–1912) in his attempts to solve the mathematics of the three-body problem, such as the orbit of the moon about the earth, about the sun, after Newton has done so for the orbit of the moon about the earth. It cannot be done analytically (meaning precisely using definite mathematics without approximation). The Chaos and Complexity theory became more systematized in the later 1970s. Mathematical models and simulations involving discrete (nonrandom) functions produced results that were unpredictable, although not random. As discussed in greater detail in Chapter 10, Chaotic and Complex Systems, Statistics, and Far-from-Equilibrium Thermodynamics, the unpredictability could appear as irreproducibility with the inference of error. However, this need not be the case in that the results are valid, even if unpredictable. The emergence of the Complexity and Chaos theory necessitates a reinterpretation of what constitutes irreproducibility.

As alluded to previously, one possible means to reduce (but not eliminate) the risk of irreproducibility is to assure rational and logical rigor in the computer programs written for modeling and simulations. However, advances in machine learning in Artificial Intelligence demonstrate that robust computational modeling and simulations are possible even when the rationality and logic of the computational algorithms are unknown, unknowable, or not proceeding from some rationale or logic (Hutson, 2018).

Consider backing a tractor—trailer to a loading dock that takes long periods of training to become proficient. Further, even experienced drivers find it difficult to coach novices to success. A neural network was programmed to simulate the situation and then develop a program that could learn to back a trailer onto a dock starting in any position in which it is physically possible to succeed. The neural network consisted of a three-layer neural network for the emulator and whose output became the input layer to the truck emulator program (Fig. 9.1) (Nguyen and Widrow, 1990).

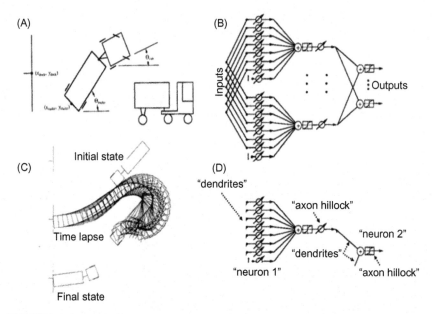

FIGURE 9.1 Schematic representation of a neural network solution to the computational problem of backing a tractor–trailer (A) to a loading dock (C). The neural network consists of 25 processing units arranged in two layers (B). The output layer drives a virtual tractor–trailer, the position of which is sent to the input units. Input to each unit (D) is adjusted by a synaptic weight that experiences can alter through the digital equivalent of dendrites. The synaptic weights modify the influences of the neurons in the prior layer on the neurons in the next layer. Weighted inputs onto the next layer neurons are summed and submitted a mathematical operation—a sigmoid function to limit the magnitude of the outputs—which is analogous to an action potential-initiating segment. Thus, the summed inputs are translated to outputs in a specific range. Actual output of the last layer is compared to input. Any difference is an error signal, which propagates back through the neural network to change the synaptic weights between the input to the output neurons. Unlike biological neuronal action potentials, the digital equivalent of the output neuron generates an output that ranges from a value of zero to one. Nonetheless, the concept is the same. Source: *From Nguyen, D.H., Widrow, B., 1990. Neural networks for self-learning control systems. IEEE Contr. Syst. Mag. 10 (3), 18–23.*

These methods are inherent in what are called evolutionary algorithms, of which genetic algorithms are a variant (Colbourn and Rowe, 2009; Dubey et al., 2016; Gerlee et al., 2011; Heberlé and de Azevedo, 2011; Hiss et al., 2010; Pizzuti and Rombo, 2014; Spirov and Holloway, 2013; Street et al., 2013; Tragante et al., 2014; Worzel et al., 2009). In genetic algorithms, algorithmic mechanisms interact to construct higher order mechanisms. The algorithms undergo "mutations" until the interactions produce a higher order mechanism that best approximates the goal. Importantly, the "mutations" need not follow from any rationale or logic and may be random. The enormous power of modern computers can sort rapidly through the consequences of any particular mutation.

One might argue that these machine learning modeling and simulations only appear to lack rationale or logic. The presumption is that with further research and knowledge, the rationale or logic eventually will be understood, restoring faith in deterministic systems. Indeed, this is the motivation behind Explainable Artificial Intelligence (xAI). Concerned about automatous decision-making by computational AI systems, the European Union passed laws instilling the right to explanation, whereby a person can demand that the AI system explain its decision (Goodman and Flaxman, 2016). For example, the citizen could demand a causal explanation, in a sense demanding conceptual reproducibility (see Chapter 6: Causation, Process Metaphor, and Reductionism). With extrapolation to biomedical research, is reproducibility in the area of machine learning still going to demand a "right to explanation" as a test of conceptual reproducibility in the broad sense?

Alan Turing demonstrated that this may not be possible. In other words, a computer program may solve a problem without anyone being able to prove that the computer program could do so, that is, provide a demonstrable explanation as a test of reproducibility. Turing developed the concept of the Universal Turing Computing Machine that could solve any computable problem. Indeed, the traditional von Neumann computing architecture that is the basis for most digital computers can be considered a variation on the Universal Turing Computing Machine. This seemed to solidify the faith that the power of computational algorithms should be able to account for any and all human endeavors, such as biomedical research, even including human behavior, including those attributed to mental states, transcendental perceptions, and meaning.

The seed of doubt can be found in Turing's own paper, "On Computable Numbers, with an Application to the Entscheidungsproblem" in 1937 (also see Turing, 1938). That paper formalized the notion of a computer algorithm from which the Universal Turing Computing Machine would evolve. However, the object of the paper was to demonstrate a mathematical problem that could not be solved—the Halting Problem. The problem asked whether it can be determined beforehand whether any mathematical procedure to solve a problem, including proofs and derivations, will arrive at a true and final conclusion in a finite period of time. This question was part of a wider agenda by David Hilbert to provide an absolute rigorous formalization of mathematics, that is, all mathematical knowledge could be derived from a set of self-evident axioms and rules for combining the axioms. An affirmative answer to the Halting Problem would have provided confidence in Hilbert's agenda. Turing demonstrated that no computer algorithm can assure beforehand that any algorithm will terminate in a solution in a finite time. In other words, a computer program may be able to solve a problem in a finite period of time, but there is no way to know by examining its computer algorithm. What does this mean for biomedical research increasingly dependent on artificial intelligence?

THE NECESSITY OF BIOLOGICAL CONSTRAINTS AND THE FALLACY OF CONFIRMING THE CONSEQUENCE

Even in mathematical and computational modeling, the notion of productive irreproducibility is foundational. Often, mathematical hypotheses are proven by denying them and then demonstrating that the denial of the hypothesis results in a contradiction or absurdity—the mathematical equivalent to irreproducibility. For example, a mathematical proof could start with a very complex statement that has the appearance of being true, through a process of deconstruction and deriving the argument or algorithm entailed in the complex statement demonstrates that 2 does not equal 2, then the argument or algorithm is wrong and the original statement can be abandoned, hence productive irreproducibility. Turing demonstrated that no computer program can tell ahead of time whether another computer program will halt on a final correct solution to a computable problem by demonstration that denial of the claim, a computer program could, resulted in an absurdity or self-contradiction. To the knowledge of this author, invocation of productive irreproducibility has not yet taken hold in mathematical and computational modeling. However, the aforementioned discussion demonstrates that productive irreproducibility is critical to any form of knowledge and the methods used to gain that knowledge—including computational and mathematical modeling. However, if only by reference to biological questions, mathematical and computational modeling and simulation are heir to all the injudicious uses of logical fallacies as biological or carbon-based research.

Since Turing's work on the Universal Turing Computing Machine, remarkable advances have been made in computational methods. Turing's computational machine depended critically on complete *a priori* specification of the algorithm. As such, any algorithm is only as good as the prior knowledge that informs the algorithm. With the development of neural network computing, prior specification of the algorithm, in terms of the specifics of the computational problem instead of in the operations of the neural network computer, is no longer necessary as described earlier and in Fig. 9.1. Rather, these neural network systems are capable of self-learning in order to solve the computational problem. A training set of causes and effects is submitted repeatedly to the neural network computer, which, by successive approximations, changes its internal computational structure to produce the correct effects from a set of causes. Then the trained neural network computer is given a new and novel set of causes for which it produces the correct effects. The neural network comes to simulate the processes that translate biological causes to biological effects. It is hoped that once simulated, the structure of the neural network can be dissected and analyzed to provide hypotheses as to the corresponding biological processes, but not always. At the very least, it provides evidence that, in principle, the biological problem is solvable and often the mathematical or computational simulations provide metaphors by which to view the biological problem.

The remarkable power of computational algorithms (given or learned), along with the dramatic increase in computer processing power, is both a blessing and a curse. It is increasingly likely that computer algorithmic modeling and simulations can create successful transfer functions between the computational analogues of the biological causes (input) and the biological effects (output). Like other mathematical optimizing procedures, such as correlational analyses described in Chapter 5, The Logic of Probability and Statistics, a solution will be found even if it is only idiosyncratic to the dataset used to develop the simulations. Application to any subsequent dataset is at risk for irreproducibility. Further, a great many different algorithms can be constructed to simulate the same biological process, as will be discussed later (Prinz et al., 2004). In these circumstances, selecting one or even a subset from the set of all possible algorithms and inferring from any one simulation to the actual biological problem sets one up for the Fallacy of Limited Alternatives and, consequently, for the risk of irreproducibility (see Chapter 4: The Logic of Certainty Versus the Logic of Discovery).

Prevention of the Fallacy of Limited Alternatives can be secured by an active consideration of all alternatives. However, the number of alternatives may become so great that this approach to prevention is not feasible. The other approach is to increase the biologically determined prior constraints on the computational modeling and simulations to exclude as many alternatives as possible.

The immense power of computational modeling and simulation greatly risks begging the question or the Fallacy of *Petitio Principii*, where the anticipated conclusion is embedded in the premises such that the result is a logical tautology, a self-fulfilling prophecy, and, consequently, scientifically uninformative. The question-begging nature in one sense may be because the premise, the computational simulation, is so powerful that it admits of any conclusion. In other words, algorithms of great power can demonstrate the predicted outcome but only because the algorithms have great power and not because of a unique relationship between the computational hypothesis and the computational predictions. In future experiments, increasing computational power, both in hardware and in concept, will require even more vigorous biological constraints to reduce the risks of unproductive irreproducibility.

The question-begging power of computational modeling and simulations means that the biological target for explication must be chosen with care. For example, if a theory or hypothesis is chosen as the target for computational modeling and simulation and the theory or hypothesis is wrong, and hence at risk for irreproducibility, the computer explication or solution still may succeed in that circumstance, a runtime error, but any attempts to generalize beyond the exact circumstance will be at risk for conceptual or broad irreproducibility. Modeling and simulations of biologically irreproducible biological claims will only result ultimately in irreproducible modeling and simulations. The successful computational modeling and simulation of demonstrably false theories and hypotheses are common problems.

SETTING THE STAGE

As discussed in Chapter 5, The Logic of Probability and Statistics, statistics deal with internal validity. Applications of statistics to computational modeling and simulations are not typical. In mathematical modeling generally, the analyses usually do not result in a range of outputs for which the notion of a Central Tendency and variance would be applicable. Similar concerns apply to computations where the algorithms are constructed previously and explicitly. Neural network modeling and simulations may produce a range of outputs for the same inputs, in which case Central Tendency and variance may be relevant; however, once the learning phase is complete, outputs tend to become precise (as opposed to accurate) where Central Tendency and variance again may be uninformative.

Another approach is to conduct Monte Carlo simulations where the values of input variables can be selected randomly from a range of values and the consequence determined. As the consequence varies with the variations in the values of the input variables, measures such as Central Tendency and variance may be interesting. Indeed, comparisons between the statistical descriptors of the values of the input variables and the outcome could provide important insights into the intervening transfer functions between the input and the output. However, caution is required when dealing with computational models and simulations that are inherently Complex or Chaotic. Slight variations in the value of the input variables may produce great differences in the consequence, demonstrating the dependence on initial conditions in Complex and Chaotic systems (see Chapter 10: Chaotic and Complex

Systems, Statistics, and Far-from-Equilibrium Thermodynamics). Indeed, such an experience might suggest that the underlying computational modeling and simulations are Complex or Chaotic. If so, then traditional statistical analyses that depend on the Large Number Theorem may not be valid as the statistical descriptors need not converge on a single value with increased numbers of simulations.

COMPUTATIONAL MEANINGFULNESS

Just as meaningfulness in clinical and basic biomedical research is important, so too is it to computational modeling and simulations. Meaningfulness in computational modeling and simulation relates directly to the purposes and goals to be achieved. One goal may be to demonstrate that a set of biological processes can achieve the biological phenomenon where the biological processes and the outcome biological phenomenon have computational analogues. Success for this goal depends greatly on the degree the modeling and simulation computational processes relate to the analogues of the biological processes, for example, Epistemic Distance and Epistemic Degrees of Freedom. In this case, there is a tension. If the dynamics of the actual biological processes were known exactly and explicitly, then what is served by the computational modeling and simulations? The results of the computational modeling and simulations would be tautological and trivial (philosophically speaking).

Computational modeling and simulations contribute more when the actual biological processes are unclear. For example, a gap can occur between two linked biological processes where the Principle of Causational and Informational Synonymy is violated (see Chapter 6: Causation, Process Metaphor, and Reductionism). Gaps can be bridged by extrapolation using the mathematical or algorithmic properties of the analogues of the biological processes and looking for convergence. However, the risk for question begging, hence tautology, is directly proportional to the Epistemic Distance and Epistemic Degrees of Freedom in the gaps between the biological processes and the computational analogues. The Epistemic Distance and Degrees of Freedom may differ between a self-learned algorithm and one that is a specified *a priori*. Nonetheless, the mathematical and computational algorithmic extrapolations may provide insight or, at the very least, hypotheses that can be explored biologically.

Another goal may be to explore the computational consequences of interactions among more fundamental mathematical and algorithmic operations based solely on the properties of the fundamental mathematical and algorithmic operations. Assuming that the fundamental mathematical and algorithmic operations have biological relevance, then it is possible that the computational consequences may have biological relevance. This notion is analogous to the concept of Logical and Mathematical Realism where nature is just such that logical and mathematical operations truly reflect nature and faith in such a notion is seen in the synthetic Geometrical Method.

SPECIFIC FEATURES TO LOOK FOR IN MATHEMATICAL AND COMPUTATIONAL STUDIES

1. What is the purpose of the mathematical or computational study?
   a. Demonstration of the feasibility of an anticipated biological theory or hypothesis.
      i. Is the biological theory or hypothesis rigorous or demonstrated reproducible?
   b. If the purpose is demonstration of potential consequences of the modeled and simulated component biological processes that are causal to an unanticipated biological phenomenon:
      i. Are the component biological processes rigorous or demonstrated reproducible?
2. What is the compliance with the Principles of Causational and Information Synonymy?
   a. Between the component biological processes:
      i. What is the Epistemic Risk in terms of Epistemic Distance and Epistemic Degrees of Freedom?
   b. Between the analogous computational models and simulations:
      i. What is the Epistemic Risk in terms of Epistemic Distance and Epistemic Degrees of Freedom?
3. Have alternative mathematical models and algorithmic procedures been considered and are the biological constraints sufficient to adjudicate between the alternatives?
4. For self-adapting (learning) algorithms, what is the nature of the training set and how similar is the training set to the novel problem to be solved?
   a. If the training set and novel prospective problem differ, how great is the Epistemic Risk of the Fallacy of Four Terms?
      i. What are the Epistemic Distance and Epistemic Degrees of Freedom between the training and the problem sets?
5. Are there stochastic simulations, such as Monte Carlo simulations, to reflect biological variability appropriately?
6. Do the mathematical and computation modeling and simulations demonstrate Complexity or Chaos?
   a. If so, have the appropriate descriptors been analyzed?

CASE STUDIES

Two cases studies in computational simulation are provided. The first is the work of Hodgkin and Huxley in modeling the mechanisms that underlie the neuronal action potential. This case is presented to demonstrate a robust and critical approach to mathematical modeling, as noted in the many caveats that were explored and discussed in the relevant publications. The second case demonstrates the computational power of modeling neural systems. As will be seen, many computational solutions for the same biological problem or phenomenon are possible, thus raising serious questions as to the biological plausibility of any particular computational modeling solution.

Case 1. The Hodgkin and Huxley model of the neuronal action potential (changes in the electrical potential [voltage] across the cell membrane of the neuron during activation—analogous to an electrical pulse of information). This case is not presented as an example of irreproducibility of a mathematical modeling and simulation but rather as an example of a well-conducted study that highlighted potential problems that could lead to unproductive irreproducibility. In a series of papers in 1952, Hodgkin and Huxley constructed a mathematical model of the action potential in the giant squid neuron's axon. The action potential represents activation of a neuron, and the sequence of action potentials encodes and conveys information in the nervous system. Their work is fundamental to the understanding of neurophysiology; the mechanisms by which some anesthetics work; and neurological disorders associated with abnormalities of the ionic conductance channels, among others. Hodgkin and Huxley were awarded the Nobel Prize in Physiology or Medicine in 1963. The history of these experiments is well described in A. McComas' "Galvani's Spark: The Story of the Nerve Impulse" (2011).

Through a number of experiments, significant biological knowledge was obtained regarding the time course of the action potential. It became clear that sodium (Na^+) and potassium (K^+) ions (carriers of electrical charges that determine the voltage across the cell membrane of the neuron) were important. However, the roles of these ions could not be appreciated in the natural state. A number of experiments had to create artificial conditions in order to control for various influential variables, such as voltage through voltage clamp experiments. However, given the artificial nature of the biological experiments, it was not possible to elucidate the biological mechanisms involving Na^+ and K^+ ions.

Hodgkin and Huxley (1952) wrote:

> The discussion in Part I shows that there is little hope of calculating the time course of the sodium and potassium conductances [flows of Na^+ and K^+ ions through the neuronal membrane that creates the action potential] from first principles [author – note their use of the term "first principles" is one kind as discussed above]. Our object here is to find equations which describe the conductances with reasonable accuracy and are sufficiently simple for theoretical calculation of the action potential and refractory period [time period following the action potential where the neuron is not excitable]. For the sake of illustration *we shall try to provide a physical basis for the equations, but must emphasize that the interpretation given is unlikely to provide a correct picture of the membrane* [emphasis added].

As noted, Hodgkin and Huxley could not create a mathematical model based on first principles, that is, an explicit knowledge of the underlying mechanisms. The only resort was to model the phenomenon—the waveform or the change in the electrical potential across the neuronal membrane over time during the action potential. Most importantly, they used a mathematical model borrowed from work on cable theory developed for the first transatlantic cable. In this sense, the cable was mathematized, in the same sense as in synthetic geometry, but as will be seen, it was extended to include ions known to be important in the action potential. However, the exact manner in which the ions were important was subsequently determined by mathematical manipulation and solution.

Hodgkin and Huxley recognized that inference from the mathematical description of the action potential to the underlying mechanisms is very difficult and while such modeling may provide some hypotheses for future study, the mathematical modeling and simulations and any inferences derived therefrom were not proof of the underlying biology—an admonition that modern computer scientists/biologists often fail to appreciate and find relevant to reproducibility in computational biomedical research. This is, at the very least, an implicit recognition of the potential for the Fallacy of Four Terms should their modeling results be conflated with notions of the actual biology. If the model results became a mechanism that "stands in" for an invoked biological mechanism in the major, minor, or bridging term, there would be the risk of the Fallacy of Four Terms and, consequently, risk for irreproducibility.

Viewing the neuronal axon as a conductor, the models could borrow from engineering in preparation for the laying of the transatlantic cable. It was known that the electrical current lost through the cable (I) could be determined by knowing the capacitance of the cable (C_m) and the rate of change in the voltage that represented the electrical signal ($\frac{dV}{dt}$). To that, Hodgkin and Huxley added a term, I_i, to represent the electrical current associated with the movement of Na^+ and K^+ through the neuronal membrane, resulting in the following equation:

$$I = C_m \frac{dV}{dt} + I_i. \tag{9.1}$$

Hodgkin and Huxley then modeled the electrical current related to ion flows (I_i) in terms of flows of Na^+ and K^+ and another term that would stand in for other ions involved in the action potential (referred to as leakage current). The resulting equation is shown here:

$$I = C_m\frac{dV}{dt} + g_K(V_m - V_K) + g_{Na}(V_m - V_{Na}) + g_l(V_m - V_l), \qquad (9.2)$$

where V_m is the membrane voltage, g_K is the K^+ conductance, V_K is the reversal potential for K^+, g_{Na} is the conductance for Na^+, V_{Na} is the reversal potential for Na^+, g_l is the conductance for the leakage current, and V_l is the reversal potential for the leakage current.

The variables for proportionality were added to curve fit Eq. (9.2) to the observed biological data to produce Eq. (9.3):

$$I = C_m\frac{dV}{dt} + g_K n^4(V_m - V_K) + g_{Na}m^3h(V_m - V_{Na}) + g_l(V_m - V_l) \qquad (9.3)$$

When solved for the variables, n, m, and h, the result was that the model approximated the biological action potential waveform obtained from the giant squid axon very closely. The resulting variables and the mathematical structures generated speculation as to why the particular form they took.

Particularly intriguing is n, which was to the fourth power (n^4) in the term related to the electrical current due to K^+. Hodgkin and Huxley speculated and later Hodgkin and Keynes (1955) suggested that n related to the probability of the K^+ channel being occupied by K^+ ions and that four ions were necessary to be in the channel if additional ions were to traverse the channel to move to the extracellular space. Interestingly, Hodgkin and Keynes modeled the flow of K^+ ions mechanically using blue and silver steel balls (Fig. 9.2). The intracellular and extracellular compartments across the neuronal membrane were modeled by two compartments connected through an opening. In one configuration, a tunnel of a certain length connected the two compartments and none (a tunnel with 0 length) in the other configuration. The apparatus was shaken and the movement of the blue and silver balls was measured. It was demonstrated that as the tunnel increased in length, it took longer for the blue balls to move into the compartment originally containing the silver balls. It was found that the tunnel between the two chambers had to be occupied by several blue balls in order for a blue ball to move into the other compartment. As this was a stochastic process (based on shaking of the apparatus), it took longer for the blue balls to get into the other chamber compared to no tunnel. The blue balls were meant to be analogues to K^+ ions, and the requirement for the channel to be occupied by four K^+ ions provided an example of

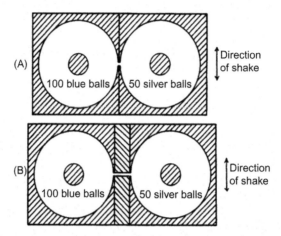

FIGURE 9.2   Diagram of mechanical model. (A) Two flat compartments were separated by a narrow gap. (B) The gap width was increased by spacers. The sides and bottoms of the compartments were made of aluminum; the top was Perspex (clear acrylic sheet). The diameter of the circular compartments was 74 mm, and that of the balls 3 mm. The drawing is roughly to scale. The central blocks of aluminum helped randomize the movement of the balls. Source: *From Hodgkin, A.L., Keynes, R.D., 1955. The potassium permeability of a giant nerve fibre. J. Physiol. 128 (1), 61−88.*

the slower time course of the changes in the K^+ conductances during the action potential. The model-derived hypothesis subsequently was verified biologically, earning Roderick MacKinnon the Nobel Prize in 2003.

Several important points need to be noted. Hodgkin and Huxley wrote of their modeling that,

> [For] both "membrane" and propagated; the conduction velocity; the impedance changes during the spike; the refractory period; ionic exchanges; subthreshold responses; and oscillations... This is a satisfactory degree of agreement, since the equations and constants were derived entirely from "voltage clamp" records, without any adjustment to make them fit the phenomena to which they were subsequently applied (Hodgkin and Huxley, 1952).

In other words, the actual biological experimental results were chosen without modification to impose biological constraints in order to ensure biological plausibility of the resulting mathematical models and therefore broad or conceptual reproducibility. They also wrote, "The agreement must not be taken as evidence that our equations are anything more than an empirical description of the time-course of the changes in permeability to sodium and potassium. An equally satisfactory description of the voltage clamp data could no doubt have been achieved with equations of a very different form, which would probably have been equally successful in predicting the electrical behavior of the membrane"—a tacit recognition of the Fallacy of Limited Alternatives.

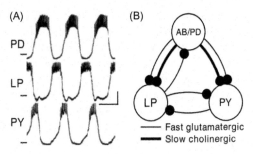

FIGURE 9.3 Biological pyloric rhythm and pyloric circuit architecture. (A) Pyloric rhythm recorded from *Homarus americanus* with intracellular electrodes. Scale bars: 1 sec and 10 mV; horizontal lines: −60 mV. (B) Schematic of a simplified version of underlying circuit (see text for explanation). Under control conditions, the pyloric network generates a triphasic rhythm with bursts occurring in the order PD-LP-PY. All synapses in circuit are inhibitory. Source: *From Prinz, A.A., Bucher, D., Marder, E., 2004. Similar network activity from disparate circuit parameters. Nat. Neurosci. 7 (12), 1345–1352.*

The Hodgkin and Huxley model has survived intact for over 65 years. This attests to its technical reproducibility. Further, the model has been extended to create a number of different neurons, other ionic conductance channels, and other species. This attests to the conceptual reproducibility. One cannot but think that in large part this is related to the inherent logic of the modeling and simulations with prior due consideration of potential logical errors.

Case 2. Multiplicity of models to achieve the same outcome—the Inverse Problem. Prinz et al. (2004) applied computational modeling and simulations to the neuronal activity patterns (rhythms) generated in the lobster stomatogastric ganglion (Fig. 9.3), which is a collection of neurons that control the pylorus of the lobster version of the stomach, in a sense studying stomach churning. The lobster stomatogastric ganglion rhythm generator consists of about 12 neurons consisting of four types: anterior burster (AB), pyloric dilator (PD) neurons, lateral pyloric (LP) neuron, pyloric (PY) neurons, ventricular dilator neuron (VD), and inferior cardiac neuron (IC). In the computational simulations, the AB and PD neurons were combined into a signal model neuron (AB/PD) and a single model neuron was used to represent the other types (Fig. 9.3). The result was a reduced preparation *in silico* that would "stand in" for the biological stomatogastric ganglion, putting the simulation at risk for the Fallacy of Four Terms.

Prinz et al. (2004) recognized the risk of the Fallacy of Four Terms, writing: "These simplifications, along with our use of single-compartment model neurons and simple synaptic dynamics, mean that the model networks studied here are best thought of as semi-realistic models of the biological pyloric circuit that allow us to examine the larger issue of how critical synapse strength is to network dynamics."

Note that the term "semi-realistic model" is a sort of double entendre saying the model is not realistic but it is realistic in some sense. The problem is that the final inference from a statement of "semi-realistic" likely will be taken in one of two ways, as relativistic as the model truly reflects the biology and hence has credibility or that it does not reflect the biology, thus without credibility and is dismissed. While neither outcome seems the intent of the authors, there is sufficient vagueness to the term to allow "having one's cake and eating it too." The potential fallacy is shown in Argument 9.1 that follows:

ARGUMENT 9.1

Major premise: *The biological pyloric mechanism* (major term) *contains neurons* (bridging term).

Minor premise: *The model* (minor term) *contains neurons* (bridging term).

Conclusion: *The model* (minor term) *is the biological pyloric mechanism* (bridging term).

The one term duplicated to produce a net of four terms is *neurons*. Clearly, *neurons* in the major premise are not exactly the same as *neurons* in the minor premise. In the major premise there are more neurons in more complicated arrangements in the *neurons* of the major premise than are the neurons and arrangements in the *neurons* of the minor premise. The risk of irreproducibility based on the Fallacy of Four Terms depends on the Epistemic Risk in relating *neurons* in the major premise to *neurons* in the minor premise. The Epistemic Risk can be broken down into Epistemic Distance and Epistemic Degrees of Freedom. The Epistemic Distance is reflected in the reduction from at least 12 neurons in the biological *neurons* to 3 neurons in the model *neurons*. If synapses are considered, there is an even greater reduction in the complexity. Epistemic Degrees of Freedom can be seen in the qualitative differences between biological neurons and model neurons. Most obvious is the combining of two specific biological neurons, the AB and PD neurons, to a single hybrid model neuron, AB/PD.

Prinz et al. (2004) took steps to mitigate the risks of a Fallacy of Four Terms. First, they constrained the range of possible outcomes of the model by imposing biologically inspired constraints. These constraints limited the range of outcomes of the model, that is, simulated pyloric rhythms, to a range obtained empirically from a study of rhythms generated in the stomatogastric ganglia of 99 lobsters (*Homarus americanus*). Presumably, the constraints on the range of outcomes of the model would translate to constraints on the input parameters, synaptic strengths between model neurons, and the conductance properties of model neurons. The inference is that the biologically inspired constraints of the model outputs would translate to biologically inspired constraints on the synaptic strengths and conductance properties with the hope that the resultant synaptic strengths and conductance properties would be biologically relevant.

Prinz and colleagues took further steps to mitigate the risks of a Fallacy of Four Terms by changing the major term, *biological pyloric mechanism*, to the major term, being *tunable networks*, such as networks of properties intrinsic to individual neurons or genetic networks within cells. Indeed, the purpose of the research was to demonstrate that tunable networks, such as the intrinsic electrical properties in individual neurons and genetic networks, both demonstrated empirically, could be generalized to neural networks as such as may be seen in the lobster stomatogastric ganglion. Thus, the actual syllogism is shown in Argument 9.2:

ARGUMENT 9.2

Major premise: *The tunable network mechanism* (major term) *contains neurons* (bridging term).

Minor premise: *The model* (minor term) *contains neurons* (bridging term).

Conclusion: *The model* (minor term) *is the tunable network mechanism* (bridging term).

Note that Argument 9.2 is the research hypothesis where the modeling is intended to demonstrate a truth or falsehood. The argument's logic relates only to whether the conclusions of the actual modeling and simulations are valid, that is, the degree of certainty. Thus, a valid syllogistic argument can lead to a false conclusion in which one can have confidence using modus tollens and thus a sense of productive irreproducibility. In that case, one knows to look at the validity of the premises to find the source of a false conclusion. However, if the argument is invalid, there is no way to go further, hence unproductive irreproducibility.

The modeling goes on to demonstrate a synonymy between the model and *tunable network mechanism* with the synonymy established by the fact that variations of synaptic weights and conductance properties can still produce a restricted range of pyloric rhythms. The condition for establishing the synonymy is by a metaphor to a previously determined *tunable mechanism* such as networks of properties intrinsic to individual neurons or genetic networks within cells. The outcome of the modeling and simulations demonstrates that a wide range of synaptic strengths and conductances can converge onto a narrow range of pyloric rhythms. The modeling establishes the possibility of such networks to do so, although it is not necessarily that such networks converge onto a single type or motif. A sense of the Large Number theorem as might be applied to look for a convergence of different networks to a Central Tendency of networks likely would have failed. Had the model failed, then it is likely that such networks necessarily cannot converge.

Argument 9.2 is more abstract and hence at some distance to the biological Argument 9.1. Within the "abstract" universe of Argument 9.2, the argument likely is valid and the demonstration of the conclusion to be true vouchsafes the truth of the premises. The biological inference depends on extrapolating from the "abstract" universe of Argument 9.2 to the biology of Argument 9.1. The extrapolation depends on how the reader evaluates the Epistemic Risk in terms of the Epistemic Distance and Epistemic Degrees of Freedom described earlier. Note that the importance of the extrapolation is that it provides metaphors by which to create future biological and computational investigations.

Using the immense processing speed of modern digital computers, Prinz and colleagues simulated 20,250,000 versions of the model, of which 4,047,375 (or 20%) met the criteria for pyloric-like rhythms. The versions covered a wide range of synaptic strengths, model architectures, and conductance properties. The findings support the authors' notion of tunable networks and, when viewed in other systems, such as in individual neurons and genetic systems, argue that such tunable networks may be a general biological property. However, this demonstrates a critical problem in computational biomedical research in particular and research in general. The issue is the Inverse Problem and the Fallacy of Confirming the Consequence.

Consider the situation where a researcher selected only 1 of the 4,047,375 versions and the subsequent simulation experiments met the criteria for pyloric-like rhythms. The research would have successfully simulated the biological pyloric rhythm and likely concluded that the model selected actually represented the biological mechanism. In truth, any of the other 4,047,374 versions also would have been successful and would equal claims to represent the biological mechanisms. There is no way within the simulation of a single version to know whether it represents the actual biological mechanism. And if another experimenter selected a single but different model from the 4,047,375 valid versions, would not the difference between the two experiments be an example of unproductive irreproducibility?

Such a mistaken claim would be an example of the Fallacy of Confirming the Consequence, which would be of the form *if specific model A implies biological phenomenon B is true and biological phenomenon B is true, then specific model A is true*. However, consider the experiment established by the argument *if specific model C implies biological phenomenon B is true and biological phenomenon B is true, then specific model C is true*. How does one reconcile the situation where *specific model* ***A*** *≠ specific model* ***C***? Would such a situation imply that one or the other demonstration is irreproducible?

The problem of inferring specific causal mechanisms when any number of them can produce the same outcome is an example of the Inverse Problem. In the research of Prinz and colleagues, there were 4,047,375 causal mechanism, all of which would produce the pyloric-like rhythm. The scientist is left with a difficult choice: are there as many causal mechanisms as there are variations in the model or is there some overarching common mechanism? If the former position is adopted, then there are at least 4,047,375 different causal mechanisms. If the latter, then the charge is to find the commonality among the possible mechanisms, which is problematic.

SUMMARY

Mathematical and computational modeling and simulations have been critical to science, particularly since Galileo and Newton. But the importance evolved from quantitative descriptions to mechanistic accounts with credibility borrowed from the presumption of the analytic truths of mathematics originating in the notion of the Geometrical Method. The remarkable advances in mathematical analyses and computational power tremendously increased the likelihood that some biological question or problem can be recast as a mathematical or computational problem and a solution can be found, even if the solution is biologically implausible. The problem is that solutions will be taken as a biological insight and potentially used to plan future research. However, such future research based on biologically implausible mechanistic solutions likely will increase the risk of unproductive irreproducibility in the broad or conceptual sense. The remarkable power and the diversity of mathematical and computational modeling and simulations, such as demonstrated by Prinz et al. (2004), raise serious issues as to what constitutes reproducibility.

The ubiquity of the use of digital electronic computers also risks irreproducibility in the narrow sense of replications. This form of irreproducibility lies in the electronics of the computer, thus making its recognition difficult. These flaws are not syntactic and thus unlikely to be detected at the time of compiling or interpreting software code into machine language. Rather, these are likely to result in runtime errors, but again the recognition of runtime errors is difficult.

Chaotic and Complex Systems, Statistics, and Far-from-Equilibrium Thermodynamics

LIMITATIONS OF TRADITIONAL STATISTICS

A major theme in concerns regarding reproducibility in biomedical research centers on the appropriate use of statistics. Many cases of irreproducibility were detected on the basis of differing statistical results. In many of those cases, as argued in this book, statistical errors relate to the necessary use of logical fallacies that lead to the need for statistics. Statistics essentially are methods used to increase the certainty of probability syllogisms that were introduced to increase the certainty of the partial syllogism that was necessary to increase the utility of syllogistic deduction but at the expense of certainty (see Chapter 5: The Logic of Probability and Statistics).

Statistics' appeal of certainty is internal to the experiments and observations, meaning that statistics is an epistemic device. Statistics only determine the boundaries of possibilities, those being probabilities, based on the data in hand. It does not attest to the reality or actuality (ontology) of the statistical inferences, such as the Central Tendency. For example, there is nothing in principle that would prevent one from finding the Central Tendency of every mammal on the planet, but how would that Central Tendency be meaningful or informative as to the nature of reality?

As an internal test to lend certainty, there must be underlying principles that govern the appropriate use of statistics for drawing inferences. These principles, as will be discussed in detail, include the Large

Number Theorem, Central Limit Theorem, and ergodicity. But what happens if the portion of reality under investigation does not comport with these principles? Note that one can still conduct statistical operations, such as finding the Central Tendency and variance, but any confidence regarding inferences to ontology will be only an illusion. The illusion will burst when attempts at replicating the experience fail, in either the local, narrow or the broad sense.

Chaotic and complex systems frequently do not comport with the necessary assumption of the principles just described. For example, the Central Tendency does not need to converge or settle on a single value with increasing sample size. Such systems may have low ergodicity, making sampling very problematic, and thus the application of traditional statistical analyses fraught. Challengingly, the very nature of the reality of concern to the biomedical researcher is chaotic and complex. In many ways, the application of traditional statistical analyses will not work, with the results risking unproductive irreproducibility.

It is important for biomedical researchers to appreciate the centrality of Chaos and Complexity in biological systems. Such appreciation first is at the descriptive level, how do Chaotic and Complex systems manifest, for example, the dependence on initial conditions and unpredictability? However, it also is important to understand the origins of manifestations of Chaos and Complexity and how the appearance (manifestations) may convey a false tractability and thus false confidence in the analyses. The consequence is that biomedical research, dependent on the "scientific" appearance that does not comport with the chaotic and complex underlying mechanisms, increases the risk of conceptual or broad irreproducibility.

To understand the manifestations, it will be necessary to introduce thermodynamics, particularly the Second Law of Thermodynamics and how the second law applies to information. As will be seen, the information in the system is what biomedical researchers are dealing with, and thus the Second Law of Thermodynamics supervenes on the very currency of their experiments. To be sure, there has been a sense of limitations since the origin of modern statistics. There has been an implicit recognition of the limits of traditional statistics, in a way reflecting the chaos and complexity of living systems. However, these issues have not found much play in discussions of biomedical research reproducibility.

As will be discussed, the notion of information is complex and some may find it counterintuitive, yet it is central to all science. One measure of information is by contrast to randomness—information is a departure from randomness. As the laws of physics apply to randomness, particularly through the Second Law of Thermodynamics, they also apply to the inverse of randomness, information, including the kind of information central to biomedical research. As will be demonstrated, physical laws drive systems to randomness. Life goes against the stream, although still subject to the same laws. This requires something unique about living processes. A tension exists among the forces regarding randomness, loss of information, and dynamics in the living to maintain information. That tension will affect the use of traditional statistics, for example, in the notion of biological variability as will be discussed.

Another counterintuitive sense of information is the notion of information as a physical entity. At its fundamental level, information is like a bit of something. One knows it is something because it takes energy to change it, much like it takes physical energy to turn a coin over from heads to tails. In other words, information and a change in information have physical (thermodynamic) costs that must be accounted for in any living organism. The cost of information and the rules for managing costs are found in physical laws, particularly the Second Law of Thermodynamics as it applies to Information. These topics are the focus of this chapter.

RESISTANCE TO STATISTICS

Perhaps one of the greatest and most unsung "heroes" in biomedical research is statistical science. Yet, the reticence of clinicians and early biomedical researchers to fully embrace statistics may be a reflection of some vaguely held sense of the fundamental limitations of statistics. While the use of statistics in biomedical research is ubiquitous, statistics were not welcomed in early biomedical research (dating early biomedical research to the development of the Cell Theory in the early 1800s). It is difficult to imagine biomedical research today without statistics. Indeed, recent concerns over the problem of reproducibility led to a greater emphasis on statistics, particularly the appropriate use thereof. Early advocates for the use of probability and statistics were opposed by physicians and biologists alike, particularly the advancing discipline of physiology (Matthews, 1995). Physicians were skeptical that the population-based approaches of probability and statistics, which in the early 1800s were successfully applied to public health issues, could be applied to the individual patient. There is some truth to that concern, as discussed in Chapter 4, The Logic of Certainty Versus the Logic of Discovery and the Fallacy of Four Terms. Perhaps more importantly, many physicians saw efforts at quantification and analysis as making medicine too scientific and denigrating physician intuitions.

Laboratory-based biomedical researchers initially opposed probability and statistics because of the great diversity of biological phenomena and the difficulty of subsuming the variety in an economical set of probability and statistical descriptors, such as the Central Tendency. The most significant concern, perhaps, was that probability and statistical correlations do not imply causation. Indeed, one of the main purposes of biomedical research is to attribute causation to the phenomena observed, thereby obtaining an economy of explanations of the variety of phenomena. In other words, because the scientist could see some common causation operating in a variety of phenomena, one did not have to invoke a separate mechanism for each variation in the phenomena. The economy of causal explanation allowed avoidance of the Solipsism of the Present Moment (nothing is generalizable beyond the immediate moment).

The distinction between phenomenological correlation and causation held by laboratory-based biomedical researchers is on shaky ground, philosophically (Hume, 1748). A philosophical school, called Operationalism, believed all measurements involved in experimentation rest on a set of fundamental or primitive operations that are axiomatic or self-evident, that is, not needing further examination or debate. Further, relationships exist between the fundamental or primitive operations and constitute the rules of inference for combining the primitive operations into the complexity involved in the actual experimentation. However, every attempt to develop such a reductionist schema to root experimentation to some anchor fails. What are left are purely relationships among observables, none of which are fundamental. In other words, explications of observations to some common principle, such as a specific form of causation, can only be in terms of other observations that likewise require explications from other observations and so on indefinitely, even if the observables require the use of telescopes, microscopes, oscilloscope, and every other form of scope.

The presumption of physiologists in the 1800s was that a causal understanding of the underlying biological principles was sufficient to understand the phenomena of the individual. One did not have to abstract the principles for the sample or ensemble. The notion was that "data speak for themselves," and if statistics were necessary to come to any understanding, the situation was desperate. As Mark Twain wrote, "Figures often beguile me, particularly when I have the arranging of them myself; in which case the remark attributed to Disraeli would often apply with justice and force: 'There are three kinds of lies: lies, damned lies, and statistics'" (Chapters from My Autobiography, North American Review, Project Gutenberg).

LARGE NUMBER THEOREM

Statistics obtain its validity internally. In research, statistical hypothesis testing has less to do with the ontological reality (where the hypotheses actually are real and shown to be true) than whether the "numbers add up" in a statistical sense. The remark by Twain regarding the self-constructed beguiling figures is prescient. The key to internal validity is predictability. The results of a specific experiment must be predictive of any repeated experiment. The results from studies of a model, such as a rodent standing in for a human, must be predictive of similar manipulations in humans. A sample of subjects must be predictive of the population. The latter prediction scenario is the basis for the Large Number Theorem.

The Large Number Theorem holds that the Central Tendency, such as the mean (average) or median of a sample of observations, will converge onto a single value that can be taken as the Central Tendency. For example, the mean (average) of a set of n observations may fluctuate for small n. As the number of observations increases, the fluctuations become less and finally settle to a range of fluctuations that is acceptable and some "middle" value in that range is taken as the true value of the Central Tendency.

The Central Limit Theorem holds that the Central Tendency in a set with large numbers of elements will be the same as the Central Tendency of the Central Tendencies of multiple smaller sets. For example, consider a set of 100 observations whose mean (average) is $\bar{x}$. Next, the set of 100 observations is broken into 10 sets of 10 observations. Each of the sets of 10 observations would have mean (average), $\bar{x_1}$, $\bar{x_2}$, ... $\bar{x_{10}}$. Each of these means that $\bar{x_1}$, $\bar{x_2}$, ... $\bar{x_{10}}$ may be different but the mean of these means would be similar to $\bar{x}$ for the set of 100 observations. The advantage of the Central Limit Theorem is that the variance of each smaller set may not be distributed normally and thus unable to take advantage of statistical methods that derive power from an assumption of normal distribution. However, the variance of the set of Central Tendencies from each smaller set becomes distributed normally with an increasing number of smaller sets. The normal distribution allows the application of more powerful statistical techniques.

These theorems depend on the concept of ergodicity (Fig. 10.1). If the data or observations are random by being distributed evenly in space (a uniform random distribution), then the ergodicity is high and one only needs to sample a relatively small volume of the space to obtain a representative sample, as would be the case for each fisherperson sampling the fish in the lake in Fig. 10.1A. If, however, the fish are distributed randomly but not evenly (for example, a Poison distribution), the two fisherpersons will have very different experiences. The result is that statistical inferences for fisherperson 1 in Fig. 10.1B will not be the same as for fisherperson 2, even though the fish are distributed randomly in the lake.

As will be discussed, ergodicity is central to the Second Law of Thermodynamics and, importantly, living biological systems have low ergodicity. Thus, the Large Number Theorem and the Central Limit Theory will be difficult when applied to living biological systems,

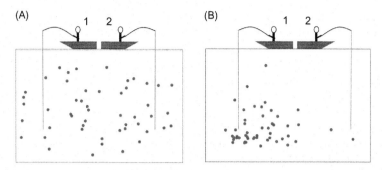

FIGURE 10.1   Schematic representation of the effects of ergodicity. (A) Fish are distributed evenly in the lake—high ergodicity. (B) Fish are concentrated under fisherperson 1. The experience of the two fisherpersons in condition A will be the same and considered reproducible. Their experiences will be different under condition B and the experiments considered irreproducible. *Reproduced from Montgomery Jr., E.B., 2019. Medical Reasoning: The Nature and Use of Medical Knowledge. By permission of Oxford University Press, USA (In Press).*

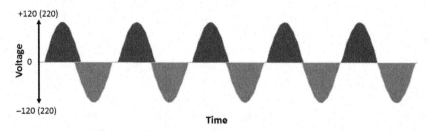

FIGURE 10.2   Sine wave representation of line voltage in an AC household current. The maximum and minimum are plus and minus 120 V, respectively. Note that the area above zero (corresponding to the amount of positive electrical current) is equal to the area below zero (amount of negative current). Thus, the mean (average voltage) over the five cycles of electrical current is 0.

necessitating the invention of other methods to achieve what traditional statistics hoped for.

BUT WHAT IF THE LARGE NUMBER THEOREM DOES NOT HOLD?

Consider that alternating current (AC) coming out of an electrical power line. Typically, the voltage swings (as a sine wave) between a positive 120 V to a negative 120 V (Fig. 10.2). The mean (average) voltage over 1/60th (1/50th in certain parts of the world) of a second coming out of an AC electrical outlet is 0. Also, the average voltage over $1,000,000 \times 1/60$ seconds is still 0. Does this mean one will not receive an electrical shock by sticking his or her finger into an electrical outlet? Obviously not, as for an instant the voltage is +120 or −120 V. As demonstrated by the fact that the mean (average) voltage over $1,000,000 \times 1/60$ seconds is still 0 indicates that the Large Number Theorem, just described, is no help in representing the nature of the electrical voltage coming from the AC electrical outlet. There are other measures that serve the same general descriptive purpose for such oscillating data, such as Root Mean Square. For a sample point of the voltage every 1 ms, it makes no difference how many cycles are included in the sample, whether it is 1 or an infinite number, as there is no convergence on a voltage other than 0 with increasing numbers of cycles sampled.

If one were to sample over 1.5 cycles, the mean voltage may not be exactly 0. However, if one were to measure over 1,000,000.5 cycles, the mean voltage may not be exactly 0 but would become vanishingly small. Thus, in a sense, the Large Number Theorem does contribute to an outcome, that is, the mean (average) voltage converges on 0 in the limit of an infinite number of samples. However, the result is still meaningless for informing as the nature and thus the danger of touching a line voltage in an AC electrical outlet.

As will be discussed, it is quite likely that the very large percentage of biological systems operate as oscillators, using the term "oscillator" in its widest notion and evidenced in periodic functions (repetitive changes in some variable over time). Biological systems will do so because of their intrinsic physics, particularly as related to the Second Law of Thermodynamics. (Originally, the field of thermodynamics was developed from studies of steam engines in the advent of steam-powered locomotives and manufacturing machines. However, thermodynamics is relevant to every machine, ranging from electrons to biology.) When Chaos and Complexity are discussed later, the concept of a limit cycle will be introduced. Limit cycles produce periodic phenomena and are ubiquitous in complex living organisms whose dynamics include mechanisms that are antagonistic to each other, for example, negative feedback loops or homeostatic mechanisms.

The Gaussian biological variability ubiquitous in observations of biological systems may be evidence of an underlying oscillator, in this case a dampened oscillator operating in an energy field (Norrelykke, 2011; Huang and Batelaan, 2013). Consider a pendulum as shown in Fig. 10.3. A forcing function, energy, is used to displace the pendulum. Afterward, inertia and the gravitational field act on the pendulum to sustain its motion. In the case of a dampened pendulum (oscillator), friction dissipates the energy to dampen the oscillations. If one plots the distributions of the different amplitudes, a probability density distribution is created that has a Gaussian distribution with a mean (average of 0).

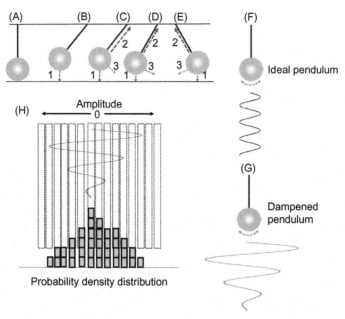

FIGURE 10.3 Schematic representation of a simple harmonic oscillator consisting of a weight (pendulum) suspended by a cable. (A) The system is at equilibrium (rest). (B) The pendulum is displaced to the left. As a consequence, potential energy is created (translated from the force that displaced the pendulum). It is denoted by vector 1. The potential energy creates force, gravity, to move the pendulum downward (B). The cable creates a restraining force that will not permit the pendulum to fall in a straight line relative to gravitational force (C). A resultant force is created (vector 3) from gravitational and restraining forces that move the pendulum toward the equilibrium position at rest (D). The pendulum gains kinetic energy from the acceleration and creates a force that moves the pendulum past the equilibrium point (E). The consequence is generation of a more potential energy gravitational force that, when combined by the restraining force, moves the pendulum in the opposite direction. And the process repeats. If no energy is lost—due to friction, for example—the pendulum will continue to sway back and forth. Over time, it will begin to trace a sine wave whose phase is 90° (F). (H) The effect of friction producing a dampened oscillator. The amplitudes of the swings diminish over time. (G) A Probability Density Distribution is shown. The amplitudes in each bin, represented by the long rectangles, are represented in the height of the distribution. This creates a Gaussian distribution.

The periodic or oscillatory nature of many biological systems is clear by observation, such as circadian rhythms, heart rate, and gastrointestinal peristalsis. However, nearly every homeostatic system keeping certain states constant, such as blood electrolytes, displays at least dampened oscillation, even if not readily observable. Thus, biological variability is not just a description of phenomena but also may reflect the underlying unobserved dynamics of the components, sometimes referred to as the system's microstates. For example, in a thermo-dynamic system such as a container of gas, the microstates are the momentums of each molecule of the gas (velocity and mass). This is central to biology, and a great deal of thermodynamics involves how macrostates, such as phenomena observable to biologists, are manifest from the microstates. For example, a biological macrostate may be the behavior of an animal while the microstate may be the dynamics of oxy-tocin molecules in the animal's brain. In the thermodynamics of a con-tainer of gas, the macrostates would be temperature and volume. What the biologist measures, typically the macrostates, can be understood from the thermodynamics of the underlying microstates; hence the rele-vance of the Second Law of Thermodynamics.

Another example of phenomena where ergodicity fails and the Large Number and Central Limit Theorems become problematic relates to dynamics that display hysteresis. An example is opening a clasp or pocket knife where it takes a relatively large force to open the knife but a point is reached where the knife "snaps" into the fully open position. The same process happens in reverse. In this case, hysteresis involves dynam-ics where there is considerable resistance to move from one state to another but any transitions are very fast so the system appears to exist in only two or a few states. In the case of the clasp or pocket knife, the knife is in one of two states, open or closed, the vast majority of the time.

Often in biology there are transitions between different states, for example, a state of rest and a state of displacement, such as activation. The system moves on a "path" between the two states; however, the "path" leading to displacement is not the same "path" taken to rest. Consider the hypothetical example shown in Fig. 10.4. A set of pro-cesses move the condition from *state V* to *state W*. A different set of pro-cesses move the condition from *state W* to *state V*. This is an example of hysteresis. Suppose a researcher is studying such a system without realizing that the system displays hysteresis and measures parameter y when parameter x is x_1. The measured values of y will be different, y_1 and y_2, depending on which process is acting, for example, moving from *state V* to *state W*, which would have results around y_1. Alternatively, if the active process is moving from *state W* to *state V*, values of y will be around y_2. These different values, if obtained in the same experiment, would increase the variance of y's greatly, decrease the statistical power, and increase the risk of a type II error. Alternatively, one experimenter may be operating under conditions favoring the path from *state V* to *state W* while another experimenter's condition favors the path from *state W* to *state V*. The results will be very different, raising the concern of irreproducibility. Biological sys-tems displaying hysteresis are frequently encountered.

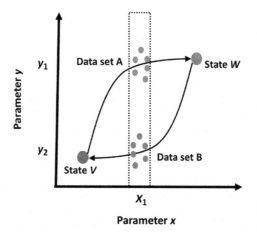

FIGURE 10.4 Schematic representation of a system displaying hysteresis. The trajectory through the parameter space (x, y) when moving from *state V* to *state W* is different when moving from *state W* to *state V*. Measures of parameter y when parameter x is x_1 will be very different, depending on which process is active or dominant. For example, filled circles at the top in the middle represent a set (A) of observations of y at x_1 under conditions that favor the transition from *state V* to *state W*, while set B would arise should the transition from *state W* to *state V* be favored. Note that when just provided the observations where x is x_1 does not inform which transition is being favored. Two experiments exactly the same but with a bias for one transition in experiment 1 and bias for the other transition in experiment 2 would have results considered irreproducible. This situation is more likely to occur in Complex or Chaotic systems, which probably involves the majority of biological conditions.

Any system that alternates between specific states, such as sleep and wakefulness, is periodic and can be described in terms of oscillators. Such states are referred to as metastable states. In Chaos and Complexity theory, many such oscillatory activity can be described by "limit cycles" (Strogatz, 1994). An example of a limit cycle may be fluctuations in the population of predators and prey in a local environment (May, 1972). The number of predators may change depending on the amount of prey. The relationship is likely oscillatory based on negative feedback of the predator on prey and the prey and the positive feedback of prey on predators. However, there may be a number of other factors, and if the effects are nonlinear, the oscillatory pattern to the predator—prey relationship may not be seen readily. Indeed, it could look random because of the complexity. However, if the prevalence of predator, P, is assessed repeatedly over time, t_i, resulting in P_i for t_i, the distributions of P_i plotted may look random, suggesting no structure (information) in the time course of predator prevalence (Fig. 10.5). However, if P_i's are plotted where the value of P_i is on one axis and the prevalence at the next measurement is P_{i+1}, a phase plot is constructed (Fig. 10.5A). The

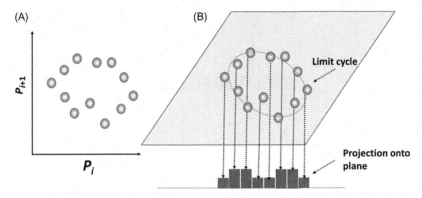

FIGURE 10.5 Hypothetical example of a system displaying a limit cycle. In this case, the model is a predator−prey relationship. At different times, the prevalence of a predator in a given location is determined. At first, it may appear that the variation in the prevalence is random due to the complex nonlinear relationships among the many factors that influence predator prevalence. (A) A phase space (also called a Return Map or Poincare plot) plots the prevalence of the predator at a given time, P_i, versus the prevalence at a subsequent time, P_2. A pattern (structure) emerges showing a circle called a limit cycle. Many times, the ability to detect a limit cycle depends on the perspective, particularly if there are a relatively large number of dimensions. Consider the situation where the limit cycle is confined to a single plane (either in the actual dimensions of the dataset or in some recombined set of dimensions, such as a principal component space as a result of Principal Component Analysis that reduces the dimensionality but with hybrid dimensions). When viewed from a perspective orthogonal to the plane containing the limit cycle, the distribution of data will appear flat (platykurtosis).

phase plot often can detect a relationship between successive measurements, P_{i+1}, that vary according to the previous measurement P_i (as often is the case in nonlinear systems). If one were just to look at the distribution of each measure it may look random. As can be seen, the distribution in the phase plot is not random, it has structure, in this case what is called a limit cycle. There is information in the changes on predator prevalence over time and thus some deterministic mechanisms underlying predator populations. What would it mean to calculate the mean (average) of a limit cycle (Fig. 10.5B)?

Other types of transitions between states may not be periodic and often are referred to as bifurcations (Fig. 10.6). For example, *state A* can suddenly change to *state B* or *state C*. To which *state A* changes to (*state B* or *C*) is unpredictable in Chaotic and Complex systems. With repetitions, a single *state A* can be associated with some mixture of *states B* and *C*. A hypothetical example is shown in Fig. 10.6. Consider a weight placed atop a thin pillar that eventually buckles with increasing weight, *x*. Restricting the hypothetical to a two-dimensional plane, the pillar can be seen as buckling to the left or right, as measured by *y*. The results of repeated experiments can be plotted as the position *y* versus weight *x*.

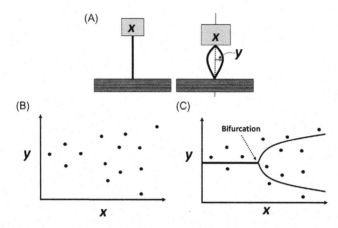

FIGURE 10.6 Schematic representation of a system that demonstrates bifurcations (transitions) between states. (A) A weight is supported by a pillar. The pillar remains upright and straight as weight (x) initially is increased. At a certain weight, the pillar from the single perspective is seen as buckling to the left or right. The direction of the buckling is unpredictable. (B) The weight and corresponding distance of the buckling (y) are plotted. At first impression, it may seem as though no relationship exists between weight, x, and buckling, y. Indeed, the correlation coefficient likely is near 0. However, the physics clearly is one of a relationship between x and y. Indeed, an equation of the form $\dot{y} = xy - y^3$ describes this process.

If one were only to see the plot (Fig. 10.6B), one might assume that no consistent relationship exists between position y and weight x, yet it is reasonable to expect one. This would be an example of conceptual irreproducibility. However, the relationship between y and x that appears random in Fig. 10.6B is clearly related by a model of the bifurcation function in Fig. 10.6C (what is called a pitchfork bifurcation).

Clearly, the application of traditional statistics that depend on the Large Number Theorem to these types of systems is problematic and potentially misleading, particularly in regard to reproducibility. The application of traditional statistics to biological questions asked of inherently complex and chaotic dynamics, in this case, periodic, oscillatory, and bifurcating processes or those with hysteresis, may give rise to an impression of irreproducibility inappropriately. It is the measure, not the measured, that is irreproducible. The challenge will be to develop appropriate statistical tests, perhaps based on scalar descriptors of Chaos and Complexity, such as the Lyapunov exponent, Kolmogorov entropy, and Kolmogorov complexity. The first challenge is the recognition of the need to do so, that is, recognizing that the biology under investigation is sufficiently Chaotic or Comples. Perhaps in the future, scientists and statisticians will check the data for Chaos and Complexity before checking for a normal or Gaussian distribution in deciding which statistical tests to perform, which typically is (should be) done.

EQUILIBRIUM AND STEADY-STATE CONDITIONS

Many of the examples given here represent dynamics or changes between states. Historically, the dynamics have been hard to analyze and have been inferred by allowing the dynamics to settle on relatively stable states. These states are referred to here as metastable states to indicate their potential for change. One way experimentally to achieve metastable states is to allow the system time to come to equilibrium. An example would be a reversible chemical reaction where A combines with B to produce AB determined by rate k_1. The rate at which AB generated through k_1 is determined by the relative concentrations of A and B. Importantly, there is an "opposing" reaction where AB dissociates into A and B and is determined by rate k_2 and the relative concentrations of AB determine the rate of conversion to A and B. Eventually, some A's and B's are consumed to produce AB until the concentrations of A, B, and AB, mediated by k_1 and k_2, reach a metastable state at which point the system is said to be in equilibrium—the rate of formation of AB from A and B is equal to the rate of formation of A and B from AB. The final concentrations of A, B, and AB are used to determine k_1 and k_2 to infer the dynamics.

The unchanging concentrations of A, B, and AB at equilibrium or steady state do not mean the dynamics are static. Rather, molecules or atoms A and B are forming AB constantly and AB is forming A and B constantly. The system's dynamics only appear to be static or unchanging. The concentrations of A, B, and AB are macrostate descriptors and not indicative of what is happening at the microstate level. In many ways, this is analogous to the fact that the macrostate descriptor Central Tendency, such as the mean, may not reflect any of the data (microstates) in the sample. The macrostate measure of Central Tendency loses information of the microstates, which will be seen as a consequence of the Second Law of Thermodynamics as applied to Information. Indeed, one of the great achievements of modern physics (statistical mechanics) has been the ability to relate macrostates (such as what is observable) to microstates (the mechanisms). In this sense, the macrostate is an epistemic tool used to understand the microstates, the latter constitutes the ontology. The key here is that biomedical research predicated on measure of macrostates that do not comport with the underlying microstates will be at risk for irreproducibility in both the narrow and the broad sense.

BIOLOGICAL MACHINES, THERMODYNAMICS, AND STATISTICAL MECHANICS

Biological systems can be conceptualized as machines that use energy, for example, heat, to perform work. It was the intent to understand machines that led to advances in thermodynamics, particularly related to steam-powered machines. A steam locomotive uses heat to boil water (low entropy) to produce steam (high entropy), which increases pressure to increase the volume of the cylinder that pushes the piston to turn the wheels (work). It takes energy to power the assembling of bricks scattered about (high entropy) into a house (low entropy). It takes energy (from glucose metabolism, for example) to assemble proteins (low entropy) from amino acids floating about in the cytoplasm (high entropy), and the energy expended creates heat (the heat of metabolism). Thus, the Laws of Thermodynamics are as relevant for any biological process, including those subject to biomedical research, as they are to a steam engine.

The science of thermodynamics owes much of its start to the scientists and engineers attempting to make steam engines more efficient at the dawn of steam-powered locomotion and industries. The ratio of the amount of energy required, representing an expense in the amount of coal bought and consumed, is relative to the movement of pistons that move the train, for example, that creates profit. In an idealized steam engine, a change in energy produces a change in heat or work done. For the steam engine, the amount of energy put into the locomotive, ΔU, is the amount of heat, Q, minus the work done, W, according to the following equation:

$$\Delta U = Q - W, \tag{10.1}$$

where heat, Q, is related to the temperature, T. From Eq. (10.1), it is clear that not all the energy goes into work, W, but that some is expended by Q, which is heat in classical macrolevel analysis. The question is what is the nature of Q? The value of Q appeared to vary with T according to

$$\Delta S = \frac{Q}{T}, \tag{10.2}$$

where ΔS is the change in entropy and is thought of as wasted energy. Later, the notion of wasted energy was related to the state of disorder in the system or entropy. Given the concerns of steam pressure and volume within the pistons, the energy that had to be added, ΔU, becomes

$$dU = TdS - PdV. \tag{10.3}$$

In an idealized gas, P is pressure, dV is the change in volume that moves the piston of the steam locomotive, and dS is the differential form of ΔS or change in entropy. Through experimentation, it was found that dS in a closed system could not be less than 0. In other words, an operating closed system, such as a machine isolated from any exchange of energy or matter with its surrounding environment, entropy could only increase or stay the same and $\Delta S \geq 0$. This is the Second Law of Thermodynamics. Consider the situation of a container of water that has a lid. The water is at 100°C and heat, Q, is added to produce water in the case form of steam, yet the steam is also at 100°C. It takes 2257 kJ (kilojoules) or 539 kcal (kilocalories) to convert 1 kg of liquid water to water vapor without any change in temperature. From Eq. (10.2), there is no change in T, although there was a change in Q; to maintain the equality, ΔS has to increase.

From the Second Law, there can be no net decrease in entropy. So, what about cells that are creating proteins (low entropy) from amino acids floating about in the cytoplasm (high entropy)? Within the cell there appears to be a net decrease in entropy, $\Delta S = S_{protein} - S_{floating\ amino\ acids}$. Somewhere else, there has to be a greater increase in entropy linked to the process that creates proteins from floating amino acids. In this case, it is energy in the form of ATP coming from glucose metabolism. Glucose is a highly organized structure (low entropy) but it is converted to CO_2 and H_2O (higher entropy) and is dissipated to the environment, creating even greater entropy. There also is radiation of heat into the environment that increases the entropy of the environment according to Eq. (10.2). (In this case, the environment is enormously large relative to the cell so that heat transfer does not increase the temperature, T, of the environment.) The net effect is an increase in entropy in the coupled system—protein synthesis linked to glucose metabolism with a net increase in entropy.

The physicist Ludwig Boltzmann reconceptualized thermodynamics to consider the systems at a microscopic level, such as atoms in an ideal gas. The past descriptors are of classical thermodynamics, such as temperature, T; pressure, P; and volume, V. T became the mean (average) kinetic energy of the atoms, P the mean force generated by collisions of the atoms on the wall of a container, and V the space through which the atoms moved. These microscopic levels of description create the realization that T, P, V, and heat, Q, are macrolevel descriptors and are consistent and thus valid internally—at the macrolevel. Yet, they do not account for the dynamics at the fundamental atomic level. Further, for the description at the macrolevel to hold, certain assumptions about the microlevel are required.

One of the key assumptions in extrapolating from micro- to macrolevel descriptors was that the microstate in any one part of the system had to have an equal probability of being in any other part of the system; in other words, ergodicity, as represented in Fig. 10.1 and Eqs. (10.4) and (10.5). Consider a volume of an ideal gas, where there has to be an equal probability of an atom being in one part of the volume as in any other part of the volume, as otherwise one part of the container will experience a different pressure from others. Similarly, the probabilities of the kinetic energy of a group of atoms in one part of the volume are the same at any other part of the volume, as otherwise one part of the volume will experience a different temperature (in violation of the First Law of Thermodynamics at equilibrium). At the microlevel, entropy S is a function of the probability of a particle, such as an atom of an ideal gas, being in any part of the entire volume and is given by the following equations:

$$S = k_B \ln W \tag{10.4}$$

$$W = N! \prod_i \frac{1}{N_i!}, \tag{10.5}$$

where S is entropy, k_B is the Boltzmann constant, W is a measure of probability, and $N!$ is the number of microstates. For N particles, there are $N!$ permutations of microstates. Thus, thermodynamics was to be understood in the statistical properties of the atoms or other particles. As will be discussed, there are systems that do not operate at equilibrium and thus violate the assumptions of equal probability throughout the system. Biological systems particularly operate outside of equilibrium.

The claim advanced here is that the macrolevel descriptors are analogous to sample statistical descriptors, such as mean (average), median, and mode. Like macrolevel descriptors, T, Q, P, and V, described earlier, that do not capture the microlevel dynamics, there may be situations where these macrolevel descriptors actually are misleading and could lead to misconceptions of reproducibility. This is most likely the case when the biological systems are operating far-from-thermodynamic equilibrium.

It is important to note that not all metastable states represent equilibrium conditions where the mechanisms are linked, for example, in the manner in which A and B become AB and by which AB becomes A and B. However, the two processes need not be linked and still reach a metastable state (static and unchanging macrolevel state). The dynamics of these two systems will be very different. Slight changes in one "direction" in the system where the opposing systems are linked are likely to result in changes to a similar degree but in the opposite "direction" in the system because of negative feedback. Uncoupled system nonequilibrium metastable states may display marked changes, even in the face of slight changes in the initial conditions. These marked changes are particularly likely if the dynamics within each process is nonlinear. Such systems are prone to Chaos or Complexity.

The thermodynamic notion of dynamics in a system can be abstracted. One method is plotting in phase space, such as a Poincare

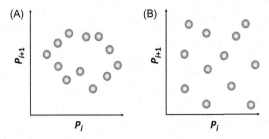

FIGURE 10.7 Schematic representation of the distribution of microstates in a phase space defined by the position of the microstate at parameter P_i versus the microstate at the subsequent measure p_{i+1}. As can be seen in A, the sequential values of the microstate form a limit cycle. The probability of a microstate value is not equal throughout the phase space as it is in B. The situation in B would be demonstrative of an equilibrium condition. Situation A is not at equilibrium and indeed may be far from equilibrium as described in the text.

plot (also called a Return Map), as seen in Fig. 10.5. The dynamics of a system variable, such as a point moving in space, can be assessed by plotting the present location, x_{i+1}, on one axis and the just prior position, x_i (Fig. 10.7). Such a plot may demonstrate "structure" and thus is not random, although it still may be Chaotic or Complex. Alternatively, it may be random and have no structure. One way to visualize such systems is as a particle that moves through space. The position of the particle over time would represent some value of some measure such as kinetic energy. At thermodynamic equilibrium, the particle would have an equal probability of being at any position anywhere in the space.

A particle representing the state at thermodynamic equilibrium, such as by its position, x_i, would have an equal probability of being anywhere in the space—the ergodic hypothesis. This allows sampling of only part of the system to gain insight into the entire system. This is analogous to studying a sample from the population of a biological system to make inferences that are generalized to the population, a risk for the Fallacy of Four Terms. However, systems not at equilibrium may violate the ergodic hypothesis, and the probability in any region of the system may not be the same as any other. Thus, sampling in only one region may not be reflective of the entire system but rather idiosyncratic to the subregion of the space explored. If a subsequent investigation (experiment) explores a different subregion, the results may be very different, leading to false inferences due to the Fallacy of Four Terms.

Biological systems generally operate far-from-thermodynamic equilibrium, which then conveys interesting properties that affect any research attempt to understand those biological systems (Jaeger and Liu, 2007). Such systems require energy to move the system out of equilibrium. For the swinging pendulum, the initial energy or forcing function moves the pendulum from its resting or equilibrium state (hanging vertically) to one extreme from which it is released to begin its swinging motion. As described earlier, the machine, the body included, involves paired processes: one that may decrease entropy, building proteins from amino acids or a house from bricks, and another that increases entropy, such as bodily or building decay. The first requires some payback for decreasing entropy that is increasing entropy elsewhere. Typically, the time course of the opposing processes is asymmetric, allowing the body and building to persist for some time before the opposing process reduces them to dust. During the time the body is maintained, the systems are not in equilibrium; indeed, they are far from equilibrium. This requires energy; for humans, this means food and oxygen, for buildings, this requires the expenditure of energy by the maintenance staff.

The input of energy and the nonlinear nature of the interactions within and between the opposing properties create the conditions for magnified and complex dynamics. Such amplification is manifest by a dependence on initial conditions in which very small changes have very large disproportionate effects to a small instigating change in the initial conditions, for example, Avalanche theory. Consider the situation where small grains of sand can be added to a pile that continues to grow in a regular and incremental manner. The effects are proportionate to the size of each grain of sand added. At a certain point, the addition of a single grain of sand is sufficient to cause the pile of sand to collapse dramatically (for a pile of sand). Now there is a very large and disproportionate effect with the addition of a single grain of sand. It is important to note that this system of adding grains of sand is a far-from-equilibrium state from the standpoint of physics and thus has the potential for an exaggerated response, the avalanche (Jaeger and Liu, 2007).

Systems far from equilibrium will produce very different effects from the same or nearly the same starting cause, called the initial conditions. [As an aside, it is important to note that repetition of an experiment is highly unlikely to have exactly the same initial conditions because of uncertainty in the sense of physics, including measurement and quantum uncertainty, rounding errors, and digitization noise in binary computer operations, for example (see Chapter 9: Case Studies in Computational Biomedical Research).] These different sources of variance in the initial conditions do not cancel but are, at least, additive. This is true even if no difference in the initial conditions is observed. It is very difficult to assume that because they are not observed, due to the measuring system, that there are no differences in the initial conditions.

Many biomedical experiments consist of comparing the initial conditions to some final conditions produced as a result of the experimental manipulation. The dependence on initial conditions may involve relatively small variances in the control condition, s_1 in Eq. (10.6), typically the initial conditions, but very large variances in the conditions or the state resulting from the experimental manipulation on the initial conditions, s_2. Consider the student's t test for unequal variances, the t statistic, which is used to determine whether a threshold for statistical significance has been crossed, is inversely related to

$$\sqrt{\left(\frac{s_1^2}{n_1} + \frac{s_2^2}{n_2}\right)}, \tag{10.6}$$

where s_1 and s_2 are unbiased estimators of the variance for the control, s_1, and the experimental sample, s_2, respectively. The result could be a marked difference in the variances that combined with unrecognized insufficient sample sizes that could increase the type I error (finding a difference when a difference does not truly exist by incorrectly rejecting the null hypothesis) or a type II error (not finding a difference when a difference truly exists by incorrectly accepting the null hypothesis) (Zimmerman, 2004). An inflated type I error rate increases the probability that should the experiment be repeated, it would be found irreproducible.

Thermodynamic systems far from equilibrium can self-organize, for example, into limit cycles, as described previously. One interesting example is the formation of a snowflake from water vapor. Two samples of water vaper, each of which will form a snowflake, can be in very close physical proximity, thereby suggesting very close initial conditions. However, the two snowflakes formed likely will be very different, other than sharing a six- or eight-point configuration. If one were to compare the two snowflakes, they would be very different and the pattern would be unpredictable. How is one to apply statistics to quantitate the differences between the two snowflakes so as to determine whether the two snowflakes were derived from the same sample, as would be expected given the assumption of close initial conditions?

Complex systems can bifurcate between two different metastable states. Consider the Lorenz attractor (Fig. 10.8). Fig. 10.8 shows changes in an analysis result from computational modeling of turbulent airflow, such as in weather. Turbulent airflow is an example of thermodynamic systems that are far from equilibrium. Lorenz developed a complex set of differential equations that he ultimately was able to simplify (Fig. 10.8). The equations were solved for x, y, and z at a certain time, t. The values of x, y, and z solved at t were incremented over a range. In this context, t can be considered time and the plot of x, y, and z over t produces a trajectory (Fig. 10.8). As can be seen, there is structure in the trajectory, for example, the series of results appear to trace a butterfly

Lorens equations

Plot of solutions for x and z at time t as t goes from 0 to →∞

$$\frac{dx}{dt} = \sigma(y - x),$$

$$\frac{dy}{dt} = x(\rho - z) - y,$$

$$\frac{dz}{dt} = xy - \beta z$$

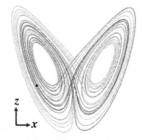

FIGURE 10.8   (Left) Lorenz equations. (Right) An example of a Lorenz attractor. When solved for different values of time, the points trace out a trajectory. In this case, the trajectory has a butterfly wing pattern. The value of any single iteration of the computational solution could be anywhere in the trajectory, that is, in one or the other wing. It would not be possible to predict. Source: *By Dan Quinn, https://en.wikipedia.org/wiki/Lorenz_system*

wing pattern. The pattern can be represented as points that are the solutions of x and z for any time t. Consider if the attractors described for the Lorenz equations were to apply to a biomedical experiment. What this suggests is that in one experiment the samples of x and z could exist in one wing. In a second experiment, it is possible that the samples of x and z could exist in the other wing. Traditional statistical hypothesis testing may have very different results between two experiments, suggesting a failure of reproducibility.

The pitchfork bifurcation shown in Fig. 10.6 is another hypothetical example. Applying a traditional correlation analysis, the experimenter might conclude that no relationship exists between increasing weight and displacement of the pillar (Fig. 10.6B). However, as shown by Fig. 10.6C, a relationship is that of the general form $\dot{y} = xy - y^3$, where y is the displacement of the pillar and x is the weight. Note the nonlinearity of the describing equation.

RECOGNITION OF COMPLEXITY IN BIOMEDICAL RESEARCH

It would not be unreasonable to suggest that there is a paucity of biologically based compared to biologically inspired published studies in Complexity and Chaos. The majority of papers published demonstrating such self-organizing systems, such as attractors, are derived from computational models applied to the biological concept rather than biological data. This paucity is striking given the thermodynamic considerations described previously and the high expectation for complexity and self-organization in biological systems. Direct studies of biomedical experimental results generally use descriptors of the degree of Complexity, such as Lyapunov exponents, Kolmogorov entropy, and Kolmogorov complexity. However, these are used much in the manner of traditional summary population statistics, such as mean (average), median, and variance.

Analyses in terms of limit cycles have been applied to biomedical questions but typically in situations where the periodic (cyclical) nature of the phenomena already is appreciated, such as epidemics, predator—prey relationships, and circadian rhythms. It is very likely that the great majority of biological phenomena do involve Complexity structure, such as limit cycles and strange attractors, if only because of the linked thermodynamic processes that increase information (reduce entropy)—decrease information (increase entropy) that tend to produce such structures are ubiquitous in biological phenomena. Differences in the rates of the opposing processes, specifically those that decrease and those that increase entropy, have a high potential to produce periodic phenomena. If the timescale of the phenomena is sufficiently slow, the phenomena will not be perceived as periodic, therefore not suggesting underlying Complexity dynamics, such as limit cycles and strange attractors.

Critical to recognizing Complexity in a particular biomedical research is to first recognize its possibility. Every good scientist or statistician knows to check the original dataset for normalcy prior to selection of the appropriate parametric or nonparametric statistical test. Similarly, good scientists and statisticians should check the original dataset for Complexity or Chaos. There are two general approaches. One is to look for familiar patterns in the dataset that are suggestive of Complexity and Chaos, and the second is to change the "perspective" by which the original dataset is viewed. This could be done by viewing the actual data from a novel perspective if it is not immediately obvious or by using mathematical and computational modeling. The difficulty with the latter is that it presupposes some prior knowledge, presumptions, or assumptions.

An example of a pattern that may suggest complexity is shown in Fig. 10.6B as might occur in a correlation analysis. As can be seen, data shown in Fig. 10.6B may suggest no relationship between x and y, but there is. The distribution of data in Fig. 10.6B demonstrates heteroscedasticity where the variance in subregions along either the horizontal or the vertical axis is not equal across the region. Thus, heteroscedasticity, for which there are statistical tests, may suggest bifurcations in an underlying Complex or Chaotic system. For typical probability density functions (distributions), a multimodal distribution could suggest bifurcations, limit cycles, or some other attractor. Platykurtosis, where the probability density distribution is relatively flat, may suggest an underlying limit cycle where the nature of the experiments projects the points in the limit cycle onto a surface that forms the plane containing the limit cycle (Fig. 10.5).

In Complex systems, such as far-from-equilibrium thermodynamic states as described earlier, the state may be only in a subregion of the phase space for a certain period of time and then bifurcate to another subregion of the phase space. The Large Number Theorem would not hold in such a circumstance and the ergodic hypothesis would fail. Thus, a failure of convergence to a Central Tendency may be an indicator of a complex system.

As with any analysis, it is best to start by visualizing data. The problem is that in Complex systems, the structure that might suggest complexity-based analyses may be difficult to discern. Visualizing in multidimensional space can also be difficult, although there are algorithms for unsupervised cluster analysis in n dimensions (Montgomery et al., 2005).

SUMMARY

Traditional statistical analyses assume the Large Number Theorem where measures of the Central Tendency, such as mean (average) and variance, converge on a single value with increasing sample size. However, there may be many biological systems where the Large Number Theorem does not apply, and traditional statistical analyses are likely to result in irreproducibility, which may be incorrect. This is very likely to occur in biological systems because these living systems operate far from thermodynamic equilibrium. Even when the biological systems appear to be in a steady state, they still may be operating far from equilibrium. Such nonequilibrium steady states require energy expenditure and generate nonlinear relationships among the variables measured in an experiment, such as unpredictability and dependency on initial conditions. The potential for the biological system being investigated should first be assessed as whether Complexity or Chaos may be playing a significant role.

11

Epilog: An Ounce of Prevention...

As the old adage says, "an ounce of prevention is worth a pound of cure." Irreproducibility is a serious problem in biomedical research, if only because of a loss of confidence. Indeed, journal editors and research administrators are making efforts to prevent irreproducibility by strong measures to ensure prevention of fraud, transparency in methods, data sharing, and statistical robustness. These are important and necessary measures but they are insufficient. As hopefully demonstrated in this book, the injudicious, but necessary, use of logical fallacies in experimentation may be a cause and one that may be preventable. This book demonstrates how an understanding of the most basic level of logic in propositional and syllogistic deduction can explain unproductive irreproducibility and hence point to measures to reduce unproductive irreproducibility and increase the probability of reproducibility. This book attempts to show how logical fallacies can be used judiciously. It is likely that more advanced logics, particularly predicate and fuzzy logic, would be even more helpful to biomedical researchers.

Perhaps encouraging a stronger grounding in logic in those interested in biomedical research would be helpful, such as in college and university courses. Unfortunately, this is not likely, as most courses in logic are traditionally provided by philosophy departments. Usually, these courses focus on valid logical forms, and fallacies are only mentioned as to be avoided. Yet, as hopefully made clear in this book, valid propositional and syllogistic deductive logics are "truth preserving," not "truth generating." The latter is the focus of biomedical researchers. The importance of the judicious use of logical fallacies can be combined with valid logic to provide an expanded field of logic, what this author describes as evolutionary logic. Advancing the field of evolutionary logic in biomedical research will require considerable additional scholarships and research.

Reproducibility in Biomedical Research
DOI: https://doi.org/10.1016/B978-0-12-817443-2.00011-2 **243**

HYPOTHESIS GENERATION

Most of the discussion regarding reproducibility in biomedical research proceeds from the experiment already framed, particularly the hypothesis, and executed. Clearly, the effort of biomedical experimentation is to translate the experimental hypothesis into scientific knowledge. Thus, an experiment with good hypotheses is more likely to be reproducible or productively irreproducible. But how do biomedical researchers arrive at good hypotheses? This has not been a subject of vigorous scholarship or research, although its importance may be becoming clearer (Montgomery, 2018). What strengthens the Scientific Method, in its modus ponens-like application, is converting the Fallacy of Confirming the Consequence of the form *if a implies b is true and b is true, then a is true* to the valid form of *if and only if a implies b is true and b is true, then a is true*. Clearly, all reasonable alternatives to hypothesis *a* are critical in meeting the *if and only if* criterion and further demonstrate the importance of the skill of hypothesis generation.

A number of mechanisms give rise to hypotheses but an important method is by metaphor (Montgomery, 2018). Metaphors are a variation of the Fallacy of Pseudotransitivity, which is derived from the valid form, the Principle of Transitivity (see Chapter 4: The Logic of Certainty versus the Logic of Discovery and Appendix A: An Introduction to the Logic of Logic). Metaphors are made even more powerful as sources of hypotheses by replacing the state-of-being linking verbs of the logical form with action verbs such as *cause* and *effect*. But the question becomes what gives rise to a specific metaphor, which, by its novelty, means that there must be some prior notion of entities and relations relevant to the metaphor.

Among the relations that can give rise to metaphors are the logical fallacies addressed in this book. These are referred to as formal fallacies. As the name suggests, there are informal fallacies, such as arguing from: (1) ignorance (Argumentum ad Ignorantiam); (2) positions of authority or inferiority (Argumentum ad Verecundiam); (3) personal attribution (Argumentum ad Hominem); (4) appeals to what was held in the past (Argumentum ad Antiquitatem); (5) beliefs in common or by consensus (Argumentum ad Populum); (6) false dichotomizations, (7) strawman arguments; (8) where the experimental question presupposes the results (Petitio principii [begging the question]); and (9) where something is discovered after an effect is thought to exist before the effect (post hoc ergo propter hoc) (www.csun.edu/~dgw61315/fallacies.html). While it may be uncomfortable to biomedical researchers that such arguments from informal fallacies play a role in science, there is considerable historical examples where they do (Kuhn, 1962).

Many of these informal fallacies are rhetorical devices whose purpose is suasion, either persuasion or dissuasion. As these rhetorical devices are used in biomedical scientific argumentation, the science is heir to the consequences of using these fallacies. Here, argumentation is taken in its widest context, including experimental design, but, perhaps more importantly, for justification of a particular experiment, inferences derived, and their more general relevance. Interestingly, in tracing the evolution of publications in the *Philosophical Transactions of the Royal Society* from 1665 to 1800, Bazerman described four stages (Bazerman, 1997, pp. 169–86). The most recent and likely persistent one was an intent to persuade. Indeed, some have argued for a more vigorous use of rhetoric in scientific reporting, if only for public consumption (Olson, 2015). However, it is not clear whether biomedical researchers can vaccinate themselves from their own use of such rhetoric. The degree to which the use of informal fallacies and rhetorical devices affects unproductive irreproducibility awaits further scholarships and research.

INDUCTION

Inductive logic is often offered as a counter to deductive logic and as somehow different than deductive logic. Often, induction is looked to as the source of scientific hypotheses, as well as evidence in translation of the hypothesis to scientific knowledge. However, it is not clear that induction is all that different from deduction, particularly syllogistic deduction, and that induction avoids many of the same problems that confront deduction (Montgomery, 2018). The point here is that biomedical researchers are unlikely to find safe harbor in their research by appealing to induction.

A fuller discussion of the potential risk for unproductive irreproducibility based on the use of induction is beyond the scope of this effort and awaits further scholarship and research. However, a brief mention serves a purpose here. Typically, induction begins with enumeration, for example, a large number of ravens are looked at, all ravens seen are black, and the induction (inference) is made that *every raven is black*. Reasoning from seeing all ravens to be black is that all ravens are black whether or not any other ravens are seen in the future. Further, this reasoning leads to empirical demonstration (experimentation) by purposely seeking out other ravens to demonstrate that they too will be black. How a particular set of animals is collected under the set *ravens* while others are excluded is an open question but clearly there must be some prior criteria that allow the collection of a set of birds to constitute the set *raven*.

The issue of induction is critical in biomedical research. For example, a large number of patients with Parkinson's disease are examined upon their death and all have loss of dopamine in the brain. The induction is made that *all patients with Parkinson's disease have dopamine depletion*. A method used to assess dopamine content in the brain of living patients has been developed and, when applied to living patients clinically diagnosed as having Parkinson's disease, approximately 10% did not have evidence of dopamine depletion. One interpretation is that the induction must be false—not all patients with Parkinson's disease have dopamine depletion. However, the scientists instead claimed that those patients without dopamine depletion do not have Parkinson's disease, rather these patients have some other disease in order to maintain the induction that *all patients with Parkinson's disease have dopamine depletion*. Note that the clinical criteria used to collect these patients into the set that was studied were exactly the same for those who subsequently demonstrated dopamine depletion and those who did not. Yet, the fact used to make such a statement—demonstration of no dopamine depletion on scans—was the result of the experiments. In other words, the nature of the experiment was changed to maintain the induction but only in retrospect. This is invalid and is an example of the informal fallacies of circular reasoning.

Induction by enumeration, as demonstrated earlier, falls to the Fallacy of Induction. It cannot be known that somewhere unbeknownst to the scientist a raven exists that is not black. Rather, the induction that *all ravens are black* bets that any subsequent experience (experiment) will demonstrate that the newly seen bird is a black raven. However, as discussed in Chapter 4, The Logic of Certainty Versus the Logic of Discovery, it is impossible to know whether every raven in the universe, the population of interest, is assessed; consequently, one must assume that only a sample has been assessed. Immediately, the Fallacy of Four Terms is confronted. The risks of the Fallacy of Four Terms can be mitigated, although not removed, by the application of probability and statistics.

Note that the Fallacy of Induction has a corollary that only a single exception from an induction is sufficient to prove the induction false. It takes only a single nonblack raven to prove the induction *all ravens are black* false. Further, the corollary presupposes the Principle of the Excluded Middle, where every raven in the set of *ravens* must be black and cannot contain any nonblack ravens. Further, the set of *nonblack birds* cannot contain a raven. It is the adherence to the Principle of the Excluded Middle that allows for the certainty of Mill's Methods of Induction with the exception of the Method of Concomitant Variation (analogous to correlation).

Mill's methods were an attempt to formulate a logic based on induction and, importantly, to establish a notion of causality. In this regard, the Method of Concomitant Variation is the weakness, as evidence of the only rarely followed admonition that correlation is note causation (well discussed by the philosopher David Hume in his "An Enquiry

Concerning Human Understanding," published in 1748). Consider the situation of scanning the brain for dopamine depletion in patients with Parkinson's disease. As those without dopamine depletion constituted only approximately 10% of the patients, the other patients will dominate the correlation analysis and suggest a statistically significant correlation between dopamine depletion and Parkinson's disease. This would be presumed as evidence that all patients with Parkinson's disease have dopamine depletion. But what about those patients without dopamine depletion, do they have some different form of Parkinson's disease?

The case of some patients with Parkinson's disease having dopamine depletion while others do not would violate Mill's other methods of induction, particularly the Method of Agreement (see Chapter 4: The Logic of Certainty Versus the Logic of Discovery). A stronger logic can be based on the notion of necessary and sufficient conditions for making a valid claim. A necessary condition is one that must be present for an effect or entity to be observed. The condition, however, may not be sufficient alone to produce the effect or entity. A sufficient condition is one that will produce an effect or entity, although this may not be necessary as other conditions could produce the effect or entity as well. The value of constructing the experiments and their interpretation as necessary and sufficient conditions is that one can apply the Principle of the Excluded Middle. Thus, it takes only a single observation where the absence of a presumed necessary condition was still associated with the presence of the effect or entity to disprove the condition as necessary. Further, it takes only a single observation where the presence of a presumed sufficient condition was not associated with the presence of the effect or entity to disprove that the condition is sufficient. A condition, such as hypothesis or experimental manipulation, that correlates with the prediction of the outcome but is demonstrated as neither sufficient nor necessary, as described earlier, would be epiphenomenal. There are numerous examples where a correlation between variables was found but was epiphenomenal. How this may relate to reproducibility in biomedical research is a subject for future scholarship and research.

A form of deductive logic that exploits the Principle of the Excluded Middle is predicate logic. This logic contains variables, such as *a* and *b*, that are quantified. The existence of quantifiers such as *there exists an a such that if a implies b is true, then b is true*, written as $\exists a$ (*if a implies b and a, then b*). In this case, *a* may be any single observation for which the rest of the logical statement is true, that is, where *a implies b is true* and there is at least one instance where *a is true* resulting in *b* being true. There may be simultaneously other instances where *a* is false, but the single instance of *a* being true is sufficient. Note that the same *a* needs not be true in every instance, but at least one instance must have *a* as true. In this case, *a* is analogous to a sufficient condition.

Another existence quantifier is of the form *for all a such that a implies b is true and a is true, then b is true*, written as ∀*a* (*if a implies b and a, then b*). This argues that every case *a* must be true for *b* to be true and thus *a* is a necessary condition. Thus, the argument relating Parkinson's disease to dopamine depletion can be written as ∀*a* such that *a* (Parkinson's disease) *implies b* (dopamine depletion) would be false as there may be cases of Parkinson's disease not associated with dopamine depletion. However, the argument written as ∃*a* such that *a* (Parkinson's disease) *implies b* (dopamine depletion) would be true.

Just as in propositional logic, complex arguments can be constructed in predicate logic. However, biomedical research experimentation can be recast as predicate logic and how this may affect reproducibility is a subject for future scholarships and research. However, the distinction between correlation and necessary and sufficient conditions argues that predicate logic, and its variations, may provide stronger support for reproducibility and productive irreproducibility.

DISCOUNTING PHILOSOPHY

It is hoped that this book demonstrated the value of logic to help ensure the advancement of biomedical research by increasing the probability of reproducibility and productive irreproducibility through the judicious use of logical fallacies and other variations on valid logical forms. The probability syllogism is derivative of the partial syllogism and is the basis for statistical analyses. The practical syllogism empowered the remarkable corpus of knowledge related to causality to help generate new hypotheses by the Fallacy of Confirming the Consequence and the Fallacy of Pseudotransitivity. It is the resulting hypotheses that create the potential for new knowledge. These and other variations on valid logical forms allow a much wider sense of philosophy to make important contributions to biomedical research. For example, the philosophical discipline of Mereology, the relationship between parts and wholes, is critical to the reductionist basis of modern biomedical research. Likewise, other fields of philosophy can assist the biomedical researcher, such as epistemology and ontology, which together are metaphysics.

To be sure, many biomedical researchers may not appreciate the potential contributions of philosophy. Rather, many developed a more intuitive knowledge of experimental design during an apprentice-like experience in the laboratory and believe these to be sufficient. Any value these experiences have for understanding the judicious and injudicious use of logical fallacies presupposes that the mentor had explicit knowledge of logic that informed her experiences. In the absence of such explicit understanding, it is unlikely that the mentee's will be spared the same deficit. Unfortunately, the crisis of unproductive irreproducibility is evidence that more needs to be done, philosophically.

Yet there may be more involved. A great many scientists appear to have an aversion to philosophy so as to reject any notion that philosophy may play a role in the advancement of biomedical science. As will be seen, many of those scientists actually were "closet philosophers" embracing exactly what they would reject in philosophers.

As will be discussed here, perhaps the earliest and most significant challenge to modern science occurred with the founding of the Royal Society in 1660 with the wholesale rejection of medieval scholastic natural philosophers, although scientists retained the name of natural philosopher until 1834 when William Whewell coined the term "scientist." The rise of modern science as a contrast to the scholastic natural philosophers is discussed in greater detail later. Modern scientists, in rejecting the medieval scholastic natural philosophers in the 1600s, "threw the baby out with the bath water" and lost the benefit of rigorous logic that philosophical knowledge provides. In this case, the words of the eminent neuroscientist Santiago Ramón y Cajal are telling when he wrote:

> Instead of elaborating on accepted principles, let us simply point out that for the last hundred years the natural sciences have abandoned completely the Aristotelian principles of intuition, inspiration, and dogmatism. The unique method of reflection indulged in by the Pythagoreans and followers of Plato (and pursued in modern times by Descartes, Fichte, Krause, Hegel, and more recently at least partly by Bergson) involves exploring one's own mind or soul to discover universal laws and solutions to the great secrets of life. Today this approach can only generate feelings of sorrow and compassion—the latter because of talent wasted in the pursuit of chimeras, and the former because of all the time and work so pitifully squandered" (Ramón y Cajal, 1999).

Yet, Ramón y Cajal then lapses into metaphysics comparable to any philosopher, writing:

> Knowing the conditions under which a phenomenon occurs allows us to reproduce or eliminate it at will, therefore allowing us to control and use it for the benefit of humanity. Foresight and action are the advantages *we obtain from a deterministic view of phenomena* [emphasis added] (Ibid., at p. 2).
>
> *Nature is a harmonious mechanism* where all parts, including those appearing to play a secondary role, cooperate in the functional whole [emphasis added] (Ibid., at p. 17).
>
> If a supreme intelligence knew all the mysterious explanations linking all phenomena in the universe, there would be one single science instead of many different sciences. *The frontiers that appear to separate fields of learning, the formal scaffolding of our classification scheme, the artificial division of things to please our intellects—which can only view reality in stages and by facets—would disappear completely* [emphasis added] in the eyes of such an individual (Ibid., at p. 55).

The next significant challenge to modern science came with the advent of postmodernism that held that scientific progress was just as "messy" and complicated as any other knowledge disciplines. Forces in everyday human life, such as political, economic, social, and psychological, were just as much factors in the scientific enterprise as they were in any other human endeavor. By the time of the postmodernists attacks, science in general and biomedical research in particular had become institutionalized, such as in the National Institutes of Health (NIH). Many of its constituents held vested interests in continuance, indeed expansion, of the biomedical research enterprise. In a sense, modern science was immunized against the criticisms of the postmodernists. Nonetheless, many of the criticisms of modern scientists by the postmodernists were important but easily dismissed—particularly by scientists, such as Sokal and Bricmont, taking "cheap shots" at the postmodernists (Sokal and Bricmont, 1999). While academics, particularly philosophers, are easy targets for satire, the issues raised early by Nietzsche in the late 1800s and Thomas Kuhn (1962) in the early 1960s were honest attempts to understand the complicated nature of the scientific enterprise by thoughtful and committed persons through historical evidence. It does raise concerns about how scientists hold themselves accountable and to whom are they accountable to, particularly relative to scholars who do not proceed from the basic assumptions taken for granted by biomedical researchers.

RISE OF SCIENTISM

Scientism, or the faith that science is somehow privileged and will answer all the important questions and lead to good health, justice, and prosperity, is a dangerous notion. Indeed, the recent revelations of unproductive irreproducibility in biomedical research are jarring because of the implicit faith of Scientism. The danger is that Scientism may lead to complacency and real reform left stillborn, at least until the next round of jarring revelations. How did scientists get to Scientism?

One could argue reasonably that modern science emerged as a reaction to prior natural philosophy that was thought dominated by the medieval scholastic natural philosophers. Many early scientists, such as Galileo, Newton, and Ramón y Cajal, ridiculed these natural philosophers and their attempts to understand nature by first philosophy, that is, assuming a set of axioms and rules of inference to construct human knowledge. Importantly, early modern scientists assumed that these axioms and rules of inference used by the natural philosophers were taken as "received knowledge" delivered by studying and arguing from ancient texts, particularly those of the ancient Greek philosophers Plato and Aristotle. The presumption or conceit of the modern scientist was that there was no attempt by the natural philosophers to establish the axioms and rules of inference from empiric evidence. This is unfair, as a great many medieval philosophers did engage in empiric research (Hannan, 2011), even Aristotle (Leroi, 2014). However, the presumption is evidenced in the motto of the Royal Society (established in 1660), *Nullius in verba*, or "Take nobody's word for it."

In many ways, the animosity on the part of early scientists, such as Galileo, may have had more to do with the reactions of the theocracy to the emerging modern science than in reaction to the natural philosophers. Indeed, the pope at the time of Galileo was quite supportive of the emerging modern science. Certainly, there were difficulties when modern science began to contradict the notions of nature as were adopted by the Christian church from the natural philosophy of Aristotle, Ptolemy, and others. However, it has been argued that the relationship became most troubled by the question whether God was constrained by the very laws of nature that God was alleged to have created (and discovered by scientists, not theologians). Galileo held that it was the case and for that was placed under house arrest. Others like Descartes and even Robert Boyle looked the other way and still found some way to give theology its due while persisting in science. Thomas Hobbes was not so lucky. His natural philosophy did not exclude a god, but he did not find it necessary to evoke a god in understanding the natural world. Hobbes was roundly ostracized as an atheist and denied scientific honors, such as membership in the Royal Society, that were freely bestowed on lesser natural philosophers.

One purpose of the Royal Society was to provide opportunities for public discussion and demonstration of new ideas regarding natural philosophy—although who constituted the "public" was restrictive. Thus, at least in the minds of the early modern scientists, they introduced experimentation and empiric observation as the *modus operandi* of the new science. However, never far and lurking unspoken was the metaphysics presumed by the physics of the modern scientist. What was interesting about Robert Boyle's air pump, which became the focal point of many scientific arguments, was not the technical nature of the air pumps (Shapin and Schaffer, 1985), but rather the physical, metaphysical, and religious ramifications of a void. Could God or Nature condone a void? Indeed, the notion of space being an empty void was still unsettled until the Michelson–Morley experiment by Albert A. Michelson and Edward W. Morley in 1887 relating to the question of whether the void of space actually is filled by an "ether" (Michelson and Morley, 1887). Even today's physics still appears unsettled as the nature of a void that is thought to constitute the vast percentage of space (Kraus, 2012; Musser, 2015).

The metaphysical presuppositions of the new modern science paralleled those of the Enlightenment of the 18th and 19th centuries, a central pillar of which was exalting science and challenging previous authorities and orthodoxies. While many of the political and social promises of the Enlightenment seem to have faded, particularly after the great wars of the time—the Thirty Years War, the French Revolution, and World War I—the faith in science has not appeared to suffer much. The faith that humanity can "science" its way out of political, sociological, and psychological turmoil persists—that faith can be called Scientism.

Following the establishment of royal patronage with formation of the Royal Society in 1662, modern Scientism received a remarkable boost as the advancement of science went from a self-indulgent patronage to a governmental responsibility (Bush, 1945). Interestingly, the argument was advanced primarily on utilitarian views of the technological implications—evidence of the Bayh–Dole Act or Patent and Trademark Law Amendments Act (Pub. L. 96-517, December 12, 1980) giving universities and other institutions of higher learning intellectual property rights resulting from governmentally funded research. With such faith supported by remarkable technological spin-offs, there was little to fan the flames of self-criticism of the fundamental presumptions—those being the metaphysics of current science (see Chapter 3: Validity of Biomedical Science, Reproducibility, and Irreproducibility). Certainly, supporters and apologists for modern science would not likely favor the argument that most biomedical research rests on logical fallacies as they do.

The nature of the biomedical research enterprise has changed just as the scientific enterprise, in general, has changed. However, in the case of biomedical research, the justificatory technological spin-offs largely relate to the clinical practice of medicine; thus, progress in biomedical research, at least in part, relates to the changes in clinical medicine. Perhaps the most profound change occurred in the late 1800s with the subsequent report to the Carnegie Foundation in 1910 by Abraham Flexner that changed medical education and, consequently, the practice of medicine (Montgomery, 2018).

The relevance to biomedical research was the adoption of "scientific" medicine by the American Medical Association, as evidenced in its code of ethics of 1847. In part, the adoption of scientific medicine was an attempt to differentiate Rationalist or Allopathic medicine [the forerunners of today's medical doctors (MDs)] from the Empirics of which homeopaths were members. Indeed, the number of homeopathic practitioners outnumbered MDs in the early 1800s in the United States, particularly as a result of the deregulation of medical licensure under President Andrew Jackson. Even before the Flexner report, there was a strong movement to align previously freestanding medical schools with universities, particularly as instituted in Germany, which favored the Rationalist/Allopathic traditions.

The German model of medical education had professors as full-time employees of the university rather than independent practitioners paid on a fee-for-service basis. Further, it was expected that the medical students would obtain the large majority of their medical education in laboratories, where their professors worked. It is interesting that those European professors who influenced medicine at the time were the same ones who decried the use of statistical analyses, such as Claude Bernard as described earlier. The door was open to potential Scientism in clinical medicine, thereby supporting a notion of Scientism in biomedical research.

Sir William Osler, the paragon of the astute clinician, opposed these particular changes in medical education. It was not that Osler was opposed to "scientific" medicine. Indeed, Osler made remarkable contributions to the science of pathology and clinical–pathological correlations. Rather, Osler believed that the best location to learn medicine was at the patient's bedside, to which the medical student and professor brought a scientific perspective. It can be argued that the Flexnerian revolution resulted in a truce whereby the traditional medical schools placed students in the hands of the PhD who then turned the students over to the clinicians. Nevertheless, strains of Scientism entered the allopathic medical tradition. Indeed, one group opposed to the American Medical Association Code of Ethics in 1847 were scientists who believed that advances in science would obviate the need for ethics and, in the interim, ethics considerations would only hamper biomedical research (Warner, 1991). Then there were the Tuskegee Syphilis Study and the Willowbrook State School scandals, which led to the Belmont report (National Commission for the Protection of Human Subjects of Biomedical and Behavioral Research, 1978) mandating independent Human Subjects Research review.

The extent of Scientism in biomedical research may be evident in opposition to the establishment of an institute among the National Institutes of Health devoted to translational research, that is, focusing on translating basic biomedical research results into tools that could be used by clinicians. Opposed by basic researchers, the compromise was establishment of the Center for Translational Science. In addressing the controversies, Fang and Casadevall (2010) quoted Vannevar Bush, who wrote:

> Discoveries pertinent to medical progress have often come from remote and unexpected sources and it is certain that this will be true in the future. *It is wholly probable that progress in the treatment of...refractory diseases will be made as a result of fundamental discoveries in subjects unrelated to those diseases, and perhaps entirely unexpected by the investigator* [emphasis added].... Basic research is the pacemaker of technological progress.... New products and new processes do not appear full-grown. They are founded on new principles and new conceptions, which in turn are painstakingly developed by research in the purest realms of science (Bush, 1945).

Not mentioned by Fang and Casadevall is the fact that a great deal (if not most) of biomedical scientific knowledge was directly due to research on ill humans, animal models of human disease, or serendipity (Le Fanuref, 1999). To be sure, there was a perfunctory acknowledgment that vaccinations, as with cowpox to prevent smallpox, preceded the scientific field of immunology. But for all the praise heaped on the great basic scientific understanding of DNA and genetics, the only new treatment to relieve human suffering as a consequence of the $3 billion investment in the Human Genome Project to date has been a single treatment for cystic fibrosis, and at that only for a limited number of mutations, and one for acute lymphoblastic leukemia pediatric and young adult patients, at the time of this writing. To be sure, there has been significant progress in developing diagnostic tests, such as for tumor typing, but in the case of the latter, this has not translated into selective approaches to cancer therapies. Yet, despite what could be reasonably considered a modest return on the human genome investment, Fang and Casadevall prescribe patience.

Perhaps more striking is that Fang and Casadevall take umbrage, writing "It is somewhat discomfiting that every grant application to the NIH must now be evaluated on its practical merits, as if an obvious practical application is an essential requirement of all research (http://enhancing-peer-review.nih.gov)." In the experience of this author, such justifications in most grant proposals were "lip service." The actual mission of the NIH "is to seek fundamental knowledge about the nature and behavior of living systems and the application of that knowledge to enhance *health, lengthen life, and reduce illness and disability* [italics added]" (nih.gov). Taking Fang and Casadevall's arguments to the logical conclusion, quantum physics is fundamental to understanding human health and relieving human suffering. To be sure, it is doubtful that Fang and Casadevall would have taken their logic to its conclusion. Certainly, they would insist in some measure of reasonableness. But then, there likewise should be a restraint imposed on reasonableness in the allocation of resources to basic biomedical research.

None of this is an argument against basic biomedical research. Indeed, this author has spent over 40 years studying basic questions about the role of the basal ganglia—thalamic—cortical system in motor control. Further, it is likely that knowledge of the human genome will lead to dramatic improvements in the human condition. Rather, the concern here is the hubris of unbridled Scientism that continues to write blank checks that others must cash. It would appear that what Fang and Casadevall are advocating is to put the fox in charge of the hen house.

But the concern here is not even the practical and social issues touched on previously. Rather, it is the insular complacency in the presumptions that science rests on some unshakable ground and is not in need of critical self-analysis, such as to its logical basis. Such self-indulgence seems misplaced in view of the crisis in reproducibility in biomedical research.

SOME SUGGESTIONS

Throughout this text a number of suggestions have been made to improve the judicious use of logical fallacies. Generally, these include making the necessary logical fallacies as close to their valid counterparts, but not completely so as to avoid tautologies and therefore still provide new knowledge with reasonable certainty. For example, one should avoid the Fallacy of Four Terms by making sure that the necessary two versions of the bridging, major or minor, terms are sufficiently close. Such closeness can be estimated, even if only qualitatively, by consideration of the Epistemic Risk, which is composed of Epistemic Distance and Epistemic Degrees of Freedom. Further, the Fallacy of Confirming the Consequence, which is the modus ponens form of the Scientific Method, can be mitigated by restructuring in the general form of *if and only if the hypothesis leads to the prediction and the prediction is true then the hypothesis is true* to the extent possible. The Fallacy of Limited Alternatives can be countered, at least to some degree, by insisting that all reasonable alternatives be considered. Similarly, experiments should be designed such that if subsequently found irreproducible it will be productive in the manner of the modus tollens form of the Scientific Method. Unless these efforts become the *modus operandi* of the biomedical research enterprise, there is little hope that they can be made a matter of routine use.

While education of the biomedical researcher about the logical foundations of biomedical research is important, the educational efforts compete with other demands, such as getting papers published and getting grant funding. Therefore, the most realistic manner by which biomedical research can reduce the injudicious use of logical fallacies is by incorporating mitigating mechanisms in the requirements for publication and grant review. The greatest challenge is that incorporating these mitigating mechanisms is whether the editors and reviewers appreciate the importance. Yet, editors and reviewers generally have the same education and experiences as the biomedical researchers whose behavior is at issue. Nevertheless, suggestions can be made.

1. Call for and support of modus tollens-like experimentation. Researchers could be encouraged to: (a) undertake critical review of important theories; (b) identify dependencies on as yet unsubstantiated premises (with a careful effort not to be misled by

quasi-facts) or reasonable alternatives for theories or hypotheses that have not been excluded; (c) determine predictions that, if demonstrated, would disprove current theories or hypotheses; and then (d) construct valid experimentation proposals that demonstrate the contravening predictions by the judicious use of logical fallacies. At the very least, such critical reviews, even if the experiments were not pursued, would provide a valuable service to the biomedical research community. It is important to note that most current reviews published, at least in disciplines with which this author has experience, rarely survey the alternatives or identify the "Achilles heel" in currently accepted theories and hypotheses. It is important to avoid the tendency to describe this kind of research as "negative" in nature, thereby creating the impression that they are less valuable than so-called "positive" studies. Knowing that a theory or hypothesis cannot be true is a "positive" contribution, and epistemically, on most secure footing and least likely to be unproductively irreproducible.

2. Require robust and explicit consideration of all reasonable alternatives to the hypothesis proposed in a modus ponens form of experimentation. This is true for research reports as well as grant applications. Indeed, the authors of reports and grant applications should act as their own "devil's advocate" and, if not, then editors and reviewers should play that role. As an interesting aside, in the process of beatification in the Roman Catholic Church, a devil's advocate was formally involved in the process from 1587 until 1983, resulting in 98 persons becoming saints. Since 1983, the number has increased fivefold (editorial "Yes, but...Brainstorming is of dubious value in discussions of scientific interests, Nature 2017;545:134). One wonders what the false positive rate became. Implement rigorous and demanding criteria when an author seeks to disqualify a potential reviewer. Indeed, those with differing perspectives and opinions should be encouraged. A lesson can be taken from judges who have to decide between conflicting expert opinions. These judges are not, typically, content experts but rather are experts in the rules of evidence. These rules of evidence are remarkable in their general applicability as honed by experience, debate, and precedence. Reviewers should act accordingly, which means that reviewers need to become experts in those factors that risk the injudicious use of logical fallacies and increase the risk for unproductive irreproducibility.

3. Formally ascertain the logical structure of the experiments proposed. If the experiments are structured similarly to a syllogistic deduction, then what *stands in* for the major, minor, or bridging term that drives the conclusion? Is the term that *stands in*, for example, a sample that

is a subset of the population of interest? If so, then does the sample have sufficient numbers to help assure generalizability to the population of interest? What are the Epistemic Distance and Epistemic Degrees of Freedom between the sample and the population of interest? What can be done to mitigate the resulting Epistemic Risk? If the sample is different in kind or quality for the population of interest, for example, an animal model that stands in for a human disease with the latter being the population of interest, then what are the Epistemic Distance and Epistemic Degrees of Freedom between the sample and the population of interest? What can be done to mitigate the resulting Epistemic Risk?

4. Encourage formal consultation and review by an applied logician/ epistemologist philosopher.

ETHICAL OBLIGATIONS

There still is too much illness, disability, and premature death in this world. Biomedical researchers have an ethical obligation to their fellow inhabitance of this world as biomedical researchers have been given the enormous privilege, through their education paid by others, to practice biomedical research in particular, and science in general. Yet, biomedical research cannot be unappreciative of the sacrifices made by society in getting the scientists to the position that they can practice science and the support to continue practice. Further, the limited resources, and what appears to be unlimited questions that can be asked, require biomedical researchers to be good stewards and not waste the sacrifices of their brethren in unproductive reproducibility.

A

An Introduction to the Logic of Logic

Logic is not just a set of arbitrary and abstract rules that one learns to pass an undergraduate introductory course on philosophy and logic. Rather, logic is an attempt to understand the nature of the universe. Any rules that define a particular logic are a consequence of that effort and not the driver. Whatever rules are created merely reflect the attempt to understand reality. If the world of reality was different, it is very likely that the rules of logic would be very different. For example, mathematics is a logical system. It is hard to imagine any biomedical research effort that does not depend on mathematics, for example, the mathematics inherent in statistical analyses. To whatever degree mathematics is appreciated by biomedical researchers and scientists, any accolades to it should be given to logic as well. The changes and evolution of mathematics also reflect the evolution and changes in logic. Since the earliest days of logic in terms of syllogistic and propositional logical deduction, logic has expanded to such things as predicate logic and fuzzy logic, as well as providing the underpinnings of mathematics.

Logic is a discipline used to help understand reality. One has a sense that one understands reality when one has confidence. Indeed, the whole issue around reproducibility in biomedical research hinges on means to establish confidence. Logic originated from an attempt to define what can be known with absolute confidence. Having first defined what type of reasoning provides absolute confidence, logic proceeds to build and create extensions to increase the utility of logic to provide absolute confidence in more complex situations. Even more importantly, logic identifies those situations where absolute confidence is not possible but nevertheless logic-like devices increase the ability to gain new knowledge while preserving as much certainty as possible. These situations are called logical fallacies. By understanding what the necessary requirement for absolute confidence is, one can contrast other means of knowledge that have utility in expanding what may be known but perhaps at the expense of absolute confidence. As discussed in Chapter 4, The Logic of Certainty versus the Logic of Discovery, these situations involve the judicious use of logical fallacies.

Logic avoids the particular circumstances in constructing the methods to certainty. Logic avoids specifics such as *that black cat over there will chase the brown mouse.* Such particular circumstances are contingent on the unique circumstances. Operating only at the level of *that black cat over there* means that a separate logic is needed when addressing *the yellow cat over here.* To do so invites the Solipsism of the Present Moment, which holds that only what is perceived at this exact moment is certain and no generalizations from the past or inferences to the future are possible. Humans engaged in such solipsism are, as described by William James (1890, p.462), as the baby whose impressions of the world are "one great blooming, buzzing confusion." Rather, logic seeks truth relationships among generalities such that the understanding is not limited to *that black cat there* or *the yellow cat here,* but to *cat* or, better yet, the logical variable *a*. Thus, a logical statement can be made that *a implies a,* regardless of whether a black or yellow cat, here or there can be substituted for *a* and it is certain (to a degree) that a *black cat there implies a black cat there.*

PROCEEDING FROM WHAT IS MOST CERTAIN

What is the most fundamental statement that can be made with absolute certainty? The answer is *a is a*; the mathematical analog would be $a = a$. The logical statement, *a is a*, is referred to as a tautology, sometimes taken as a pejorative sense, which is incorrect. All one has to do is stare at that black cat over there and then an instant later stare at that black cat over there. Assuming that the sensory impressions attendant to staring at that black cat over there are the same, then the black cat at both instances is the same black cat, $a = a$, where the *a* to the left of the equal sign is the same as the *a* to the right of the equal sign. Indeed, it could not be otherwise because if they were not, then either the *a* to the left or the *a* to the right would be recognizably different and no reasonable person would hold that they are equal or the same. One of the *a*'s would not be an *a* but something different. One could repeat this experiment with any number of objects, *the yellow cat here, the tree here,* and so on and it would always be the same, $a = a$. In logic, one can replace the equal sign by the term *implies,* which is represented by the symbol →, so the $a = a$ become *a implies a* or *a* → *a.*

One can understand the logical certainty *a implies a* by use of Venn diagrams that represent sets (Fig. A.1). Two sets, represented by circles,

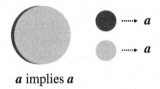

a implies *a*

FIGURE A.1   Venn diagram of the logical form *a implies* ($\rightarrow$) *a*, which is true. As can be seen, everything in set *a* to the left of the operational relation (*implies* or $\rightarrow$) is a member of the set *a* to the right. The two sets are said to be mutually wholly contained in each other.

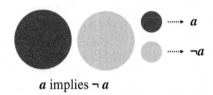

a implies $\neg$ *a*

FIGURE A.2   Venn diagram of the logical form *a implies* ($\rightarrow$) *not a* ($\neg a$), which is false. As can be seen, nothing in set *a* to the left of the operational relation (*implies* or $\rightarrow$) is a member of the set *not a* ($\neg a$) to the right. The two sets are said to be mutually exclusive of each other.

are shown where the darker circle represents the set of *a*'s that are to the left of the equal sign or the $\rightarrow$ and the lighter circle represents the set of *a*'s that are to the right of the equal sign or the $\rightarrow$. The circles in the Venn diagram are superimposed so as to be identical (in Fig. A.1); they are offset slightly so one can see that there are two circles and, thus, two sets.

From the discussion, one readily appreciates that the statement *a implies a* is true under any and every possible specific condition for *a*. One can build on this certainty to state with absolute certainty that *a implies not a*, also written as *a* $\rightarrow$ $\neg a$, where $\neg$ means *not* is false always (Fig. A.2). It turns out that the statement *a implies not a* is false has great utility. Any complex argument that can be shown to result in the claim that *a implies not a is true*, a logical impossibility or absurdity, is an invalid argument where one or more of its premises are false or one or more of its propositions are invalid. Consider a logical argument that begins with the proposition that *p implies q* and the subsequent analysis that demonstrates the conclusion to be *a implies not a is true*, which is an absurdity. If the intervening steps between the initial proposition *p implies q* and the conclusion now of the form *a implies not a* have true premises and valid propositions, then one knows the initial premise *p implies q* cannot be true and, importantly, *p implies not q* must be true. This form of proof is called *Reductio ad Absurdum*.

Reductio ad Absurdum has been used in mathematics to achieve rather remarkable findings, such as the Gödel's Incompleteness theorem, which posits that there are mathematical or logical truths that are not provable. *Reductio ad Absurdum* is also the fundamental basis for demonstrating productive irreproducibility. For example, an experiment proceeds from the hypothesis that *a implies b*, for example, hormonal use in postmenopausal women, *a*, reduces cardiovascular disease, *b* (see Chapter 6, Causation, Process Metaphor, and Reductionism). The experiment proceeds from that hypothesis to an experimentally verified prediction, lower hormone use in the experiment, *a*, implies fewer deaths in subjects of the experiment, *b*, that is, *a implies b*. However, what was found was *not b* (¬*b*), that is, no fewer deaths in subjects treated with hormones. The propositions *a implies b* and *a implies not b* both cannot be true, as otherwise there would be an absurdity. Hence the experiments are not reproducible.

The logical notions that *a implies a is true* and *a implies not a is false* can be readily appreciated from the Venn diagrams in Figs. A.1 and A.2. In these examples, the linking term, *implies*, is a state-of-being linking verb analogous to *is*. This *a implies a* is equivalent to saying *a is a*, which is absolutely true. Similarly, *a implies not a* is equivalent to saying *a is not a*, which is absolutely false. The state-of-being linking verb can also be considered as a statement of membership in a particular set, for example, *a* is a member of set *a* and *not a* (¬*a*) is not a member of set *a*. Any logical argument where the initial proposition is *p implies q* that results in a final conclusion that *a implies a* likely means that *p implies q*. The fact that *a implies a* is a tautology also means that *p implies q* is a tautology. However, this could be very useful if *p* and *q* previously were not recognized as being the same. The recognition then become a new scientific fact, although not necessarily new and novel knowledge.

Linking verbs that combine logical variables, such as *a* and *b*, need not imply identity or exhaustive membership in a particular set. For example, in Fig. A.3, in the logical premise *a implies b*, *b* is a subset with set *a*. What is true for set *a* is also true for set *b* because set *b* is contained within set *a*. Yet set *b* is not set *a*. In this case, the logical

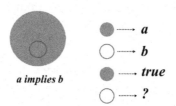

a implies b

● ┄┄► *a*
○ ┄┄► *b*
● ┄┄► *true*
○ ┄┄► *?*

FIGURE A.3   Venn diagram of the modus ponens logical form *if a implies b is true and a is true, then b is true*. The circles represent *a* and *b* and *b* is within *a*. As *a* is true, the entire area within *a* is true, which includes *b*. Therefore, *b* must be true.

proposition *if a implies b is true and a is true, then b is true* can be readily appreciated from set membership as shown in Fig. A.3. This form of propositional logic is called modus ponens. The proposition can be rephrased to say *as b is a member of set a and all members of set a are true, then all members of set b must also be true*. However, the proposition *if a implies b is true and b is true, then a is true* is not valid, just because members of *b* do not mean that all members of set *a* are true, for example, members in set *a* that are not in set *b* so that the truth of members in set *b* is not evidence that all members of set *a* are true (Fig. A.4). Note that the antecedent of the logical proposition *if a implies b is true* does not hold that *a* is true and only speaks of the relation of *a* to *b*.

As discussed in Chapter 4, The Logic of Certainty versus the Logic of Discovery, propositions of the form *if a implies b is true and b is true, then a is true* are invalid in that the conclusion *a is true* cannot be certain. This does not mean that *a* cannot be true but only that its truth or falsehood cannot be known with certainty. This invalid form is known as the Fallacy of Confirming the Consequence, where *b* is the consequence of the implication *if a implies b*. Importantly, this fallacy is one mode of the Scientific Method where *if a* (hypothesis) *implies b* (prediction) *is true and b* (prediction) *is true, then a* (hypothesis) *is true*. Because this use of the Scientific Method is invalid, results from its use cannot be taken as certain, even though the results may be true.

A variation of the Scientific Method is of the form *if a* (hypothesis) *implies b* (prediction) *is true and b* (prediction) *is false, then a* (hypothesis) *is false*. This is the valid form in propositional logic called modus tollens and is absolutely true (Fig. A.5). As discussed in Chapter 4, The Logic of Certainty versus the Logic of Discovery, this form of logic provides for productive irreproducibility. The predication (*b*) is found false, a form of irreproducibility, but it is productive because it leads to the absolute certainty that the hypothesis (*a*) is false.

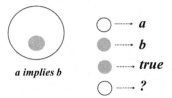

a implies b

········▶ *a*

········▶ *b*

········▶ *true*

········▶ *?*

FIGURE A.4 Venn diagram of the modus ponens-like logical variation known as the Fallacy of Confirming the Consequence and is of the form *if a implies b is true and b is true, then a is true*. The circles represent *a* and *b* and *b* is within *a*. As *a* is true, the entire area within *b* is true and that area (members) is *a*, which overlaps with *b*, is true; however, the truth of *b* bears no weight on the members of *a* that are outside of *b*. Therefore, the conclusion that *a is true* is invalid. This does not mean that *a* cannot be true but just that the truth or falsehood of *a* cannot be known with certainty.

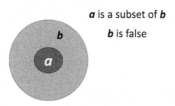

FIGURE A.5   Venn diagram of the modus tollens form of propositional logic, which is of the form *if a implies b is true and b is false, then a is false,* which is known with certainty. The circles represent *a* and *b* and *b* is within *a*. As *b* is false, at least some areas (some members of the set) within *a* are false. The Principle of the Excluded Middle holds that *a* must be true or false, but not both true and false and not neither true nor false. Consequently, *a* can only be false with certainty.

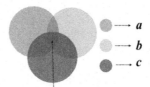

a inclusive OR b inclusive OR c

FIGURE A.6   Venn diagram of the inclusive *OR* function, which is of the form *a inclusive OR b inclusive OR c,* which is true if some combination of *a, b,* or *c* is true. The logical premise (*a inclusive OR b inclusive OR c*) is represented by the overlapping area. Note that even if *a* is true, *b* may not be true. While the area or members of set *a,* all of which are true, are also in *b,* the truth of *a* cannot prove that all of set *b* is true. Indeed, the overlap between *a* and *b* may not have any members, thus, the status of *b* is unknown. The same attends between *a* and *c* and *b* and *c.* However, it just takes one set to be true for the proposition *a inclusive OR b inclusive OR c* to be true.

Other logical operators greatly extend the utility of propositional logic such as the *OR* function, which holds that the statement *a OR b is true if either a or b* (or both in one case) *is true.* This particular case of the *OR* function is called the *inclusive OR* for reasons that will be become clear later. The *inclusive OR* operator can be represented in the Venn diagram shown in Fig. A.6. A truth table is constructed that shows the truth or falsehood of *a OR b* (Table A.1). Another form of the *OR* operator is the *exclusive OR* function, which holds that *a exclusive OR b* is true if either *a* or *b* is true but not both are true as shown in the Venn diagram in Fig. A.7 and Table A.2. As will be demonstrated, the *OR* function, and particularly confusion about whether the *inclusive* or *exclusive OR* function is implied, can lead to unproductive irreproducibility (see also Chapter 4, The Logic of Certainty versus the Logic of Discovery).

The *not* function ($\neg$) was introduced before, which holds that *if a is true, then not a* ($\neg a$) *is false* and *if a is false, then not a* ($\neg a$) *is true.* Another useful logical operator is the *AND* function, which holds that *a AND b is true only if a and b are true, otherwise a AND b is false.* The

TABLE A.1   Truth Table for the Logical Proposition *a Inclusive OR b*

a	b	a inclusive OR b
True	True	True
False	True	True
True	False	True
False	False	False

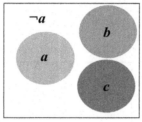

a exclusive OR b exclusive OR c

FIGURE A.7   Venn diagram of the *exclusive OR* function, which is of the form *a exclusive OR b exclusive c*, which is true if only one of the variables, *a, b,* or *c,* is true and false for any other combination. In this case, the circles represent the variables *a, b,* and *c* as well as their denial or *not* function of the variables, ¬a, ¬b, and ¬c; in other words, areas where *a* must be false, *a,* when *a* is true. If *a* is true, then all members outside of *a,* that is, ¬a, are false. As *b* and *c* lie in the area ¬a, all members of *b* and *c* must be false. The same issues apply to *a* and *c* when *b* is true and to *a* and *b* when *c* is true.

TABLE A.2   Truth Table for the Logical Proposition *a Exclusive OR b*

a	b	a exclusive OR b
True	True	False
False	True	True
True	False	True
False	False	False

associated Venn diagram is shown in Fig. A.8, and the truth table is shown in Table A.3. The *NAND* operator is interesting because virtually every digital computer operation is just combinations of the *NAND* operator, which holds that *a NAND b is true only if both a and b are false, otherwise a NAND b is false.* When one considers the power of modern

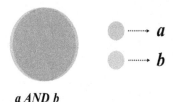

a AND b

FIGURE A.8　Venn diagram of the *AND* function, which is of the form *a AND b*, which is true if both *a* and *b* are true. Note that the circles that represent *a* and *b* are coextensive or superimposed on each other. If the members in one circle or set are false, so will the members of the other set. The only way for members of the combined set in the *AND* function to be true is if both circles contain only true members.

TABLE A.3　Truth Table for the Logical Proposition *a AND b*

a	*b*	*a AND b*
True	True	True
False	True	False
True	False	False
False	False	False

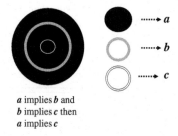

a implies *b* and
b implies *c* then
a implies *c*

FIGURE A.9　Venn diagram of the Principle of Transitivity, which is of the form *if a implies b is true and b implies c is true, then a implies c is true*. As can be seen, *b* is contained within *a* in whom all the members are true. Thus, all the members of *b* are true. Members of *c* are contained within *b* and because all the members of *b* are true, then all the members of *c* are true. Thus, *c* is true because *a* is true, hence *a implies c* is true.

digital computers, it demonstrates the incredible power of propositional logic.

　　The Principle of Transitivity is important, particularly because a variation (a fallacy) is important in biomedical research. The principle holds *if a implies b is true and b implies c is true, then a implies c*. The form can be restated that if *a* is a set wholly contained within the set of *b* and *b* is a set wholly contained within set *c*, then *a* is a set wholly contained in set *c*. Fig. A.9 shows the associated Venn diagram.

PROCEEDING TO WHAT IS NOT CERTAIN BUT USEFUL AND DANGEROUS

As discussed earlier, valid logical arguments essentially are tautologies, which eventually resolve to *a implies a*. The only novel knowledge is the recognition of what initially did not seem to be a tautology. The tautological nature of *a implies a* is easily recognized when the *implies* operator is replaced by the *equal* (=) or the *is* linking verb. Similarly, *a implies b*, if true, is a tautology because *a* is a set wholly within set *b*. Thus, whatever is true of *b* is true of *a*; in that sense, some part of *b* *is a* or some part of *b* = *a*. As was seen from the various Venn diagrams, the logical variables were considered as sets and the operators relate the various sets. The notion of a set implies membership. Thus, the logical variables, such as *a* or *b*, signify membership in set *a* or set *b*. For example, the logical variable *a* can signify a *cat* or *all cats* in the set of *cats*. Either an entity, signified by the logical variable, is a member of or is in a set or is not a member or not in the set.

An entity as described previously, signified by the logical variable *a*, is either in the set or out of set *a*, it cannot be both in and out, nor can it be neither in nor out of set *a*. This is the Principle of the Excluded Middle, also called the Principle of Bivalence. In the language of propositional logic, a logical variable is either true or false, not both true and false, and not neither true nor false. The Principle of the Excluded Middle gives propositional logic its absolute certainty. However, it is very rare in biomedical research that a hypothesis or claim comports with the Principle of the Excluded Middle. There are a number of different logics whose particular purpose is to loosen the Principle of the Excluded Middle such as predicate logic and fuzzy logic. As will be seen, the Principle of the Excluded Middle also governs syllogistic logic, and attempts to loosen the principle result in the partial syllogism that gives rise to probability and statistics (see Chapter 5, The Logic of Probability and Statistics and Appendix B).

The valid propositional logical arguments are limited to tautologies and, thus, do not generate or create new knowledge. Propositional logic is described as truth preserving, not truth generating. Clearly, the aim of biomedical research is to gain new knowledge. For that reason, the logic used by biomedical researchers departs from strictly valid propositional (and syllogistic) logic but, in doing so, becomes uncertain. The logical forms typically used by biomedical researchers are logical fallacies. The key to productive biomedical research is to use logical fallacies that, although invalid, are "close enough" to the valid forms so as to increase certainty. However, the logical fallacies cannot be too close to the valid forms or minimal new knowledge will be gained.

The modus ponens-like form of the Scientific Method is the Fallacy of Confirming the Consequence (also called abduction) represented in the Venn diagram shown in Fig. A.4. The only way to make the modus ponens form of the Scientific Method is to recast the form from *if a* (hypothesis) *implies b* (prediction) *is true, and b* (prediction) *is true, the a* (hypothesis) *is true* to the form *if and only if a* (hypothesis) *implies b* (prediction) *is true, and b* (prediction) *is true, the a* (hypothesis) *is true*. This form can be taken to say that *b* is a member of *a and only a so what is true of b must be said of a*. Alternatively, *the only set of a comprises the set of b*. Indeed, randomized controlled trials are an attempt to establish the if and only if qualifier (see Appendix B).

Rarely in biomedical research can the experiment be based on *if and only if a* (hypothesis) *implies b* (prediction) *is true, and b* (prediction) *is true, then a* (hypothesis) *is true*. More often, several possible hypotheses are considered for a hypothesis. The logical form of the experiment becomes *if (a or b or c) (competing hypotheses) implies b (prediction) is true and b (prediction) is true, then (a or c or d) is true*, ignoring for the moment that this is the Fallacy of Confirming the Consequence. Assuming that the logical fallacy is close enough to the valid logical form of modus ponens, the challenge is to determine which of the alternative hypotheses, *a*, *c*, or *d*, are true.

Typically, other experiments are done to determine which among the alternative hypotheses, *a*, *c*, or *d*, truly implicate the prediction, *b*. This is where the type of *OR* function becomes critical. If the relationship between the logical variables, *a*, *c*, or *d*, is the *exclusive OR*, then demonstrating that *a implies b* is true (within the limits of the Fallacy of Confirming the Consequence), then *c implies b* must be false and that *d implies b* must be false. Another approach would be to demonstrate that [(*a implies not c*) *AND* (*c implies not d*) *AND* (*and a implies not d*)] *is true*. In other words, *a*, *c*, and *d* are all mutually incompatible.

There are situations where the relationship is an *inclusive OR* function risks logical fallacies, thereby increasing the risk for unproductive irreproducibility. If the logical form of the experiment becomes *if (a inclusive OR b inclusive OR c) (competing hypotheses) implies b (prediction) is true and b (prediction) is true, then (a inclusive OR c inclusive OR d) is true*, ignoring for the moment that this is the Fallacy of Confirming the Consequence. If the research was to subsequently demonstrate the *a implies b is true* and were then to conclude that *c implies b* and *d implies b* were both false, the researcher would commit the Fallacy of Limited Alternatives. The truth of *a implies b* has no bearing on whether or not *c implies b* or *d implies b*. There are no mutual incompatibilities among *a implies b, c implies b*, and *d implies b*. However, demonstrating that *a implies b* is false means that either or both *c implies b* and *d implies b* must be true. When applied to probabilities, the Fallacy of Limited Alternatives is the Gambler's Fallacy.

There is a particularly pernicious form of the Fallacy of Limited Alternatives, which is when alternatives are not even considered. Just the fact that *a* (single hypothesis) *implies b is true* from demonstrating that *b is true* (within the limits of the Fallacy of Confirming the Consequence) leads some researchers to not even consider alternative hypotheses. Clearly, if alternative hypotheses are not considered, the fallacious modus ponens-like form of the Scientific Method cannot be strengthened by fulfilling the *if and only if* qualifier.

The Principle of Transitivity allows one to recognize a tautology over multiple logical variables and is of the form *if a implies b and b implies c is true, then a implies c*. In this case, *a* is realized as *c*, hence the tautology. To extend the utility of the principle, even if absolute certainty is given up, the principle can be changed to the Fallacy of Pseudotransitivity. This fallacy is of the form *if a implies b and c implies b is true, then a implies c* and is represented in the Venn diagram in Fig. A.10. The fallacy can be restated, as set *b* is wholly contained in set *a*, so that whatever may be said of set *a* can also be said of set *b*. Further, set *b* is wholly contained within set *c* and what can be said of set *c* can be said of set *b*. However, this does not mean that set *a* is wholly contained in set *c*.

Despite the uncertainty associated with the Fallacy of Pseudotransitivity, the fallacy is extremely important, indeed fundamental, to biomedical research as the source of hypotheses for biomedical research. Indeed, the Fallacy of Pseudotransitivity is the form of a

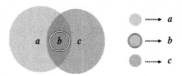

a implies b and c implies b then a implies c

FIGURE A.10   Venn diagram of the Principle of Pseudotransitivity, which is of the form *if a implies b is true and c implies b is true, then a implies c is true*, but the argument is invalid. The truth or falsehood of the premise *a implies c* cannot be known with any certainty. As can be seen, *b* is contained within *a* in whom all the members are true. Thus, all the members of *b* are true. Members of *b* are contained within *c* and those members in *b* and *c* will be true. However, there are members in *c* that may not be members of *a* so the truth of all the members in *a* cannot ascertain that all members of *c* are true. Thus, the truth or falsehood of *a implies c* cannot be known with certainty and, thus, this form of logic is invalid and is the Fallacy of Pseudotransitivity.

metaphor. For example, a clinical trial of drug X established for disease Y is tested for its effect on disease Z based on the similarity of disease Y to disease Z—typical off-label use. The logic is of the form *if disease Z (a) implies disease Y (b) and if drug X (c) implies disease Y (b), then disease Z (a) implies drug X (c)*. Just because drug X works for disease Y does not mean, with certainty, that drug X will work in disease Z. However, it is possible, and the degree of possibility depends on the degree of similarity between the pathophysiological mechanisms between disease Y and disease Z.

The Fallacy of Pseudotransitivity is invoked any time a lower animal species provides a rationale for studying a higher species. For example, the benefit of cancer treatment (c) in rodent cancer (b) suggests that cancer treatment (c) may help in human cancer (a). The rationale for the experiment is *if a* (human cancer) *implies b* (rodent cancer and the implication is appropriate because the rodent is being used in lieu of experimenting at early stages in humans) *is true and c* (cancer treatment) *implies b* (rodent cancer—demonstrated previously in preclinical trials) *is true, then a* (human cancer) *implies c* (cancer treatment).

The approaches described previously are fundamental to biomedical research. It is estimated that 20% of accepted medical treatments are off-label. The use of rodent models to develop new cancer treatments is highly successful. But note that preclinical trials have not been 100% successful, absolute certainty, because the method fundamentally is based on the Fallacy of Pseudotransitivity.

EXTENSION TO SYLLOGISTIC DEDUCTION

Another form of deductive logic is syllogistic deduction, which is of the form shown Argument A.1. The syllogism can be understood based on the Venn diagram shown in Fig. A.11. It shows that the set of *all humans* is wholly contained within the set of things that are of *things mortal*. *Socrates* is a set wholly contained in the set of *humans*. It is absolutely certain that *Socrates* then is a set wholly contained within the set of all *things mortal*. The syllogistic deduction can be rephrased as *if Socrates (a) implies human (b) is true and human (b) implies mortal (c) is true, then Socrates (a) implies mortal is true (c)*.

As in propositional logic, a valid syllogistic deduction with true premises is a tautology. No new knowledge is gained other than to recognize what can be said of humans, their being mortal, can be said of Socrates. However, there is a variation on the syllogistic deduction that increases the utility, although lessening certainty and opening the risks

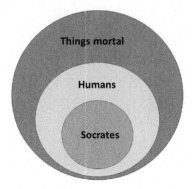

FIGURE A.11   Venn diagram of the syllogistic deduction given in Argument A.1. The major term in the major premise and conclusion is the set of things mortal. The minor term in the minor premise and the conclusion is the set Socrates. The bridging term is the set of humans. As can be seen, the set Socrates is wholly contained in the set humans, which is wholly contained in the set of things mortal. Thus, the set Socrates is wholly contained in the set of things mortal.

ARGUMENT A.1

Major premise:	*All humans* (bridging term) *are of things mortal* (major term).
Minor premise:	*Socrates* (minor term) *is a human* (bridging term).
Conclusion:	*Socrates* (minor term) *is mortal* (major term).

of unproductive irreproducibility. That variation is of the form called the Fallacy of Four Terms. For example, what if in Argument A.1, Socrates was a different kind of human, one that is immortal (Fig. A.12)? The conclusion of the argument would be false, which means that the argument is invalid or the premise is false. In this case, the bridging term, *human*, in the major and minor premises is not the same. *Human* in the major premise is of the kind that is mortal. *Human* in the minor term is of the kind that is immortal. Thus, there are four terms, the major and minor terms and two variations on the bridging term.

The Fallacy of Four Terms is ubiquitous and fundamental in biomedical research. It is invoked any time a sample is used to draw inferences to a population as in Argument A.2. It is the basis of doing research on a lower species of animals in order to hypothesize to trials in humans, as shown in Argument A.3. It is the basis for using a computer model and simulation (in silico) or reduced specimens such as tissue slices (in vitro) to create hypotheses for experiments in animals (in vivo), as shown in Argument A.4.

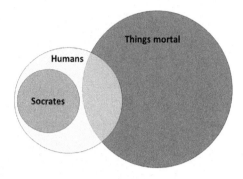

FIGURE A.12   Venn diagram of the invalid syllogism where *Socrates* is a different *human* who is not *mortal*. The major term in the major premise and conclusion is the set of *things mortal*. The minor term in the minor premise and the conclusion is the set *Socrates*. The bridging term is the set of *humans*. As can be seen, the set *Socrates* is wholly contained in the set *humans*, but the set of *humans* includes those who are *mortal* (the overlap between humans and things mortal). However, Socrates is not in the overlap and, thus, what can be said for *things mortal* cannot be said of *Socrates*. Hence, the syllogism is invalid and is called the Fallacy of Four Terms. In a sense, the set *humans* is divided into those who are mortal, overlapping with the set of things mortal and a part that does not overlap and hence is not mortal. The division of the set *humans* into two parts is what gives rise to the four terms in the fallacy.

ARGUMENT A.2

Major premise: *A sample of patients with disease **A** (bridging term) improves with treatment **X** (major term).*

Minor premise: *A population of patients (minor term) is disease **A** (bridging term).*

Conclusion: *The population of patients (minor term) improves with treatment **X** (major term).*

Conclusion: *My patient with disease **A** should improve with treatment **X**.*

ARGUMENT A.3

Major premise: *Rodents with disease **A** (bridging term) improve with treatment **X** (major term).*

Minor premise: *A population of patients (minor term) is disease **A** (bridging term).*

Conclusion: *The population of patients (minor term) improves with treatment **X** (major term).*

Conclusion: *My patient with disease **A** should improve with treatment **X**.*

ARGUMENT A.4

Major premise: *Computer model of system A* (bridging term) *demonstrates mechanism X* (major term).

Minor premise: *Humans* (minor term) *have system A* (bridging term).

Conclusion: *The population of patients* (minor term) *demonstrates mechanism X* (major term).

In each of these cases, the bridging terms in the major premises may not be the same as the bridging term in the minor premise and the arguments are Fallacies of Four Terms. In Argument A.2, the sample of subjects may not be the same as the population of concern. For example, if the sample solely consisted of middle-aged Caucasian males without other medical problems and the population of concern spans a large range of ages, races, ethnicities, genders, and comorbidities, then it is not at all clear that the bridging term *disease A* is the same between the major and the minor premises. Indeed, sampling is critical in statistics used to establish probabilities to be used in scientific reasoning (see Chapter 5, The Logic of Probability and Statistics and Appendix B, Erwin Bert Montgomery Jr.). Biomedical researchers often must go to great efforts to ensure that their sample is representative of the population of concern. Similar concerns are obtained when reasoning between species or from a model, either computation or a reduced preparation, to the whole animal.

WHERE IT GETS MORE UNCERTAIN, FROM STATE-OF-BEING LINKING VERBS TO CAUSATION

As demonstrated repeatedly here, valid deductive logic gains its absolute certainty because the premises comport with the Principle of the Excluded Middle. This principle is affected by using state-of-being linking verbs, *a is b*, or membership in sets. Unfortunately, these constructions are of relatively little utility in gaining new knowledge. Thus, even going back to Aristotle, the state-of-being linking verbs were changed to action verbs such as *cause*, of which Argument A.5 is an example and is known as the practical syllogism according to Aristotle.

The linking verb is the action verb *eat*. But Argument A.5 may not be true, as Socrates may choose not to eat, even though Socrates is hungry. One could attempt to make Argument A.5 more certain by replacing *eat* with the state-of-being linking verb *is* or *are* as shown in Argument A.6, but this clearly does not make sense.

ARGUMENT A.5

Major premise: *All humans* (bridging term) *may eat when hungry* (major term).

Minor premise: *Socrates* (minor term) *is a human* (bridging term).

Conclusion: *Socrates* (minor term) *may eat when hungry* (major term).

ARGUMENT A.6

Major premise: *All humans* (bridging term) *are when hungry* (major term).

Minor premise: *Socrates* (minor term) *is a human* (bridging term).

Conclusion: *Socrates* (minor term) *is when hungry* (major term).

Virtually every hypothesis that becomes the basis of a biomedical experiment entails the notion of cause, for example, *drug X is tested to treat disease A because drug X causes a change in the mechanisms that cause disease A*. Another example is of the form if mechanism *X* causes normal function in *organism A*, then agents that disrupt *mechanism X* should cause a change in function in *organism A* and the appropriate experiments can be conducted.

Introduction to the Logic of Probability and Statistics

This appendix is not intended to be a full discussion of probability and statistics, but rather to introduce some concepts to better understand the logic fundamental to probability and statistics. There are a great many introductory textbooks available for learning the various methods of statistical analyses. This discussion builds on some of the fundamental concepts of deduction and induction covered in Appendix A, An Introduction to the Logic of Logic. This effort is based on the premise that many cases of irreproducibility rarely involve errors in the mechanics of the statistical analyses applied. Rather, many cases are due to logical failures. There is indeed a logic to statistical analyses. Many cases fail because the recognized assumptions necessary to comply with the necessary logic for a particular statistical method have not been met. Most often these flaws are not recognized because the fundamental logic necessary for the statistical analysis is not appreciated and consequently cannot guide sound experimental statistical design.

The problem is compounded by the fact that biomedical research requires the use of logical fallacies to gain new knowledge, as traditional propositional and syllogistic deduction are incapable of generating new knowledge other than the recognition of tautologies (see Chapter 4: The Logic of Certainty versus the Logic of Discovery). However, to avoid the risk of irreproducibilities, the judicious use of logical fallacies requires that the fallacies comport with the traditional logical forms to a reasonable degree.

THE NEED FOR PROBABILITY AND STATISTICS

Traditional valid propositional and syllogistic deduction with true premises is guaranteed to produce true and valid conclusions. These logical forms do so by adherence to the Principle of the Excluded Middle, also called the Principle of Bivalence. This principle holds that a premise, proposition, or conclusion must be true or false, not both true and false, and not neither true nor false. However, there are a few situations in biomedical research where the scientific issues, hypotheses, and experimental designs conform to the Principle of the Excluded Middle. For example, consider the syllogism in Argument B.1:

ARGUMENT B.1

Major premise: *All smokers* (bridging term) *die from cancer or heart disease* (major term).

Minor premise: *John Doe* (minor term) *is a smoker* (bridging term).

Conclusion: *John Doe* (minor term) *will die of cancer or heart disease* (major term).

If the premises are accepted as true and the propositions are valid, then the conclusion is true and certain. Note that no probability is needed or, alternatively, the probability is maximal at 1 or 100%. However, the major premise, *smokers die from cancer or heart disease*, is not necessarily true, as some smokers may die of other reasons. Consequently, the syllogism is of little help and any experiment based on the syllogism is likely to be irreproducible. For example, a survey can be conducted to ascertain the cause of death in a set of smokers. One sample of smokers may just be that all died from cancer or heart disease. However, another sample may have subjects who died of other reasons.

ARGUMENT B.2

Major premise: *Some smokers* (bridging term) *die from cancer or heart disease* (major term).

Minor premise: *John Doe is a smoker.*

Conclusion: *John Doe may die of cancer or heart disease.*

In response, the partial syllogism was invented, for example, consider Argument B.2. However, the partial syllogism is of relatively little help. How would one conduct an experiment, using a survey as described previously, to determine causal factors in the death of smokers? However, if the terms *some* and *may* are replaced by probabilities, the resulting probability syllogism is much more helpful; for example, one can quantify the *some* in the partial premise to produce the probability syllogism in Argument B.3 that follows.

ARGUMENT B.3

Major premise:	*80% of smokers* (bridging term) *die from cancer or heart disease* (major term).
Minor premise:	*John Doe is a smoker.*
Conclusion:	*John Doe has an 80% chance of dying of cancer or heart disease.*

The syllogisms may be generated by a causal argument knowing that components contained within tobacco smoke contain carcinogens or compounds that promote atherosclerosis. Alternatively, the syllogism may be generated by induction, for example, every deceased smoker died of cancer or heart disease. The question is where do the probabilities come from? Most often, the probabilities are generated by enumeration. A number of observations are made, such as a sample of deceased smokers, and those that died of cancer or heart disease are identified. The ratio of those that died from cancer or heart disease to the total number of deceased smokers is determined. This ratio has values between 0 and 1, or often is represented by a percentage where the ratio is multiplied by 100.

PROBABILITY CALCULUS

Probabilities are critical in biomedical research. Often, whether or not a claim resulting from an experiment being accepted or not is based on the probability that the null hypothesis is rejected. This probability often is represented as the p value where a $p < 0.05$ means that the experimental result has less than a probability of 0.05% or 5% being due to chance and not being due to the experimental manipulation. Like propositional logic, complex chains of probability reasonings can be constructed, for example, as a logical calculus. Logical arguments use logical variables and operators that relate the logical variables. Given the truth values (whether true or false), the truth value of the logical chain can be determined. For example, in the logical *AND* function linking two entities, such as *a* and *b*, into *a AND b*, there is a mapping of the truth values of *a* and *b* onto *a AND b* to create a truth table as shown in Table B.1.

There are two logical variables, *a* and *b*, each of which can be either true or false. As can be seen, there are four combinations of truth values for both of the variables. The variables are related by the logical operator *AND* to produce the logical argument *a AND b*. The truth of *a AND b* relative to the truth values of *a* and *b* is shown in Table B.1. The logic argument *a AND b* is true only when both *a* and *b* are true. There is

TABLE B.1   Example of a Truth Table

a	b	a AND b
True	True	True
True	False	False
False	True	False
False	True	False

virtually no limit to the complexity of logical arguments. Indeed, every operation in the computer is a set of logical operators, particularly the *NAND* function where *a NAND b* is true only when both *a* and *b* are false. Similarly, logical chains of probabilities can be constructed constituting a probability calculus.

In probabilities, the proposition *a AND b* is converted to the probability of *a* and *b*, which equals the product of the probability of *a*, written as P(*a*), and the probability of *b*, written as P(*b*). The probability of the proposition *a AND b* becomes P(*a and b*) = P(*a*) × P(*b*) requiring P(*a*) be independent of P(*b*). In other words, P(*a*) does not vary with P(*b*) and P(*b*) does not vary with P(*a*). The syllogism in Argument B.1 becomes the probability syllogism in Argument B.3. The probability of John Doe dying from cancer or heart disease is the probability for any smoker and is written as P(*smoker dying of cancer or heart disease*) times the probability that John Doe is a smoker, which is written as P(*John Doe is a smoker*). The probability of John Doe dying of cancer or heart disease becomes P(*smoker dying of cancer or heart disease*) times P(*John Doe is a smoker*) written as P(*smoker dying of cancer or heart disease*) × P(*John Doe is a smoker*). If the probability of John Doe being a smoker is only 0.5 (50%), then the probability of John Doe dying of cancer or heart disease becomes 0.8 × 0.5 or 0.4 (40%).

Probability calculus is critical in biomedical research. Consider the question of whether humans identified as at risk of being detected as having gene *X* will have disease *Y*? Perhaps not all humans having gene *X* will be detected and the test for detecting disease *Y* is not 100% accurate. If the probability of the human subjects having gene *X* is 50% and the probability of detecting disease *Y* is 50%, then the chance of a human having both gene *X* and disease *Y* is 25%. The experiment is conducted and 25% of the human subjects test positive for disease *Y*. This is what would happen by chance. If the experiments resulted in 50% of the humans having disease *Y*, then the probability of having gene *X* is greater than 50%, the test for disease *Y* is more than 50% accurate, or both. Unfortunately, from the experiment alone, one cannot know. These calculations only work if the probability of detecting gene

X, $P(X)$, and the probability of the test to detect disease Y, $P(Y)$, are independent. This may not always be the case; in fact, it often is not. It may be that the technology for detecting gene X is the same as detecting disease Y and the joint probability will be 100% even though both detections are only 50% accurate.

The situation becomes more difficult if less than 100% of the humans tested have gene X. The probability of a human having gene X prior to testing for gene X is called prior probability. For example, if only 50% of the humans tested have gene X and the test is only 50% sensitive in detecting gene X, the probability of detecting gene X in the sample of humans will be 25%, even though 50% of the humans in the sample have gene X.

The probability of detecting gene X [corresponding to $P(a)$] and having disease Y [corresponding to $P(b)$] is written as $P(a \text{ and } b)$ and equals $P(a) \times P(b)$ when the probability of a, $P(a)$, is independent of the probability of b, $P(b)$. In the case where $P(a)$ and $P(b)$ are not independent, then $P(a \text{ and } b) = P(a) \times P(b \mid a)$, where $P(b \mid a)$ is the probability of b given a. In the aforementioned case, the probability of detecting gene X depends on the probability of the humans in the sample having gene X written as $P(a)$. The probability that the test can detect gene X when gene X is present is $P(b)$. Thus, the ability to detect gene X in the sample of humans, $P(a \text{ and } b)$, is $P(b \mid a)$. Note that if $P(a)$ and $P(b)$ are independent, then $P(b \mid a)$ just becomes $P(b)$ because it does not matter what $P(a)$ is.

The nonindependent probability calculus just described is critical for scientific decision-making. The concept of accounting for prior probabilities is an example of Bayesian analysis. For example, consider two conditions, disease A and test T, used to diagnose the disease. The probabilities related to A and T are not independent because if they were, then the probabilities related to T would not have anything to do with the probabilities of A and hence would not be a good test. The fact is that when A and T occur together, $A \text{ and } T$, the diagnosis is made. Therefore, fundamental to making a diagnosis is estimating the probability of $A \text{ and } T$ or $P(A \text{ and } T)$, which results in the first equation (Eq. B.1) shown here. Two equations are possible, the second is Eq. B.2. Setting Eqs. B.1 and B.2 equal and then recombining give rise to Bayes' theorem, Eq. B.3, which is critical in decision-making.

$$P(A \text{ and } T) = P(T) \times P(A|T) \tag{B.1}$$

$$P(A \text{ and } T) = P(A) \times P(T|A) \tag{B.2}$$

$$P(T) \times P(A|T) = P(A) \times P(T|A)$$

$$P(A|T) = \frac{P(A) \times P(T|A)}{P(T)} \tag{B.3}$$

Eq. B.3 can be interpreted as the probability that the subject has disease A if test T is positive, $P(A \mid T)$, which depends on the pretest probability of the subject having the disease, regardless of the test. $P(A)$, the probability of someone with disease A having a positive test, is $P(A \mid T)$, and the probability of having a positive test regardless of having disease A is $P(T)$. The probability of someone with disease A having a positive test, $P(A \mid T)$, is the sensitivity. The pretest probability of the subject having the disease regardless of the test, $P(A)$, is a measure of how common (prevalent) disease A is.

Bayesian analysis can help understand some cases of irreproducibility. Consider a clinical trial of test T for disease A as described previously. The scientist selects patients known with confidence to have disease A and finds that the probability of having a positive test T is 80% (sensitivity). In Eq. B.3, $P(T|A)$ equals 80% and $P(A)$ is 100%. Thus, it would seem that the scientist could have absolute confidence that test T is a good experimental measure of disease A.

From the experiments just conducted, the scientist cannot know the probability of a negative test T in someone without disease A, written as $\neg A$, where the $\neg$ means *not* is 70% (specificity), which would be written as $P(\neg T | \neg A)$. Thus, there is a probability of having a positive test T even in the absence of disease A written as $P(T | \neg A)$, in this case 0.3 or 30%. $P(T)$ is just the probability of T being positive, which is 80% in this case, which includes those with disease A as well as those without disease A, and would be written $P(T) = (P(T|A) \times P(A)) + (P(T|\neg A) \times P(\neg A))$, which equals 0.7 in this case. Note that $P(T|A)$ and $P(T|\neg A)$ are independent probabilities, one cannot infer one from knowledge of the other. The probability of a person having disease A with positive text T is 80%, written as $P(T|A)$. With great confidence, the experimenter goes on to test additional subjects suspected as having disease A. The test has little value as the human subjects already are known to have disease A. However, some of the humans do not have disease A. If the prior probability of having disease A in the second sample is only 50%, then the probability of having a positive text is 0.8×0.5 for those with disease A plus 0.3×0.5 for those without disease A for a combined probability, $P(T)$, of 0.7. Thus, the probability of having disease A if test T is positive is 0.057 or 57%. Perhaps test T is not a good predictor of disease A, which is discovered by subsequent repeat experiments. The result is that the studies are irreproducible.

DETERMINATION OF THE PROBABILITIES

Most often, probabilities are determined by enumeration. For example, one study may examine 100 cases of deceased persons who smoked and find in 99 of those cases the smoker died of cancer or heart disease. Therefore, the probability is 99% that a smoker will die of cancer or heart disease. But, would the same probability be obtained if another and different set of 100 deceased smokers was studied? Just because the probability was 99% in a past study does not mean it will be 99% in the future—this problem is known as Hume's Problem of Induction after the 18th-century philosopher David Hume. More likely, the risk by enumeration is likely to be different, for example, another enumeration on another set of deceased smokers demonstrated a risk of 93%.

The only condition where the Fallacy of Induction can be avoided is if every case was reviewed. In other words, if the medical histories of every patient in the universe (assuming there is only a single universe) who ever smoked were examined, then the number of those persons dying from cancer or heart disease would be sufficient to enumerate the probabilities of John Doe, who smokes, of dying of cancer or heart disease. But, how often is it possible to exhaustively study very issue of interest or concern in medicine so as to completely determine the relevant probabilities? Hence the need of statistics as an alternative to complete enumeration will be discussed later.

Enumeration over all possible subjects, the population, would provide the highest confidence. However, this generally is not feasible. Further, the Fallacy of Induction suggests that studying what is considered the entire population still lacks complete confidence. A typical induction is of the form *All ravens ever seen are black* (sample), *therefore all ravens are black* (population). However, one cannot know whether somewhere unbeknownst to the scientist, there exists a raven that is not black. Consequently, in the very large majority of cases, the biomedical researcher must assume that the subjects of experimentation are a sample and not the population, which necessarily exposes the experiments to the risk of the Fallacy of Four Terms when any effort is made to generalize from the sample of the experiment to the population of concern. Even if a complete enumeration is possible, it would be obsolete shortly after completion because new persons begin smoking and, by Hume's Problem of Induction, one cannot assume that their risk of dying from cancer or heart disease will be the same as those studied previously.

PROBABILITIES OF PROBABILITIES—STATISTICS

As described earlier, probability is central to scientific reasoning. However, there often is no outside independent "gold" standard to give absolute confidence that the probabilities the scientists use in their reasoning have absolute confidence. Consequently, the probabilities of the probabilities must be studied. For example, repeated and independent samples of the population of concern may result in multiple and different probabilities established by enumeration (counting). But, which is the "correct" probability to use in scientific thinking?

The centrality of probabilities in scientific thinking is exacerbated by the types of decisions scientists have to make. The driving force behind statistics in scientific decision-making is that the scientists, nearly always, must make dichotomous "yes" or "no" decisions. Is the hypothesis examined in the experiment true and, consequently, added as a principle or fact to the scientific corpus? Is it not true and, thus, cannot only be discarded but flagged for future scientists to avoid? Can the experimental results be taken as confidently true or false? Is there sufficient probability to continue the line of research or to abandon the line of research?

Dichotomous decisions follow naturally from deductive logic by virtue of the Principle of the Excluded Middle (see Chapter 4: The Logic of Certainty Versus the Logic of Discovery and Appendix A: An Introduction to the Logic of Logic), but at the expense of the range of problems that can be addressed scientifically. Extension of the syllogistic deduction to the probability syllogism (see Chapter 5: The Logic of Probability and Statistics) expands the range of problems made tractable by allowing some range of truth in the sense of the probability of truth. Note that the truth is still an all-or-nothing proposition, but confidence in the truth may be a matter of probability that lies between 0 and 1 or between 0% and 100%. The problem becomes how to have confidence in the probabilities now incorporated in logic?

Key to having confidence in the biomedical knowledge claimed from the experiment first requires confidence in the sample. Confidence comes from the reproducibility of the samples, and reproducibility is increased to the degree that the sample is made to represent the population of concern, thereby minimizing the risk from the Fallacy of Four Terms, within the limits of ergodicity that will be explained later. However, it is highly unlikely that every sample will be exactly the same. With repeated sampling, for example, increasing the size of the sample or the number of samples, when is a sample size or series of samples enough? Also, when the samples are all different, how is one to decide which sample is the most representative of the population of concern?

How would one go about comparing different samples, particularly when each sample has very large numbers of observations? One could extract a small set of abstractions or descriptors, such as the Central Tendency and variance, which then are thought to represent the sample. Now the comparison between samples can be effected by comparison of their abstractions or descriptors. However, this only exacerbates the concerns regarding the confidence in the samples. Now one has to have confidence that the abstractions or descriptors, such as the Central Tendency and variance, reflect each sample and that the sample or the set of samples reflects the population of concern.

A number of mathematical operations can be used to define the Central Tendency. These include the mean (average), median (the observation value where 50% of the observations are greater and 50% are less), and mode (the most common observed value). These measures of Central Tendency may be quite different for the same sample. In these circumstances, it is not necessarily clear which should be taken as the Central Tendency. The mean, $\bar{x}$, can be determined from Eq. B.4, where x_i is each observation indexed by i, which goes from 1 to n, where n is the number of observations in the sample and is determined by the following equation:

$$\bar{x} = \frac{\sum_{1}^{n} x_i}{n}. \tag{B.4}$$

The variance is a measure of the variability or dispersion of the actual observations of the sample. One method of determining the variance is given by the following equation:

$$\sigma^2 = \frac{\sum (x - \bar{x})^2}{n - 1}. \tag{B.5}$$

Often, the variability is reported as the standard deviation, σ, which is the square root of the result of Eq. (B.5). It is critical to note that for most applications of probability calculus and statistics, each value of a measure or observation in a research study is independent of the others. An example of a distribution is shown in Fig. B.1. Fig. B.1 is an interval histogram, where the vertical axis represents the number or percentage of observations with values in the ranges shown along the horizontal axis.

Historically, there have been a great many debates regarding which abstraction or descriptor represents reality or, at least, is the best for scientific purposes (Gigerenzer et al., 1989). However, the choice is critical as it is used in the probability syllogisms necessary for the interpretation of experiments. Often in biomedical research, the Central Tendency is considered the *de fault* representative of reality, although often this is a metaphysical position. Indeed, it is not unheard of for no observation

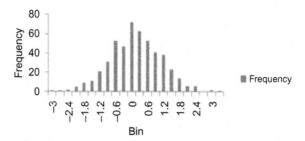

FIGURE B.1   An example of the distribution of a series of 500 observations. The display is an interval histogram with bins shown on the horizontal axis. The vertical axis represents the number of observations in each bin. In this case, the mean was 0 and the standard deviation was 1. Note that the shape of the distribution is symmetrical—essentially the same distribution on each side of the mean. Also, the frequencies of observations in the bins fall off exponentially as the bins are further from the mean on both sides. This type of distribution is termed a normal or Gaussian distribution.

to have the value of the mean. To be sure, recent practice is to constrain the interpretation of the Central Tendency by also reporting the confidence intervals that reflect the variance of the observations around the Central Tendency. For example, the effect of an experimental manipulation is reported to produce a change in a measure of the sample as of X, which represents the Central Tendency, such as the mean of changes in each subject as a result of the manipulation with a 95% confidence interval of Y and Z. The 95% confidence interval holds that 95% of the observations lie between Y and Z.

The confidence interval is designed to give a degree of confidence in the presumption that the Central Tendency reflects reality. For example, a value X of 0 would suggest that the experimental manipulation did not produce any effect. A 95% confidence interval where $Y = -0.1$ and $Z = +0.1$ would provide greater confidence than $Y = -100$ and $Z = +100$. Consider the example of where a value X of 2 would suggest that the experimental manipulation did produce an effect. A 95% confidence interval where $Y = -0.1$ and $Z = +10$ would provide less confidence than $Y = -1$ and $Z = +3$.

One approach often used to determine confidence intervals is to use the standard deviation calculated from the sample. The 95% confidence interval runs from the mean, $\bar{x}$, minus $1.96 \times$ standard deviation (sd) to the mean plus $1.96 \times$ standard deviation, often written as $\bar{x} - sd < \bar{x} < \bar{x} + sd$. Again, how much confidence can one have in a description of the mean and standard deviation? Fig. B.2 shows two distributions that all have the same mean and standard deviation, yet they are quite different. Note that the Central Tendency, for example, the mean and the standard deviation, can be calculated from any distribution as shown in Fig. B.2. The critical issue is that the mean and standard deviation do

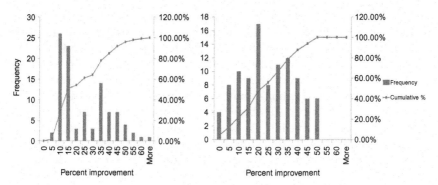

FIGURE B.2   Example of two distributions (interval histograms) of the observations of two different samples, but with same mean and standard deviation. *Reproduced from Montgomery Jr., E.B., 2019. Medical Reasoning: The Nature and Use of Medical Knowledge. By permission of Oxford University Press, USA (In Press).*

not represent reality; indeed, Fig. B.2 demonstrates this by the fact that the same mean and standard deviations are associated with very different realities reflected in the actual distributions of observations. Measures of Central Tendencies and variances are "internal" to the experiment and cannot be considered as any "external" validation. This can cause significant problems for reproducibility, particularly for Chaotic and Complex Systems, where the Central Tendency and variance are very poor abstractions or descriptors (see Chapter 5: The Logic of Probability and Statistics). For example, experiments with the same mean and variance in effects may be considered reproducible when they are not and the converse is true as well.

LEGITIMIZING ASSUMPTIONS

While the Central Tendency and variance are purely "internal" constructions that do not vouchsafe their particular "view' of reality, there are some notions that seem to provide credibility to the Central Tendency and the variance. The Large Number theorem holds that the mean of a sample will converge to a single number as the number of observations, n, increases. For example, the difference between successive calculations of the mean will tend to be 0 as the number of samples is increased in each successive calculation of the mean. This can be expressed in the following equation:

$$\lim_{n \to \infty} (\bar{x}_{n+1} - \bar{x}_n) = 0 \tag{B.6}$$

The fact the mean converges on a single number with increasing n would seem to imply that the mean is a stable and hence accurate descriptor of reality. However, such an inference may not be justified on its own. If, for example, the range of values an observation can take, x_i, is relatively limited, but the number of observations, n, is very large, the sum of all observations may not increase in proportion to the increase in the number of observations; consequently, the difference between succeeding means, $(\bar{x}_{n+1} - \bar{x}_x)$, will not increase because the contribution of x_{n+1} to the sum from which the mean is calculated, $\sum_0^n x_i$, will be less as n increases. In the limit of $n \to \infty$, the change in the mean approaches 0.

Similar concerns attend the abstractions and descriptors of the variability of the observations, such as the standard deviation, which is the square root of the variance as would be calculated by Eq. (B.5). Just as in the calculation of the mean, changes in the numerator for calculation of the standard deviation may not increase proportionately to the increase in n, particularly in distributions that are not normal or Gaussian (as explained later). In which case, the standard deviation will get smaller with increasing n, even though the actual distribution of the observations does not change. Thus, very large sample sizes, such as may be involved in what is called Big Data, may have standard deviations that are small because of the large n and not because of the nature of the observations. This could have an enormous impact on confidence of the experiments. For example, as discussed previously, the confidence interval is just that, an interval for which one can confidently conclude the Central Tendency has some meaning in reality. The 95% confidence interval ranges from the mean minus 1.96 × standard deviation to the mean plus 1.96 × standard deviation. If the standard deviation is artifactually small (less than would be the true variance), the confidence interval will be narrower and convey a false confidence in the reality of the Central Tendency.

One problem with very large data sets is demonstrating an effect, such as the change in some measure, with very narrow confidence intervals that provide a false sense of validity. It may well be the case that the calculated effect size is not meaningful or important to the scientific question, but is instead relying just on the statistical effect size, such as the mean difference affected by the experimental manipulation, which will not allow the scientist to know. Rather, the meaningful or important effect size should be stipulated prior to the experiments.

There is another implication of the Large Number theorem in the opposite direction. In order to have confidence in any statistical inference, such that an experimental manipulation does produce a meaningful change in the sample, a minimal number of observations are necessary. Fig. B.3 demonstrates the effect of the number of

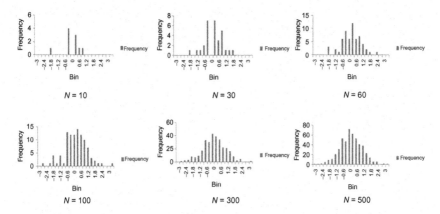

FIGURE B.3 Distributions of observations with different sample sizes. A population of 500 observation values was created with a mean of 0 and a standard deviation of 1. As can be seen, it takes 300 observations to have high confidence that the distribution resembles that of the population of 500 observations. How much confidence can be had with a fewer number of observations is problematic. *Reproduced from Montgomery Jr., E.B., 2019. Medical Reasoning: The Nature and Use of Medical Knowledge. By permission of Oxford University Press, USA (In Press).*

observations on attempting to infer what is the distribution of observations of the population of concern. A population of 500 random numbers was created using a mean of 0 and a standard deviation of 1. A series of samples of different sizes, *n*, were selected with replacement and the resulting distributions are shown in Fig. B.3. As can be seen, it takes 300 observations to have high confidence that the distribution resembles that of the population of 500 observations. As will be seen, the sample size used in experiments greatly affects the confidence that can be achieved for inferences drawn from statistical analysis of the data. An insufficient number of observations within the samples can lead to errors in the inferences drawn. Indeed, insufficient sample size is a frequent cause of irreproducibility. Errors in the inferences can go both ways, such as falsely rejecting the null hypothesis and claiming there is a difference (type I error) and falsely accepting the null hypothesis (type II error).

Knowing the mean and standard deviation of a sample allows one to determine when another observation is a member of the same population that the sample is thought to represent. For example, consider an observation with a value of 1.95, a sample with a mean of 0, and a standard deviation of 0.98. In a normal or Gaussian distribution, 95% of the observations are within 2 × standard deviation or, in this case, −1.96 to +1.96. This means that the observed value is within 95% of the range of values within the sample and, thus, the probability that the observation is not from the population is less than 5%. The *p* value, or the probability, is described as $p < 0.05$. However, if the observation was 2.0, then the odds are less than 5% that the observation is from the same population represented by the sample.

The Large Number theorem often does not hold in chaotic and complex systems, such as living organisms. Such systems, often starting from identical or nearly identical states, can have widely varying and often very distinct final states. For example, successive observed values can bifurcate to very different final states. Consider the situation of a heavy weight supported on a thin column (Fig. B.4). At some point as the weight increases, the column will buckle. Considering only a two-dimensional system, the column will buckle to the right or left. Buckling to the left creates negative numbers while to the right creates positive measures of the position of the middle of the column. A series of trials of increased weights and measuring the position of the midpoint of the column is undertaken. It is quite possible that in every trial to that time, the column buckles to the left, producing a negative mean position of the midpoint of the column. As the number of trials included in the calculation of the mean increases, there will be a convergence of the final mean in accordance with the Large Number theorem. However, it just may be that in the next set of trials, the column buckles to the right, producing positive midpoint positions. Successive means will diverge toward the right to a different final mean. The mean is considered the abstraction or descriptor of the final state. It is possible that many biomedical research experiments involve Chaotic or Complex Systems, and the presumption of the Large Number theorem presupposed by the use of the mean as the Central Tendency could lead to irreproducibility.

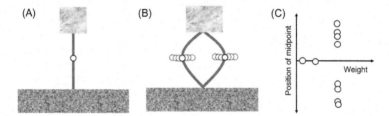

FIGURE B.4   Schematic representation of bifurcation in a complex system. A weight rests on a column (A). As the weight increases, a point is reached where the column buckles (B). In this case in two dimensions, the column can buckle either to the right or to the left. The position of the midpoint of the column is plotted versus the weight (C). At the lower weights, the position of the midpoint remains the same. However, at a certain increased weight, the column buckles and the midpoint of the column moves, either to the right or to the left. It is possible that the first trials result in buckling to the left. The distribution of the midpoints would have a normal or Gaussian distribution about the mean position to the left. If increasing numbers of trials all result in buckling to the left, the Large Number theorem holds that the mean position becomes stable at a single position. The next series of trials results in buckling to the right. Now the mean changes greatly and moves to the left and would continue to move (be unstable) as the subsequent trials continue to buckle to the left.

Brownian motion and diffusion of one liquid placed in another are examples of complex systems that do not comport with the Large Number theorem. The original description of Brownian motion by Robert Brown in 1827 was consequent to observing a grain of pollen suspended in a drop of water under a microscope. The grain of pollen moved about as a result of collisions with water molecules. The distance traveled with each collision varies, and the set of individual collisions has a normal or Gaussian distribution. The pollen moves through the water because of large, although rare, collisions. Subsequent collisions have a higher probability of being smaller and a high probability of not in the direction exactly opposite to the previous direction. In a sense, the position of the pollen grain at any one time can be considered as the sum of all the collisions prior and, thus, analogous to the mean of the collisions. If the Large Number theorem held in this case, the grain of pollen would not move, but perhaps jitter in place. The fact that the pollen moves around (just as particles of a drop of ink placed in a glass of water diffuse throughout the water) demonstrates that the Large Number theorem does not hold.

The Large Number theorem also depends on the concept of ergodicity, which presumes that the degree of variability is distributed evenly over the population of concern. Consider the situation of two fisherpersons at different ends of a long boat (Fig. B.5). In Fig. B.5A, the distribution of fish is distributed evenly in the lake. Estimating the mean number of fish caught per unit of time will be the same for both fisherpersons, provided

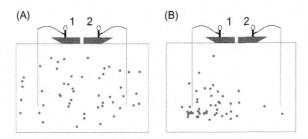

FIGURE B.5   Schematic representation of ergodicity, that is, the evenness of distributions in the population. In this case, the population is the total number of fish in the lake. Each fisherperson is sampling the population. (A) The distribution of fish is random, but distributed evenly. In this case, both fisherpersons will have samples (number of fish caught per unit time) and would be good representations of the population. The distribution of fish in A would have high ergodicity. (B) The distribution of fish is random, but not distributed evenly. The experiences of the fisherpersons would be quite different, and each sample would not be a good representation of the population. Fisherperson 1 may have an overestimate of the fish caught per unit time for the entire population, while fisherperson 2 may have an underestimate. *Reproduced from Montgomery Jr., E.B., 2019. Medical Reasoning: The Nature and Use of Medical Knowledge. By permission of Oxford University Press, USA (In Press).*

a sufficient number of fish are caught, according to the Large Number theorem. In this case, the lake displays high ergodicity, meaning one can fish anywhere in the lake and come up with the same mean number of fish caught per unit of time. However, if the distribution of fish in the lake is not distributed evenly, the mean number of fish caught per unit time by each fisherperson will be different (Fig. B.5B). The lake has low ergodicity even though each fisherperson may find higher ergodicity in the small area of the lake in which they fish.

Assume fisherperson 1 in the experiment represented in Fig. B.5 attempts to demonstrate that his experiment on the number of fish caught per unit time is reproducible. The next day, fisherperson 1 fishes from the position of fisherperson 2 previously, where fisherperson 1 will find his estimate of the number of fish caught per unit time irreproducible in the second fishing trip, even though (assuming) the total number of fish in the lake has not changed. The irreproducibility of the fishing experience is due to the lack of ergodicity in the distribution of fish in the lake and not due entirely to the nature of the experiment. The experimenter perhaps should not have assumed that the fish were distributed evenly throughout the lake or that the distribution would remain the same (stationarity).

Complex living biological systems are prone to low ergodicity based on fundamental principles of physics as discussed in Chapter 10, Chaotic and Complex Systems, Statistics, and Far-from-Equilibrium Thermodynamics. In order for biomedical researchers to minimize irreproducibility or make false claims about irreproducibility (irreproducible studies of irreproducibility), perhaps experimenters should first ask whether the biological system being studied is Chaotic or Complex, just as good scientists ask whether their set of observations are normally distributed or not, as will be discussed later. Generally, most scientists understand the importance of first knowing the distribution of their data in order to choose the appropriate statistical methods, but it is unclear how many appreciate the importance of first testing for Chaotic or Complex Systems behavior of their data.

NOT ALL DISTRIBUTIONS ARE THE SAME

Normal or Gaussian distributions of observations were shown in Fig. B.1. However, this need not be the case. Consider an experiment testing how quickly a subject can move his or her hand in response to a go cue—the reaction time. There are multiple factors that affect the reaction time, such as attention to the go cue. As the attention of the subjects varies, the reaction times will vary and become longer with less attention paid. However, there is a fundamental physical and physiological limit to how fast the subject can respond, even with maximal attention. Consequently, the distribution will be skewed to the right (Fig. B.6). Note that the mean can be calculated from any distribution, but the question becomes whether it is the true Central Tendency representing the reality of the system. In a skewed distribution, Central Tendency measures, such as mean or median, can be quite different.

The notion of the mean can be considered as a balance beam where the fulcrum is the mean. According to the Large Number theorem, the mean should become steady, which on the balance beam would only be the case if the weights on one side of the fulcrum were the same as the weights on the other side of the balance beam (Fig. B.7). However, not only are the weights important, but also the distance of the weights from the fulcrum. To determine the mean, the weights × distance to the fulcrum (which is the value of the observations in this experiment) are summed on each side of the fulcrum. The position of the fulcrum is adjusted so that the weight × distance [or just distance when unit weights (all the same value) are considered] on one side equals the sum of the weight × distance on the other side. If the unit weights are distributed in the same manner on each side of a particular point, it is relatively easy to determine that particular point, which becomes the fulcrum representing the mean. Further, if the distribution of distances is such that they cluster around a single point, again it is relatively easy

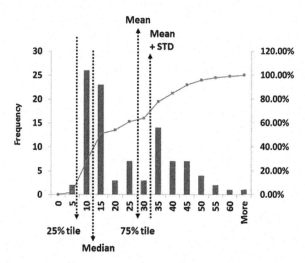

FIGURE B.6   An example of a skewed deviation to the right. As can be seen, the skewed deviation of the distribution results in different Central Tendencies when measured as the mean, median, or mode and different variances when measured as the standard deviation or percentiles (quartiles). *Reproduced from Montgomery Jr., E.B., 2019. Medical Reasoning: The Nature and Use of Medical Knowledge. By permission of Oxford University Press, USA (In Press).*

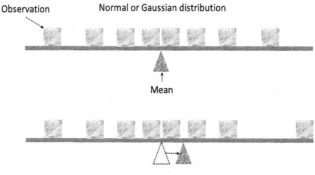

FIGURE B.7   Schematic representation of the effects of a skewed deviation on calculating the mean. The mean can be considered the position of the fulcrum in which the balance beam on which the weights rest is stable and balanced. In the case of normal or Gaussian distribution, the arrangements of the weights to the left of the fulcrum, which would cause the balance beam to rotate in a counterclockwise rotation, are balanced by the similar distribution to the right of the fulcrum. However, in the skewed distribution, the single weight that is far to the right without a counterbalancing weight on the left would cause the balance beam to rotate in a clockwise rotation. To compensate, the fulcrum, representing the mean, moves to the right.

to determine the position of the fulcrum, the mean, in which the beam will balance. However, if there is a skewed distribution, even with a minimum number of values, the balancing point has to be moved, and often a considerable distance. This situation is represented by the two balance beams shown in Fig. B.7. As can be seen, just moving one observation a greater distance from the others changes the balancing point or mean greatly.

Similar to the considerations regarding the mean being influenced by the nature of the distribution, the standard deviation is also affected. The numerator contains the expression $\sqrt{\sum(x_i - \bar{x})^2}$, which means the difference between the individual observation, x_i, and the mean, $\bar{x}$, increases exponentially with the difference as the difference is squared. A difference between x_i and $\bar{x}$ of 1 unit adds 1 unit to the sum, a difference of 2 adds 4, and a difference of 3 adds 9. As shown in Fig. B.7, observations that lie a distance from the mean can have an outsized effect on estimating the variance.

In experiments where distribution of the observations is highly skewed, one has to question the value of abstractions or descriptors, such as the mean and standard deviation described previously. As will be seen, these effects of distributions have a significant impact in attempting to test hypotheses statistically, such as an experimental manipulation applied to subjects in the experimental group and not applied to the control group. The question is posed whether the experimental group samples and the control group samples come from the same population. If not, then the null hypothesis (holding that they do come from the same population) can be rejected and the inference made that the experimental manipulation had an effect on the sample and, thus, on the population of concern. Many statistical tests, such as parametric tests, make important assumptions as to the distributions of data, such as requiring a normal or Gaussian distribution. Indeed, it is the ability to presume a normal or Gaussian distribution that gives these tests their power and, thus, preference. Use of these tests when the distribution is not normal or Gaussian can result in invalid analyses, which means that they could be misleading. This could result in true irreproducibility or false claims of irreproducibility. Accordingly, many scientists use methods, transformations, to make nonnormal distributions into normal distributions.

EXPERIMENTS CREATE DIFFERENT SAMPLES, OR DO THEY?

Most biomedical research experiments involve comparing samples, particularly where the subjects of one sample undergo an experimental manipulation, such as being given a drug and constituting the experimental group, while experimental manipulation is withheld from the second group constituting the control group. The scientist asks wither the manipulation had an effect by comparing the two groups. However, this often is not straightforward. Consider the distribution of results of the observations of each group shown in Fig. B.8. As can be seen, the two distributions overlap. Note that even if the experimental manipulation did absolutely nothing, it is highly likely that the distributions of the experimental and control groups would not be superimposable, as there would be some overlap. The question then is how much overlap can there be and the two samples still be considered different?

Another way to answer the question is to ask whether the two samples represent the same or different populations? If the two samples represent the same population, then it is likely that the experimental manipulation did not make any observable change in the subjects of the experimental group. The null hypothesis would be stated as the two samples came from the same population and failure to demonstrate that the two groups did come from different populations would be to accept the null hypothesis.

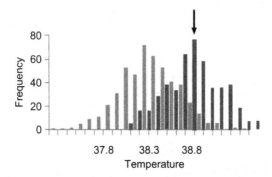

FIGURE B.8   Hypothetical distributions of the values of observations made in two samples, for example, the body temperature given drug X. One sample consists of subjects receiving an experimental manipulation (lighter colored bars to the left) and the other consists of subjects in which the experimental manipulation was withheld. The question is whether the drug caused an increase in body temperature. *Reproduced from Montgomery Jr., E.B., 2019. Medical Reasoning: The Nature and Use of Medical Knowledge. By permission of Oxford University Press, USA (In Press).*

One common test, called hypothesis testing, referring to acceptance or rejection of the null hypothesis, is the student's *t* test discussed here as for unpaired observations (subjects of the experimental group are not the same subjects in the control group) for equal sample size and equal variance, for illustrative purposes. The method calculates what is called the *t* statistic, the value of which indicated the degree of confidence that the null hypothesis can be rejected:

$$t = \frac{\bar{x}_1 - \bar{x}_2}{\sqrt{sp^2 \left(\frac{1}{n_1} + \frac{1}{n_2} \right)}} \tag{B.7}$$

The *t* statistic is determined by the differences in the means of the two groups, $\bar{x}_1 - \bar{x}_2$, experimental and control in the numerator. The denominator has the term *sp*, which is the pool variance, and n_1 is the number of observations in the experimental group and n_2 for the control group. The *sp* is similar to the variance (square of the standard deviation) for a single distribution. The *sp* is given by the following formula:

$$sp^2 = \left[(n_1 - 1)sd_1{}^2 + (n_2 - 1)sd_2{}^2 \right] / (n_1 + n_2 - 2). \tag{B.8}$$

In Eq. (B.8), sd_1 is the standard deviation of the distribution for the experimental group and sd_2 is the standard deviation of the control group. Calculating the *t* statistic also required degrees of freedom, which in this case is given by $(n_1 + n_2 - 2)$. Degrees of freedom roughly indicate the many different combinations possible from the number of subjects in the sample and the number of samples.

The t statistic is interpreted based on the degrees of freedom and is analogous to the p value described earlier. Thus, a t statistic of 1.96 indicates only a 5% probability that the two samples are from the same population and, if a 5% probability is acceptable, then the null hypothesis can be rejected.

Note that the t test involves the use of means and standard deviations, which are affected by the distributions of the observations within the sample as described earlier risking type I and type II errors. As described previously, the t test operates best when the distributions are normal or Gaussian. However, the t test may still be of help in distributions that are not too skewed, but in an interesting way. With a skewed distribution the standard deviations are likely overestimated, that is, being larger than would be a reasonable estimate of the variability. A larger standard deviation would result in a larger pooled variance, which would reduce the t statistic. This would be conservative, meaning that there is less likelihood of a type I error. However, it also means a greater likelihood of a type II error. The meaningfulness of a type I error versus a type II error—in other words, which type of error would be more tolerable—is not definable from the statistical tests, but rather only for the consequences for the aim or purpose of the biomedical research study. As mentioned earlier, there are statistical hypothesis tests that do not make assumptions as to the distribution of the values of the observations in the samples, but generally they have less power, thereby increasing the possibility of a type II error. Nonetheless, an appropriate use of statistical hypothesis testing requires a thorough understanding of the assumptions required for the most optimal use of the test.

CONTROLLED TRIALS OR EXPERIMENTS

Fundamental to biomedical research is the notion of controlled experiments. There are a number of different concepts regarding control. For example, given that complex biological systems may have reasons why a particular experimental manipulation designed to test a specific hypothesis may produce the experimental prediction, as noted in discussions of the Scientific Method, also known as the Fallacy of Confirming the Consequence (see Chapter 4: The Logic of Certainty Versus the Logic of Discovery), which is of the form *if a implies b is true and b is true, then a is true*. The experiment may also be affected by the Inverse Problem in that one cannot know which factors, such as the hypothesized cause or some other cause, produced the experimental effect. To counter the risks, the Scientific Method is restructured to *if and only if a implies b is true and b is true, then a is true*. One aspect of a controlled experiment is to ensure *if and only if a implies b*.

One way to ensure *if and only if a implies b* is by eliminating all possible competitors to *a* in leading to *b*. However, achieving the situation of *if and only if a implies b* is very difficult and, according to the Fallacy of Induction, one can never be absolutely confident that it has been achieved. Nevertheless, it often is helpful to try. For those causal factors other than *a* that are uncontrollable or unknown, one way to control for these other factors is to make as sure as possible that these other factors are present equally in the experimental and control groups. This allows the experimenter to take advantage of Mill's inductive method of agreement, which is of the form shown in Argument B.4. In this case, entities *A*, *B*, and *C* are potential confounds in studying the effects of *D*, where *D* is hypothesized to produce a novel effect, *Z*. The hypothesis is confirmed.

ARGUMENT B.4

Observation control group:	*Entities A, B, and C are associated with observations W, X, and Y.*
Observation experimental group:	*Entities A, B, C, and D are associated with observations W, X, Y, and Z.*
Conclusion:	*D is associated with Z.*

Confounds may vary over time, thereby complicating the analyses. The method of induction becomes that shown in Argument B.5. The conclusion in Argument B.5 is invalid if $time_1$ and $time_2$ are not the same. Thus, effect *Z* could be associated with entity *D*, the presence of $time_2$ or the absence of $time_1$. This problem is particularly acute in situations where the same subject is both the experimental group and the control group or where the experimental manipulation occurs after the observation in the control period. In this case, it is not possible to make $time_1$ and $time_2$ the same. One could randomly assign half the group to have the experimental manipulation at $time_1$ and the other half at $time_2$, in which case Argument B.5 becomes that shown Argument B.6.

ARGUMENT B.5

Observation control group:	*Entities A, B, C, and $time_1$ are associated with observations W, X, and Y.*
Observation experimental group:	*Entities A, B, C, D, and $time_2$ are associated with observations W, X, Y, and Z.*
Conclusion:	*D is associated with Z.*

ARGUMENT B.6

| Observation control group: | *Entities A, B, C, and time₁/time₂ are associated with observations W, X, and Y.* |

Observation control group: *Entities A, B, C, and $time_1/time_2$ are associated with observations W, X, and Y.*

Observation experimental group: *Entities A, B, C, D, and $time_1/time_2$ are associated with observations W, X, Y, and Z.*

Conclusion: *D is associated with Z.*

In this case, there can be more confidence that entity D is associated with observation Z. The method of timing the experimental manipulation is called counterbalancing.

STATISTICAL POWER, MULTIPLE COMPARISONS, AND CONFIDENCE

As discussed earlier, the number of observations per sample is critical to confidence in any inference from the statistical analyses. An insufficient number of subjects reduces the power of the analyses. Typically, power is expressed as the probability of finding a difference (rejecting the null hypothesis) at a certain level of confidence, such as a $p < 0.05$. A power of 0.8 means that there is an 80% probability of finding a statistically significant result where the result has less than a 5% probability of being due to chance. The power depends on the magnitude of the effect, for example, the difference in the means of the experimental and control samples, and the variance of the measures.

The optimal power level depends greatly on the nature of the experiments and the purpose for which the experiments are conducted. For example, preliminary and exploratory studies may use a power that is relatively low in order to avoid a type II error. Alternatively, if the consequences are more severe, then a much greater power would be most reasonable in order to avoid a type I error.

There are two general uses of power analyses. The first is pre-hoc and is used for sample size calculations when planning an experiment. One wants to have a sufficient sample size to avoid a type I or type II error. One wants to avoid the situation where the sample size is found insufficient on an interim analysis, and a secondary effort is undertaken to increase the sample size. This results in a situation of multiple comparisons, which increase the risk of a type I error, for example, an experiment in which the threshold for a statistically significant effect is set for $p < 0.05$. The researchers find that based on the sample size, the p value is 0.05 but this is not $p < 0.05$ as required before the experiments were undertaken. The investigators decide to reopen subject recruitment again, holding a $p < 0.05$ as the threshold for rejecting the null hypothesis and claiming a statistically significant result. The scientists now have engaged in multiple comparisons, one after the first study and again after the second reopened study. If the probability of a finding not due to chance is 5% for each run, then the probability of finding a statistically significant result is 0.05×2 or 0.10. The investigators may find a $p < 0.05$ on the second analysis thinking that this truly reflects the probability of rejecting the null hypothesis due to chance being 5%, but in actuality it is 10%.

The second use is post-hoc to ensure that the study just completed has sufficient power to have confidence in inferring from the outcomes. The pre-hoc use for calculating the planned sample size is based on best estimates (guesses) as to the effect size and variability of the outcome measures. But what if these estimations were wrong? In the worst-case scenario, the actual effect size was much less or the variance was much more than the estimates. In this case, the actual study may have been underpowered even though the conduct of the experiments comported with the sample size calculations pre-hoc.

The situation of multiple comparisons causing an increased risk of a type I error occurs more often when there are multiple outcome measures. For example, consider the situation where there are three outcome measures. Each may have a probability of having a statistically significant outcome due to a chance of 4.5%. However, the joint probability that at least one of the outcome measures having less than a 5% probability is based on the number of combinations, such as how many have 3 of the 3, plus 2 of the 3 plus 1 of the 3, for a total of 7 combinations. Thus, there is a 31.5% probability of any one of the outcomes being due to chance. Roll the dice enough times and one will win. There are corrections for multiple comparisons, such as reducing the p value threshold to $p < 0.025$ in the case of two comparisons.

Glossary of Concepts

This glossary contains brief descriptions and explanations of various concepts addressed in the book. As some readers may be unfamiliar with some of these concepts that make their appearance on multiple occasions, the glossary can be accessed repeatedly to clarify understanding.

Abduction A form of logic that resembles deduction but is a variant called the Fallacy of Confirming the Consequence, which is of the form *if a implies b is true and b is true, then a is true* (*see* Fallacy of Confirming the Consequence). For example, one might reason that *if dopamine depletion (a) causes (implies) parkinsonism (b) and the experimental subjects have Parkinsonism (b), then the subjects have dopamine depletion (a is true)*. However, the subjects could have Parkinsonism from any number of other causes, such as damage to the globus pallidus externa. As can be appreciated, abduction and the Fallacy of Confirming the Consequence are one instantiation of the Scientific Method.

A Priori problem of induction Typically an induction is derived from identifying a common phenomenon among a collection of particular observations to induce a general principle. For example, a large number of ravens are observed and a commonality of black color is found. The generalization is induced that *all ravens are black*. However, the question is what were the criteria that allowed the collection of a specific set of birds such that they are were of black color? Perhaps it may have been the shape of the peak, but it is unlikely that only ravens have that particular shape of beak. Perhaps it is the size of the raven, but there are many other species of birds of the same size that would not be admitted as ravens. Note that one cannot use black color as a criterion for inclusion of birds into the set of ravens. That would be invalid and an example of circular reasoning. Thus, the nature of prior criteria is fundamentally import to the induction and is problematic, hence the *A Priori* Problem of Induction.

Bayes' theorem A mathematical principle applied to probabilities. The theorem allows modifications of what may happen (posterior probabilities) based on what has happened in the past (prior probabilities). For one example, consider the case where an experimenter has *test X* to test for *condition A*, such as a disease. When the test is applied to a group of subjects known to have *condition A*, *test X* is found to be positive (abnormal) in 9 of 10 subjects, or 90 percent positive. This represents the test's sensitivity. When *test X* is applied to subjects known not to have *condition A*, 1 out of 10 subjects (10 percent) is found to be positive. This means that *test X* is correctly negative in 90 percent of those without *condition A*, which represents a specificity of 90 percent. *Test X* is going to be applied to a group of subjects, each of which may or may not have *condition A*. If 80 percent of the group has *condition A* (prior probability), *test X* will be positive in 72 percent of the subjects with *condition A* (posterior probability). Those subjects not detected are false negatives. Importantly, *test X* will be positive in two cases where the subjects do not have *condition A*. These cases will be false positives. Two studies using the same *test X* but on samples with very different prior probabilities will have very different outcomes (posterior probabilities) and could be unproductively irreproducible.

The same concept can be applied to estimations of irreproducibility based on the concept of statistical power, although the same concept can be used in estimations and have qualitative rather than quantitative questions. In the case of statistical power, the power calculation determines the probability of detecting an effect, such as the result of an experimental manipulation, in a sample with a variance (variability of the outcome measures). For example, a sample with an effect size of X percent (percent change in an outcome measure) with standard deviation Y percent and sample size n with a specified level to claim statistical significance, such as a p value, can be determined. In this case, the power is similar to the sensitivity of the test, S. However, the issue becomes how true is the sample representative of the population of concern? The sensitivity, S, would be a measure of the truth (reproducibility) of the scientific claim in the sample (experiments used)—posterior probability. However, it is possible that the scientific claim may not be 100 percent true in the population of concern (prior probability); consequently, the probability of an experiment actually being true is S modified by the probability in the population of concern, which may be less than S. One can ask what prior probability is necessary to have a sufficient probability of the scientific claim from the experiment, which reflects a sufficient probability of being true for the population of concern, hence the degree of conceptual or broad reproducibility or productive irreproducibility.

Broad reproducibility Refers to reproducibility of a general scientific claim that transcends the specifics of a particular experiment. For example, a claim about that anticancer effect of a treatment in a rodent model now tested for efficacy in humans or testing a claim from computational modeling (in silico) or reduced preparations such as tissue cultures (in vitro) to whole organisms (in vivo).

Causal syllogism A derivation from the practical syllogism, which is a variant of the deductive syllogism where the state-of-being linking verb, such *is* or *are*, is replaced by the action verb, in this case *cause* (*see* Syllogistic deduction). As few scientific observations and hypotheses are related by state-of-being verbs, other linking verbs are necessary, such as *cause*. However, while the causal syllogism has the appearance of a syllogistic deduction, it does not have absolute certainty.

Causational synonym, principle of *see* Principle of Causational Synonymy

Central limit theorem A mathematical assumption that holds that if a series of samples are taken from the population, the distribution of the means of those samples will be normal or Gaussian (*see* Distribution), even if the distribution of observed values within each sample is not a normal or Gaussian distribution. This allows the use of more powerful statistical methods. An example is use of a composite score from an instrument that has multiple variables that do not have a normal or Gaussian distribution. The composite score can be considered the mean of that subject's responses to the instrument. Taking the mean over a sample of subjects is tantamount to taking a mean of a series of means presumed to be normal or Gaussian in distribution.

Central tendency A measure thought to reflect the characteristics of a sample in a single value. This single value is thought to be representative of the sample or, in a bit of metaphysical extrapolation, represents the true nature of the sample where the observed values are variable for a variety of reasons, typically instrumental or methodological errors. In other words, if every instrument and measurement were perfect, the result would always be exactly the same (unfortunately, in complex biological systems, this is unlikely, *see* Chaos and Complexity). There are a great many different ways of calculating the Central Tendency, for example, the mean (average), the median (value where half the observations are greater and half are less), or mode (the most frequently appearing observed value). Depending on the distribution of the observed values, these different measures of the Central Tendency may be quite different, which raises the question which is the true measure?

Chaos and complexity Refers to systems whose architecture or structure can be clearly defined, such as by explicit mathematical formulas. However, the results of actions within the Chaotic or Complex systems cannot be predicted with reasonable accuracy or precision. Indeed, it is the unpredictability of Chaotic and Complex systems that is the main defining feature. Chaotic and Complex systems are highly dependent on the initial conditions, such as the starting conditions of the experiments. Thus, a set of experiments appearing to have the same starting conditions may have very different results. Such cases could be considered irreproducible, thus considered disproof of the experimental hypotheses. However, this may not be true and would be an example of unproductive irreproducibility.

Clinical meaningfulness This addresses whether an effect demonstrated in an experiment is sufficient to produce a clinically important change in a subject's well-being. For example, an effect from an experimental manipulation produces an improvement in the patient's disease, such as quality of life, that a reasonable person would consider a true benefit. An effect may be considered clinically meaningful if it results in a significant reduction in healthcare cost compared to not applying the experimental treatment. Note that clinical meaningfulness is not synonymous with statistical meaningfulness. A treatment derived from biomedical experimentation may produce a change in the outcome measure that meets the criteria for statistical significance by rejecting the null hypothesis (the result is not different than what would happen by chance) (*see* Statistical meaningfulness), but the subsequent application of the treatment does not produce a clinically meaningful change in patients.

Complexity and Chaos see Chaos and Complexity

Conceptual reproducibility see Broad reproducibility

Deduction A method of argumentation. As applied in logic, true conclusions are assured by true premises and valid propositions. However, the conclusions ultimately represent tautologies (*see* Tautology), which do not lead to new and novel knowledge. Thus, valid deductive logic is truth preserving rather than truth generating. There are a number of different types of deductive logics. The types of primary concern in this book are propositional logic, for example, *if a implies b is true and a is true, then b is true*, and syllogistic deduction, for example, *if all a's are b and c is an a, then c is a b*. Variations on the forms of deductive logic are important in biomedical research. As these variations are fallacious forms, results from their application cannot be taken with certainty (*see* Fallacies).

Distribution As used typically in statistics, distribution describes the range of observed values in a sample, for example, heights of adult males. Often the distribution is shown as an interval histogram where the heights are measured and divided in bins along the horizontal axis that reflect a range of heights. The number of observations, that is, males, with heights in that bin are reflected in the vertical axis. Perhaps the best-known distribution is Normal or Gaussian distribution. The shape of the distribution is critical, as many statistical tests are valid for only certain distributions.

Duhem−Quine thesis A logical argument is found to result in a false conclusion. At least one premise must be false or at least one proposition must be invalid. But which one? The Duhem−Quine thesis holds that, from the argument alone, one cannot know which premise, proposition, or some combination caused the argument to be invalid. The unfortunate consequence is that if one knew which false premise or invalid proposition was at fault, the experiments could be modified more easily. Rather, attempts to identify the cause of the invalid argument require importing knowledge from outside the experiment, which risks introducing bias that could lead to future experiments being unproductively irreproducible.

Epistemic degrees of freedom A component of Epistemic Risk (*see* Epistemic Risk) that relates to the degrees of freedom involved in the conceptual linkage between hypotheses and predictions in an experiment. Typically, the degree of freedom is related to how tightly the series of causal mechanism are linked, for example, within an experiment how similar is the mechanism when reasoning from a sample to the population or how similar is the model (simpler species, reduced preparations, or computational models) to the population of concern (higher species or whole organisms)? For example, changes in nervous system oxygen metabolism are used to "stand in" for neuronal activities, with the latter based on changes in the electrical potentials across the cell membrane of excitable cells. Changes in oxygen metabolism are the basis for functional magnetic resonance imaging (fMRI). However, the binding of oxygen to hemoglobin is linked to changes in the electrical potential across the cell membrane of excitable cells through a series of complex interactions. Further, changes in the oxygen binding to hemoglobin are linked to detection by fMRI through a series of complex interactions. The result is that the dynamics of changes in oxygen binding to hemoglobin are very different than those that involve changes in the electrical potentials across the membrane of excitable cells. Thus, experiments that infer changes in neuronal activities from changes in the oxygen binding to hemoglobin are at a relatively high risk of irreproducibility. This is not to say that such inferences are false but only that one cannot have absolute confidence that the changes in oxygen binding to hemoglobin detected truly reflect neuronal activities.

Epistemic distance A component of Epistemic Risk (*see* Epistemic Risk) that relates to the conceptual distance between hypotheses and predictions of an experiment. Each link in the causal chain relating the hypotheses to the predictions may be a source of variation in degree rather than in kind, increasing the risk for unproductive irreproducibility. For example, glucose metabolism in rodents may be relatively the same as humans, thereby reducing the Epistemic Distance and hence Epistemic Risk. However, oncogene expression, particularly in view of epigenetic factors, may be more complex in humans, thereby increasing the Epistemic Distance when inferring from rodent studies to humans.

Epistemic risk A conceptual tool used for assessing risks involved in the necessary use of fallacies. For example, Epistemic Risk can be used to judge hypotheses used in the Scientific Method, particularly when used in a modus ponens-like manner. Epistemic risk is composed of Epistemic Distance (*see* Epistemic Distance) and Epistemic Degrees of Freedom (*see* Epistemic Degrees of Freedom).

Epistemology A branch of the philosophical disciple of metaphysics. Epistemology relates to the theory of knowledge, such as how is knowledge obtained and how one can have certainty about what is thought to be known? This is different from ontology, which relates to questions about reality, such as objects and forces.

Ergodicity A concept related to the homogeneity of observations in a collection of observations, such as data in a database. The concept derives from the notion of how reproducible are samples drawn repeatedly from the collection of observations? An analogue to two fisherpersons is illustrated in Fig. 1.

Evolutionary epistemology A generalization from Evolutionary Logic to include other methods of epistemology that are not necessarily logical in nature (*see* Evolutionary Logic). The mechanisms and principles underlying Evolutionary Epistemology are shaped by the entities and operations in the environment that relate to the need for decision-making and the consequences of the decisions. For example, as in Evolutionary Logic, the need in the intellectual environment places a premium on certainty, with the result being the evolution of deductive logic.

Evolutionary logic this is a conceptualization of logic where the structure of the logical systems is driven by the need for certainty and the structure of the environment. For

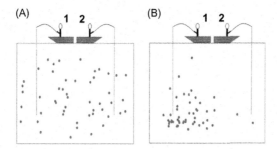

FIGURE 1 Schematic representation of the effects of ergodicity. (A) Fish are distributed evenly in the lake—high ergodicity. (B) Fish are concentrated under fisherperson 1. The experiences of the two fisherpersons in condition A will be the same demonstrating high ergodicity and considered reproducible. Their experiences will be different under condition B, thus low ergodicity, and the experiments are considered irreproducible. Source: *From Montgomery, E.B., Jr. Medical Reasoning: The Nature and Use of Medical Knowledge, Oxford University Press, in press.*

example, if success or survival in an environment places a premium on certainty, deductive logic likely will be a result. However, in the environment of biomedical research, the need for new knowledge is at a premium and the application of valid deductive logic is not going to be productive and hence will not be perpetuated. The modus ponens-like variant of propositional deductive logic has more utility, even though less certain, and has survival value as the Scientific Method.

Fallacies Forms of arguments whose conclusions cannot be taken as absolutely certain, as would be the case of deduction. This does not mean that the conclusions from the arguments cannot be true but only that the argument does not provide absolute confidence in the truth of the argument. Often, fallacies are invalid forms of valid deductive logical forms, thus giving the appearance of being valid arguments, for example, the Scientific Method, which is the Fallacy of Confirming the Consequence when used in the modus ponens-like form.

Fallacy of Affirming a Disjunct A logical fallacy of the form *if d implies (a or b or c) is true, and d is true then (a or b or c) is true.* This is valid and true. However, if one selects *a* over *b* or *c* as the explanation, one is engaging in the fallacy.

Fallacy of Composition One form of the Mereological Fallacy that relates the parts to the wholes. In this version, the properties of the parts are attributed to the whole. This approach is fundamental to Reductionism. To the degree Reductionism is valid (*see* Reductionism), the fallacy is not likely to be a risk. The test of the validity of the reduction is whether the whole can be reconstructed from the parts.

Fallacy of Confirming the Consequence A logical fallacy, hence invalid, resulting from a variant of the valid modus ponens form of propositional logic (*see* Modus Ponens). The valid form is *if a implies b is true and a is true, then b is true.* The Fallacy of Confirming the Consequence is of the form *if a implies b is true and b is true, then a is true.* The problem with the fallacy is that *b* could be true for any number of reasons other than *a*. This fallacy is a consequence of the Inverse Problem (*see* Inverse Problem) and is the basis for the Scientific Method.

Fallacy of Division One form of the Mereological Fallacy that relates the parts to the wholes. In this version, the properties of the whole are attributed to the part. This approach is fundamental to Reductionism. To the degree Reductionism is valid (*see* Reductionism), the fallacy is not likely to be a risk. The Fallacy of Division is particularly problematic when an experimental manipulation is directed at a part of an organism while the outcomes are the consequence of the remaining organism. The outcomes reflect the subsequent behaviors of the entire remaining organism. To the degree that the remaining parts adapt to the loss of the part due to experimental manipulation, the outcomes will not be a true reflection of the properties of the part affected by the experimental manipulation.

Fallacy of Four Terms A fallacy, invalid form, that is derived from syllogistic deduction (*see* Syllogistic deduction). Typically, a valid syllogistic deduction has three terms with the major and minor terms linked through a bridging term. However, the invalid form arises when there are four terms, for example, when there are two versions of the bridging term.

Fallacy of Induction A fallacy that holds that all inductions, such as *every raven observed is black, therefore all ravens are black*, are suspect, not that they are not true because one cannot know somewhere unbeknownst to the observer that a nonblack raven exists. This fallacy is different than the *A Priori* Problem of Induction (*see* *A Priori* Problem of Induction).

Fallacy of Limited Alternatives A fallacy that is of the form *if (a or b or c) implies d is true; d is true, b and c are false, then a must be true*. Regardless of the truth status of *b* or *c*, the conclusion that *a is true* is invalid as it represents the Fallacy of Confirming the Consequence. However, including *b* and *c* as hypotheses to be considered and then finding them false appears to lend confidence (falsely) that *a is true*. Further, there may be some other unbeknownst condition, *e*, that is responsible for *d*. The risk for the Fallacy of Limited Alternatives depends on the nature of the "or" function, whether it is an "*inclusive OR*" or an "*exclusive OR*." In an "*exclusive OR*," if one of *a exclusive OR b, exclusive OR c is true*, then the others must be false and if two are false, then the remaining must be true. In an "*inclusive OR*" function, any of the alternatives, *a inclusive OR b inclusive OR c*, any variable may be true and at least one must be true. For example, confusing an *exclusive OR* for an *inclusive OR* function can lead to the Gambler's Fallacy (*see* Gambler's Fallacy).

Fallacy of Pseudotransitivity A fallacy that is a derivative of the valid Principle of Transitivity, the latter being of the form *if a implies b and b implies c, then a implies c*. The Fallacy of Pseudotransitivity is of the form *if a implies b and c implies b, then a implies c*. The fallacy is the logical form of metaphors. The Fallacy of Pseudotransitivity is fundamental to forming deductive syllogisms, where *b* becomes the bridging term that links the major term, *a*, and the minor term, *c*, to form the conclusion that *a* implies *c* (*see* Syllogistic deduction).

Futility studies Experiments designed to specifically disprove a hypothesis, for example, using the logical form modus tollens (*see* Modus Tollens). Such studies utilizing the modus tollens form of the Scientific Method retain their certainty compared to use of the modus ponens-like form of the Scientific Method, which is the Fallacy of Confirming the Consequence. For example, an experiment can be constructed as *if experimental manipulation A caused parkinsonism and the actual experiment does not demonstrate parkinsonism, then manipulation A cannot be a cause of parkinsonism*. This is an example of productive irreproducibility (*see* Productive irreproducibility). Interestingly, often it is easier to demonstrate that an agent cannot be a cause of some phenomena than it is to prove that it does. Those agents in which the futility study has failed (cannot disprove the hypothesis) can be pursued further as a potential cause.

Fuzzy logic A type of logic that loosens the Principle of the Excluded Middle where state-of-being linking verbs are expressed as probabilities. In this case, logical variables may have intermediate values, for example, as ranging from 0 to 1. Solving an argument involves functors whose operations change the values, similar to athematic operations.

Gambler's fallacy A fallacy where past occurrences are thought to change the probability of future occurrences of events that are independent. For example, a gambler might believe that his luck has been so poor previously that he is due for a win. Alternatively, a gambler on a winning streak believes that he will continue to win.

Induction A form of logic in which commonalities among a set of observations are used to infer a general principle, for example, *every raven observed is black, therefore all ravens are black*. Induction is thought fundamental to science in general and to biomedical science in particular. Induction is subject to the Fallacy of Induction (*see* Fallacy of Induction) and the *A Priori* Problem of Induction (*see* *A Priori* Problem of Induction).

Induction, the A Priori Problem of *see* *A Priori* Problem of Induction

Information In its widest connotation, information is characterized by nonrandom state changes. For example, a sensible English sentence can be considered a sequence of states, corresponding to places in which letters, punctuations, and blanks are inserted—the precise order conveys information. This definition of information can be applied to the binary code in digital computers or to the sequence of base pairs in DNA. Indeed, any phenomenon that is not random, for example, has some structure, contains information.

Informational synonymy *see* Principle of Informational Synonymy

Injudicious use of logical fallacies *see* Judicious use of logical fallacies

Inverse problem This holds that one cannot determine which of any number of possible causes of the same phenomenon is the cause of that phenomenon from consideration of the phenomenon alone. When one arbitrarily chooses some causes and not others, the Fallacy of Limited Alternatives is at risk. This concept can be applied to any logical argument, such as an experiment. From the phenomenon of a failed experiment, such as irreproducible experiments, one cannot know which component of the experiment is at fault, an example of the Inverse Problem applied to the Duhem–Quine thesis (see Duhem–Quine thesis).

Judicious use of logical fallacies It is an epistemic fact of life that valid logical deductions do not lead to new knowledge other than the recognition of tautologies (*see* Tautology). In order to gain new knowledge, variations on deductive logic are used and these are known as logical fallacies. However, the fact that they are fallacies automatically means that their application does not result in absolute certainty as valid deductive logic would, thereby placing the experiments at a risk of unproductively irreproducibility. However, the fallacies can be used in a manner that maximizes the utility in gaining new knowledge while minimizing the risk for unproductive irreproducibility. This can be accomplished by using the fallacy in a manner that is close to the valid deductive form but sufficiently different so as to provide new knowledge. Such practice represents the judicious use of logical fallacies. Lack of such care in practice would not minimize the risk and would be considered an injudicious use of logical fallacies.

Large Number theorem A mathematical assumption critical to traditional statistical analysis, such as determining the Central Tendency. For example, the mean (average) of a sample is a measure of Central Tendency (*see* Central Tendency). The theorem states that the mean of a sample will stabilize or become asymptotic to a single value as the sample size increases. The presumption is that the stable value of the mean represents the true "nature" of the sample and, by inference, the true nature of the population of concern. The theorem may not be applicable in Chaotic and Complex systems, which are frequent in biological systems.

Local reproducibility This refers to a type of reproducibility within a single experiment that may have multiple trials, runs, or other replications. Indeed, statistical analyses of the result of a single experiment are based on the reproducibility within the experiment. Consider the following syllogism:

Major premise:	*Sample 1 of patients with disease A responds to treatment X*
Minor premise:	*Sample 2 has disease A*
Conclusion:	*Sample 2 responds to treatment X*

In this case, *disease A* is the bridging term. If sample 2 with *disease A* is not identical to patients that make up sample 1 in every relevant manner, then there are two bridging terms and thus the conclusion is invalid by the Fallacy of Four Terms. One bridging term is *Sample 1 of patients with disease A* and the second is *Sample 2 of patients with disease A.*

Mereological fallacy The fallacy of attributing the function of the whole to a part (*see* Fallacy of Division) or from the part to whole (*see* Fallacy of Composition).

Modus ponens A valid form of deduction, such *if a implies b is true and a is true, then b is true,* which guarantees true conclusions from true premises and valid propositions. The great value of recognizing the modus ponens form is that it will be true regardless of what is substituted for *a* or *b*. It does not matter what *a* or *b* actually is.

Modus tollens A form of deduction where a conclusion can be guaranteed to be false such as the logical form of *if a implies b is true and b is false, then a must be false.* Any experiment of this form will result in confidence that the hypothesis (*a*) is false, regardless of what specifics are substituted for *a* or *b*.

Narrow reproducibility Refers to reproducibility between multiple executions of the exact same experiment but different from the notion of local reproducibility (*see* Local reproducibility).

Negative Predictive Value The probability that some experiment will demonstrate the negative to be true. For example, an experiment is constructed to demonstrate a particular result from a measurement of an experimental manipulation thought to indicate a specific state or situation, for example, a state or situation that is abnormal. It also is important to know the probability of the measurement not showing the state or situation when an experimental manipulation was withheld, for example, the normal or natural state. The probability that a measurement will not demonstrate an altered state or situation when the manipulation is withheld describes the Negative Predictive Value of the experiment. The Negative Predictive Value incorporates the specificity of the test itself and the probability that the not altered state will be encountered in the actual experiments (*see* Specificity).

Null hypothesis A prediction used in statistical hypothesis testing that there is no difference in what is being compared. For example, as the results in experimental group are not different than the results obtained by chance or in the control group. Hypothesis testing is based on accepting or rejecting the null hypothesis. Thus, if the null hypothesis is rejected, then the experimental group results could not be obtained by chance, at a predefined level of probability.

Null study A null study is one whose result is indeterminant, not a confident demonstration of a difference in the experimental outcome (a positive study) or not a confident demonstration of no difference in the experimental outcome (a negative study). Typically, a null study is one where no statistically significant difference is found between the experimental and the control group where a prior sample size calculation was not done to demonstrate a reasonable probability of detecting a difference if a difference were to exit.

Ontology A branch of the philosophical disciple of Metaphysics. Ontology relates to the theory of reality, such as what is real as opposed to the theory of knowledge or how knowledge comes about (*see* Epistemology).

Partial syllogism A variation syllogistic deduction that is less certain but often has greater scientific utility. Consider the valid syllogism *all persons who smoke will develop cancer or heart disease; John Doe smokes, therefore John Doe will die of cancer or heart disease.* If the premises of the syllogism are true, it is certain that *John Doe will die of cancer or heart disease.* Yet, it is a reasonable possibility that a smoker dies in an accident. The partial syllogism of the form *some persons who smoke will develop cancer or heart disease; John Doe smokes, therefore, John Doe may die of cancer or heart disease* can be created from the valid syllogism and more accurately reflects reality even though it cannot be certain that *John Doe will die of cancer or heart disease.*

Population of concern Refers to all subjects, conditions, circumstances, and instances for which an experimentally derived claim is to be generalized to. For example, the effectiveness of *treatment A* for Parkinson's disease can only be tested on a limited number of patients, the sample, yet the findings are thought applicable to every patient with Parkinson's disease, present and future.

Positive Predictive Value The probability that some experiment will demonstrate that positive to be true. For example, an experiment is constructed to demonstrate a particular result from a measurement of an experimental manipulation thought to indicate a specific state or situation, for example, a state or situation that is abnormal, hence a positive test. The Positive Predictive Value incorporates the sensitivity of the test itself and the probability that the altered state will be encountered in the actual experiments (*see* Sensitivity).

Practical syllogism A syllogistic deduction (*see* Syllogistic deduction) with true premises provides a certainty of conclusions based on the Principle of the Excluded Middle (*see* Principle of the Excluded Middle) and the state-of-being linking verbs among major, minor, and bridging terms. Yet, seldom are the premises of an experiment related by state-of-being linking verbs. Indeed, most biomedical experiments are based on action linking verbs, for example, and particularly important, is when *cause* is substituted for *is* or *are*, thus creating a practical syllogism. Note that the greater utility of the practical syllogism comes at the cost of certainty.

Principle of Bivalence *see* Principle of the Excluded Middle

Principle of Causational Synonymy This holds that whatever the mechanisms are in a causal agent that creates an effect, those mechanisms must be present in the effect, a concept dating back to the ancient Greeks, although not often fully appreciated. For example, consider my hand moving water. The electrons in the outer orbits of the atoms that make up the surface of my hand repel the electrons in the outer orbit of the atoms that make up water. Thus, a complete explanation of why my hand can move water ultimately must include a discussion (whether implicit or explicit) of the electrons involved. Further, the electrons involved in the cause (my hand) are of the same kind as in the effect (water). An experiment whose hypothesis is not causally synonymous with the predictions likely is at risk for unproductive irreproducibility because a failure to ensure causational synonymy increases the Epistemic Distance and Epistemic Degrees of Freedom (*see* Epistemic Distance and Epistemic Degrees of Freedom).

Principle of the Excluded Middle A fundamental principle of deductive logic and induction where a premise must be either true or false and cannot be both true and false or neither true nor false. This principle ensures certainty even if it limits applicability.

Principle of Informational Synonymy A more general form of the Principle of Causational Synonymy (*see* Principle of Causational Synonymy) based on a definition of causation as effects on the nonrandom states within the experiments. This relationship between cause and effect can be interpreted as changes in the information state in the cause affecting information states in the effect. The Principle of Information Synonymy goes further based on the application of the Second Law of Thermodynamics. The second law holds that in any closed system (a system whose content is not affected by things external to the system), entropy can only increase. Entropy is a measure of randomness. In the case of the Principle of Informational Synonymy, information is considered the inverse of entropy and hence can only decrease in any closed system. Thus, when considering the interactions of an experiment, cause, such as an experimental manipulation, and effect, the outcome, the information content in the effect cannot be greater than the information content of the cause. Just as entropy increases in any physical process that is irreversible, information is lost with any manipulation of information that is irreversible, for example, defining a set of observations from the Central Tendency. Given just the Central Tendency, it is impossible to reconstruct the contributing observations. It is possible that the loss of information in biomedical experimentation could contribute to unproductive irreproducibility.

Probability Relates to the chance of a specific event or circumstance being true (or false) or present (or absent), typically by enumeration. In a given set of observations with n members, the number of observations of a specific type, s, is determined. A ratio of s to n, or s/n, constitutes the probability. The ratio can be multiplied by 100 to convert to a percent probability. Interestingly, probabilities are based on the past but are used to predict the future. Thus, it does not necessarily follow that future experiences will be the same as probabilities determined from the past (this is known as Hume's Problem of Induction). The uncertainty creates risks for irreproducibility, where an experiment demonstrated on probability and a subsequent replication resulted in a different probability.

Probability syllogism A derivation from the partial syllogism (*see* Partial Syllogism) where the state-of-being linking verb is replaced by a conditional or qualitative verb such as *may* or *some*. The probability syllogism replaced the conditional or qualitative verb with a probability.

Process metaphor Most metaphors involve target and source domains of the general form, *a is to b as c is to d*, where *c is to d* is known, the source domain, and is the source of inferences as to the claim *a is to b* which is the target domain. For example, a metaphor may be *a cell is to a tissue as a citizen is to a town*. The relationship of a citizen to the town is the citizen is a component of the town and the collective activities of the citizens constitute the activities of the town. By virtue of the metaphor, one comes to infer the cells are constituents of tissues and that the collective activities or functions of the cell become the activities or functions of the tissue. In these cases, the source domain is definite, *c is to d* or that citizens are constituents of town. In the Process metaphor, the source domain is not a definite entity but rather a process. For example, the velocity of an object is to some finite value as time interval is reduced toward 0. In other words, as velocity is the change in distance divided by the change in time, one cannot measure the instantons velocity as the time change would be 0 and dividing any number by 0 is indeterminant. The approach is taken to measure the change distance of an interval of time, which produces a non-0 change in time that allows calculation of the velocity. A process is then engaged where the time interval is reduced and the trend as this process continues points to what would be the instantaneous velocity. The instantaneous velocity is the target domain of a metaphor where the source domain is the processive reduction of the time interval ever closer but never reaching 0. The Process metaphor is fundamentally critical to Reductionism in biomedical research.

Productive irreproducibility An experiment or set of experiments found irreproducible. However, this in itself is not a problem; indeed, science advances by providing previous scientific claims irreproducible. Thus, while a failed experiment may be considered irreproducible, it still can make a significant scientific contribution. The contribution lies in the fact that the hypotheses pursued in the failed experiments can be dismissed from future considerations, provided that one has confidence that the experimental outcomes are valid. If the experiments are invalid, one cannot have confidence in the results and the hypotheses cannot be dismissed readily. This would be an example of unproductive irreproducibility.

Quartile range A measure of the variability of a set of observations. Typically, the range is measured by the value in which 25 percent of the observations are less than the value of the 25th percentile and the value where 75 percent of the observations are greater than the 75th percentile.

Reductionism A conceptual approach based on reducing the complexity of the scientific question to simpler and, presumably, more tractable questions. For example, for many reasons, it may not be possible to explore a scientific hypothesis by experimentation on humans. Consequently, the experiments are done in reduced or simplified experimental conditions, such as lower species, tissue cultures, or computational modeling. The key metaphysical presumption is that these reduced models standing in for more complex organisms, such as humans, truly reflect the same concern in the more complex organism. This may not be the case. There are several forms of Reductionism. Methodological reductionism uses reduced or simplified measures, for example, using a simpler biomarker as an indicator of a more complex biological state or circumstance. Theoretical reductionism reduces the complexity of the hypothesis or theory, for example, having a model less complex than the behaviors observed as standing in for the mechanisms of the more complex biology. Ontological reductionism is the notion that a simpler organism has the same nature or reality as a more complex organism, for example, glucose metabolism in a rodent is exactly the same as glucose metabolism in a human. However, the true test of Reductionism is the ability to reconstruct more complex from simplified experimental observations. In fact, it may be the presumption of Reductionism that leads to broad or conceptual irreproducibility (*see* Broad reproducibility).

Reproducible irreproducibility Based on the experience that multiple runs or trials of other replications within an experiment or in repletion of exactly the same experiment are unlikely to give exactly the same results. Thus, a set of trials, runs, or other replications with the same experiment or between exactly the same experiments will not be exactly reproducible. The critical question is to what degree the irreproducible results are still reproducible with respect to the intent of the research. This concept is fundamental to the use of variance (variability) in statistical analyses.

Sample It may not be feasible to test every subject of concern in an experiment, for example, a trial of *treatment A* for patients with Parkinson's disease. Consequently, *treatment A* is tested on a smaller collection of patients with Parkinson's disease and this collection is a sample. The significance of the experiments is important to the degree that the sample, the relatively small number of patients with Parkinson's disease actually subject to the experiment, can be generalized to the population of concern, such as all patients with Parkinson's disease, present, future, and possibly past.

Scientific Method A procedure intended to generate new knowledge attributed to Francis Bacon in the early 17th century. Typically the procedure involves (1) identify or formulate a problem, the answer to which generates new knowledge; (2) review current relevant information; (3) formulate a hypothesis that likely could be an answer; (4) convert the hypothesis to a testable hypothesis by deriving expectations that can be subjected to testing; (5) demonstrate whether the expectations result; (6) if the expectations do not result, re-evaluate and reformulate the hypothesis; and (7) if the predictions are demonstrated, the hypothesis is supported. The procedure that includes eventuality 7 is the modus ponens-like form of the Scientific Method and is invalid. This means that should step 7 be followed, that step does not ensure the true or falsehood of the hypothesis and is known as the Fallacy of Confirming the Consequence (*see* Fallacy of Confirming the Consequence). The steps through 6 are valid, that is, if the predictions are not found, then the hypothesis must be false with certainty. This form is the modes tollens form of valid propositional logic (*see* Modus Tollens).

Second Law of Thermodynamics as Applied to Information The Second Law of Thermodynamics states that entropy, a measure of randomness, cannot decrease in any closed system. Entropy can be considered inversely proportional to the information (structure) contained in the system. Thus, the Second Law of Thermodynamics as Applied to Information means that in any closed system, information cannot increase, only decrease. Further, any process within the closed system that is irreversible will result in a loss of information (for an example, *see* Reductionism).

Sensitivity The probability that a test will demonstrate some state or condition when the state of condition actually exists. For example, consider a test to detect Parkinson's disease. The sensitivity of the test is the percentage of subjects known to have Parkinson's disease and with the test results thought indicative of the presence of Parkinson's disease, such as abnormal or "positive" test results. Note that this is a different concept than Positive Predictive Value (*see* Positive Predictive Value).

Solipsism of the present moment Solipsism refers to the notion that all one can know is what is immediately present in consciousness. To claim to know otherwise invokes appeals to recollections of the past or predictions of the future that are liable to misconceptions. These misconceptions are not present in what one is immediately conscious of as there is nothing else that could be compared to arrive at the claim of misconception. Clearly, such solipsism means that the past cannot be a guarantee of the future and consequently any appeals to any generalized knowledge derived from past is suspect.

Specificity The probability that a test will demonstrate the absence of some state or condition when the state or condition actually does not exist. For example, consider a test to detect Parkinson's disease. The specificity of the test is the percentage of subjects known not to have Parkinson's disease and with test results thought not indicative of the presence of Parkinson's disease, such as a normal or "negative" test result. Note that this is a different concept than Negative Predictive Value (*see* Negative Predictive Value).

Standard deviation A measure of the variability of a set of observations. Typically, the standard deviation is the square root of the variance.

Statistical significance An outcome from an experiment that meets the probability threshold, such as the p value, sufficient to reject the null hypothesis (the result is not different than what would have happened by chance). Note that statistical significance is not synonymous with clinical meaningfulness (*see* Clinical meaningfulness).

Syllogistic deduction A form of logical deduction, consisting of major and minor premises and a conclusion, for example. The premises and conclusions contain major, minor, and bridging terms. The general structure is:

Major premise:	*All humans* (bridging term) *are mortal* (major term)
Minor premise:	*Socrates* (minor term) *is a human* (bridging term)
Conclusion:	*Socrates* (minor term) *is mortal* (major term)

If the premises are true, then the conclusion has absolute certainty. The verbs linking the terms are state-of-being verbs, *are* and *is*, which imply status such as membership. Thus, this form of syllogistic deduction can be understood from Set Theory where the major term (things *mortal*) is the universal set. The bridging term (*human*) is a subset wholly contained within the universal set (things *mortal*). The minor term (*Socrates*) is a subset wholly contained within the set of the bridging term. Consequently, the minor term (*Socrates*) must be a subset wholly contained within the universal set (things *mortal*).

Tautology A situation in which a scientific claim is true by definition or by virtue of being identical. By definition, the claim *All unmarried men are bachelors* is true by definition. No new knowledge is gained other than the recognition that a bachelor is an unmarried man.

Thermodynamic equilibrium Generally refers to any physical system in flux where two or more opposing processes interact to arrive at a constant state. Historically, the processes involved energy and entropy, such as in heat and work, hence the term "thermodynamic." For example, a process involving A leads to increased B while another process involving B leads to decreased A, the interactions will arrive at a balance between these two processes manifest by stable amounts of A and B. The stable amounts of A and B would be called a steady-state condition. Note, steady-states can arise from processes that are not in equilibrium.

Type I error Generally associated with the inappropriate rejection of a null hypothesis. This is interpreted as two sets of observations being statistically significantly different when in reality they are not different.

Type II error generally associated with the inappropriate acceptance of a null hypothesis. This is interpreted as two sets of observations being not statistically significantly different when in reality they are different.

Unproductive irreproducibility *see* Productive irreproducibility

Variance even as the exact experiment is repeated, the result likely will not be exactly the same, as there will be variability. Variance is a means to measure the variability. There are a number of different measures of variability, for example, variance, standard deviation, or quartile range.

Bibliography

Aird, F., Kandela, I., Mantis, C., 2017. Reproducibility project: cancer biology. Replication study: BET bromodomain inhibition as a therapeutic strategy to target c-Myc. eLife 6, e21253. Available from: https://doi.org/10.7554/eLife.21253.

Arikha, N., 2007. Passions and Tempers: A History of the Humours. Harper Collins, New York City, NY.

Arrowsmith, J., 2011. Trial watch: phase II failures: 2008–2010. Nat. Rev. Drug Disc. 10, 328–329. Available from: https://doi.org/10.1038/nrd3439.

Atamspacher, H., Massen, S. (Eds.), 2016. Reproducibility: Principles, Problems, Practices, and Prospects. John Wiley & Sons, New York, NY.

Atmanspacher, H., Demmel, G., 2016. Methodological issues in the study of complex systems. In: Atmanspacher, H., Massen, S. (Eds.), Reproducibility: Principles, Problems, Practices, and Prospects. John Wiley & Sons, New York, NY (Chapter 10).

Bailey, D.H., Borwein, J.M., Stodden, V., 2016. Facilitating reproducibility in scientific computing: principles and practice. In: Atamspacher, H., Massen, S. (Eds.), Reproducibility: Principles, Problems, and Prospects. John Wiley & Sons, New York, NY, pp. 205–232.

Baker, M., 2012. Independent labs to verify high-profile papers. Nature. Available from: https://doi.org/10.1038/nature.2012.11176.

Baker, M., 2016. 1,500 scientists lift the lid on reproducibility. Nature 533 (7604), 452–454. Available from: https://doi.org/10.1038/533452a.

Bálint, S., Bitter, I., Czobor, P., 2015. Neurobiological correlates of cognitive flexibility in ADHD: a systematic review of the literature. Psychiatr. Hung. 30 (4), 363–371 (in Hungarian).

Barker, R.A., Barrett, J., Mason, S.L., Björklund, A., 2013. Fetal dopaminergic transplantation trials and the future of neural grafting in Parkinson's disease. Lancet Neurol. 12 (1), 84–91. Available from: https://doi.org/10.1016/S1474-4422(12)70295-8. Review.

Barton, J., Bain, C., Hennekens, C.H., Rosner, B., Belanger, C., Roth, A., et al., 1980. Characteristics of respondents and non-respondents to a mailed questionnaire. Am. J. Public Health 70 (8), 823–825.

Bazerman, C., 1997. Reporting the experiment: the changing account of scientific doings of the Philosophical Transactions of the Royal Society, 1665-1800. In: Harris, R.A. (Ed.), Landmark Essays on Rhetoric of Science: Case Studies, Vol. 11. Hermagoras Press, Mahwah, NJ.

Begley, C.G., Ellis, L.M., 2012. Raise standards for preclinical cancer research. Nature 483, 531–533.

Bell, J.L., DeVidi, D., Solomon, G., 2001. Logical Options: An Introduction to Classical and Alternative Logics. Broadview Press, Peterborough, Ontario, Canada.

Bernatowicz, A.J., 1958. Teleology in science teaching: professors and textbooks alike make uncritical use of teleological and anthropomorphic language. Science 128 (3336), 1402–1405.

Bertler, Å., 1961. Occurrence and localization of catecholamines in the human brain. Acta Physiol. Scand. 51, 97.

Bickle, J., 2007. Ruthless reductionism and social cognition. J. Physiol. Paris 101 (4–6), 230–235. Available from: https://doi.org/10.1016/j.jphysparis.2007.11.009.

Birkmayer, W., Hornykiewicz, O., 1961. Der L-Dioxyphenylalamin (=Dopa)-Effekt bei der Parkinson-Akinese [The L-3,4-dioxyphenylalanine (DOPA)-effect in Parkinson-akinesia]. Wien Klin. Wochenschr. 73, 787–788 (in German).

Bizzarri, M., Palombo, A., Cucina, A., 2013. Theoretical aspects of systems biology. Prog. Biophys. Mol. Biol. 112 (1–2), 33–43. Available from: https://doi.org/10.1016/j.pbiomolbio.2013.03.019.

Boccia, M., Dacquino, C., Piccardi, L., Cordellieri, P., Guariglia, C., Ferlazzo, F., et al., 2017. Neural foundation of human moral reasoning: an ALE meta-analysis about the role of personal perspective. Brain Imag. Behav 11 (1), 278–292. Available from: https://doi.org/10.1007/s11682-016-9505-x.

Bochner, S., 1966. The Role of Mathematics in the Rise of Science. Princeton University Press, Princeton, NJ.

Brown, P., Marsden, C.D., 1998. What do the basal ganglia do? Lancet 351 (9118), 1801–1804.

Burgess, J.P., 2009. Philosophical Logic, Princeton Foundations of Contemporary Philosophy. Princeton University Press, Princeton, NJ and Oxford.

Bush, V., 1945. Science, the Endless Frontier, a Report to the President, from the Director of the Office of Scientific Research and Development. U.S. Government Printing Office, Washington, DC.

Button, K.S., Ioannidis, J.P., Mokrysz, C., Nosek, B.A., Flint, J., Robinson, E.S., et al., 2013. Power failure: why small sample size undermines the reliability of neuroscience. Nat. Rev. Neurosci. 14 (5), 365–376. Available from: https://doi.org/10.1038/nrn3475. Epub 2013 Apr 10. Erratum in: Nat. Rev. Neurosci. 2013 14 (6), 451.

Byson, B., 2003. A Short History of Nearly Everything. Black Swan (UK), Broadway Books, New York City, NY.

Carlsson, A., 1959. The occurrence, distribution and physiological role of catecholamines in the nervous system. Pharmacol. Rev. 11 (2, Part 2), 490–493.

Carlsson, A., 1971. Basic concepts underlying recent developments in the field of Parkinson's disease. In: McDowell, F.H., Markham, C.H. (Eds.), Recent Advances in Parkinson's Disease. F.A. Davis Company, Philadelphia, PA (Chapter 1).

Chen, J., Li, Z., Lv, Y.F., Li, C.L., Wang, Y., Wang, R.R., et al., 2015. Empathy for pain: a novel bio-psychosocial-behavioral laboratory animal model. Sheng Li Xue Bao 67 (6), 561–570 (in Chinese).

Colbourn, E.A., Rowe, R.C., 2009. Novel approaches to neural and evolutionary computing in pharmaceutical formulation: challenges and new possibilities. Fut. Med. Chem. 1 (4), 713–726. Available from: https://doi.org/10.4155/fmc.09.57. Review.

Collins, F.S., Tabak, L.A., 2014. NIH plans to enhance reproducibility. Nature 505 (7485), 612–613.

Collins, H., 2016. Reproducibility of experiments: experimenters' regress, statistical uncertainty principle, and the replication imperative. In: Atamspacher, H., Massen, S. (Eds.), Reproducibility: Principles, Problems, Practices, and Prospects. John Wiley & Sons, New York, NY (Chapter 4).

Coppieters, I., Meeus, M., Kregel, J., Caeyenberghs, K., De Pauw, R., Goubert, D., et al., 2016. Relations between brain alterations and clinical pain measures in chronic musculoskeletal pain: a systematic review. Pain 17 (9), 949–962. Available from: https://doi.org/10.1016/j.jpain.2016.04.005.

Cornish-Bowden, A., Cárdenas, M.L., Letelier, J.C., Soto-Andrade, J., 2007. Beyond reductionism: metabolic circularity as a guiding vision for a real biology of systems. Proteomics 7 (6), 839–845. Review.

Cotzias, G.C., 1966. Manganese, melanins and the extrapyramidal system. J. Neurosurg. 24 (Suppl 1), 170–180.

Cotzias, G.C., Papavasiliou, P.S., Miller, S.T., 1964a. Manganese in melanin. Nature 201, 1228–1229.

Cotzias, G.C., Papavasiliou, P.S., Vanwoert, M.H., Sakamoto, A., 1964b. Melanogenesis and extrapyramidal disease. Fed. Proc. 23, 713–718. Review.

Cotzias, G.C., Van Woert, M.H., Schiffer, L.M., 1967. Aromatic amino acids and modification of parkinsonism. N. Engl. J. Med. 276 (7), 374–379.

DeLong, M.R., 1990. Primate models of movement disorders of basal ganglia origin. Trends Neurosci. 13, 281–285.

Design of the Women's Health Initiative clinical trial and observational study, 1998. The Women's Health Initiative Study Group. Control Clin. Trials 19 (1), 61–109.

Dubey, A.K., Gupta, U., Jain, S., 2016. Epidemiology of lung cancer and approaches for its prediction: a systematic review and analysis. Chin. J. Cancer 35 (1), 71. Available from: https://doi.org/10.1186/s40880-016-0135-x. Review.

Ehringer, H., Hornykiewicz, O., 1960. Verteilung von Noradrenalin und Dopamin (3-hydroxytyramin) in Gehirn des Menschne und ihr Verhalten bei Erkankungen des extrapyramidalen Systems [Distribution of noradrenaline and dopamine (3-hydroxytyramine) in the human brain and their behavior in diseases of the extrapyramidal system]. Klin. Wochenschr. 38, 1236–1239 (in German).

Evans, J.R., Mason, S.L., Barker, R.A., 2012. Current status of clinical trials of neural transplantation in Parkinson's disease. Prog. Brain Res. 200, 169–198. Available from: https://doi.org/10.1016/B978-0-444-59575-1.00008-9. Review.

Fahn, S., 2015. The medical treatment of Parkinson disease from James Parkinson to George Cotzias. Mov. Disord. 30 (1), 4–18. Available from: https://doi.org/10.1002/mds.26102.

Fahn, S., Oakes, D., Shoulson, I., Kieburtz, K., Rudolph, A., Lang, A., et al., 2004. Levodopa and the progression of Parkinson's disease. N. Engl. J. Med. 351 (24), 2498–2508.

Fanelli, D., 2018. Opinion: is science really facing a reproducibility crisis, and do we need it to? Proc. Natl Acad. Sci. USA 115 (11), 2628–2631. Available from: https://doi.org/10.1073/pnas.1708272114.

Fang, F.C., Casadevall, A., 2010. Lost in translation—basic science in the era of translational research. Infect. Immun. 78 (2), 563–566. Available from: https://doi.org/10.1128/IAI.01318-09.

Feldherr, C.M., Akin, D., 1999. Signal-mediated nuclear transport in the amoeba. J. Cell. Sci. 112 (Pt 12), 2043–2048.

Feyerabend, P., 1975. Against Method. Verso Books, New York City, NY.

Feyerabend, P., 2010. Against Method: Outline of An Anarchistic Theory of Knowledge. Verso, London.

Filion, M., 1979. Effects of interruption of the nigrostriatal pathway and of dopamine agents on the spontaneous activity of globus pallidus neurons in the awake monkey. Brain Res. 178, 425–441.

Flexner, A., 1910. Medical Education in the United States and Canada: A Report to the Carnegie Foundation for the Advancement of Teaching (PDF), Bulletin No. 4. The Carnegie Foundation for the Advancement of Teaching, New York City, NY, p. 346.

Garijo, D., Kinnings, S., Xie, L., Xie, L., Zhang, Y., Bourne, P.E., et al., 2013. Quantifying reproducibility in computational biology: the case of the tuberculosis drugome. PLoS One 8 (11), e80278. Available from: https://doi.org/10.1371/journal.pone.0080278.

Gentili, C., Cristea, I.A., Angstadt, M., Klumpp, H., Tozzi, L., Phan, K.L., et al., 2016. Beyond emotions: a meta-analysis of neural response within face processing system in social anxiety. Exp. Biol. Med. (Maywood) 241 (3), 225–237. Available from: https://doi.org/10.1177/1535370215603514.

Gerlee, P., Basanta, D., Anderson, A.R., 2011. Evolving homeostatic tissue using genetic algorithms. Prog. Biophys. Mol. Biol. 106 (2), 414–425. Available from: https://doi.org/10.1016/j.pbiomolbio.2011.03.004.

Gigerenzer, G., Swijtink, Z., Porter, T., Datson, L., Beatty, J., Krger, L., 1989. The Empire of Chance: How Probability Changed Science and Everyday Life. Cambridge University Press, Cambridge.

Gilbert, S.F., Sarkar, S., 2000. Embracing complexity: organicism for the 21st century. Dev. Dyn. 219 (1), 1–9. Review.

Goodman, B., Flaxman, S., 2016. European Union regulations on algorithmic decision-making and a "right to explanation". arXiv 1606, 08813v3 [stat.ML].

Graham, D.J. 2004. Risk of acute myocardial infarction and sudden cardiac death in patients treated with COX-2 selective and non-selective NSAIDs. <http://www.fda.gov/downloads/Drugs/DrugSafety/Post-marketDrugSafetyInformationforPatientsandProviders/ucm106880.pdf>.

Greenfield, J.G., Bosanquet, F.D., 1953. The brain-stem lesions in Parkinsonism. J. Neurol. Neurosurg. Psychiatry 16 (4), 213–226.

Grodstein, F., Manson, J.E., Stampfer, M.J., 2001. Postmenopausal hormone use and secondary prevention of coronary events in the nurses' health study: a prospective, observational study. Ann. Intern. Med. 135 (1), 1–8.

Gross, P.R., Levitt, N., Lewis, M.W. (Eds.), 1997. The Flight from Science and Reason. Annals of the New York Academy of Sciences.

Gruener, A., 2015. The effect of cataracts and cataract surgery on Claude Monet. Br. J. Gen. Pract. 65 (634), 254–255. Available from: https://doi.org/10.3399/bjgp15X684949.

Hamani, C., Mayberg, H., Stone, S., Laxton, A., Haber, S., Lozano, A.M., 2011. The subcallosal cingulate gyrus in the context of major depression. Biol. Psychiatry 69 (4), 301–308. Available from: https://doi.org/10.1016/j.biopsych.2010.09.034.

Hannan, J., 2011. The Genesis of Science: How the Christian Middle Ages Launched the Scientific Revolution. Regnery Publishing, Inc, Washington, DC, US.

Heberlé, G., de Azevedo Jr., W.F., 2011. Bio-inspired algorithms applied to molecular docking simulations. Curr. Med. Chem. 18 (9), 1339–1352. Review.

Hiss, J.A., Hartenfeller, M., Schneider, G., 2010. Concepts and applications of "natural computing" techniques in de novo drug and peptide design. Curr. Pharm. Des. 16 (15), 1656–1665. Review.

Hodgkin, A.L., Huxley, A.F., 1952. A quantitative description of membrane current and its application to conduction and excitation in nerve. J. Physiol. 117 (4), 500–544.

Hodgkin, A.L., Keynes, R.D., 1955. The potassium permeability of a giant nerve fibre. J. Physiol. 128 (1), 61–88.

Holtzheimer, P.E., Husain, M.M., Lisanby, S.H., Taylor, S.F., Whitworth, L.A., McClintock, S., et al., 2017. Subcallosal cingulate deep brain stimulation for treatment-resistant depression: a multisite, randomised, sham-controlled trial. Lancet Psychiatry 4 (11), 839–849. Available from: https://doi.org/10.1016/S2215-0366(17)30371-1.

Horrigan, S.K., 2017. reproducibility project: cancer biology. Replication study: the CD47-signal regulatory protein alpha (SIRPa) interaction is a therapeutic target for human solid tumors. eLife 6, e18173. Available from: https://doi.org/10.7554/eLife.18173.

Horrigan, S.K., Courville, P., Sampey, D., Zhou, F., Cai, S., 2017. Reproducibility project: cancer biology. Replication study: melanoma genome sequencing reveals frequent PREX2 mutations. eLife 6, e21634. Available from: https://doi.org/10.7554/eLife.21634.

Houart, G., Dupont, G., Goldbeter, A., 1999. Bursting, chaos and birhythmicity originating from self-modulation of the inositol 1,4,5-trisphosphate signal in a model for intracellular Ca^{2+} oscillations. Bull. Math Biol. 61 (3), 507–530.

Hu, C., Di, X., Eickhoff, S.B., Zhang, M., Peng, K., Guo, H., et al., 2016. Distinct and common aspects of physical and psychological self-representation in the brain: a meta-analysis of self-bias in facial and self-referential judgements. Neurosci. Biobehav. Rev. 61, 197–207. Available from: https://doi.org/10.1016/j.neubiorev.2015.12.003.

Huang, W.C.-W., Batelaan, H., 2013. Dynamics underlying the Gaussian distribution of the classical harmonic oscillator in zero-point radiation. J. Comput. Methods Phys. Available from: https://doi.org/10.1155/2013/308538. Article ID 308538.

Huang, H., Watts, R.L., Montgomery Jr., E.B., 2014. Effects of deep brain stimulation frequency on bradykinesia of Parkinson's disease. Mov. Disord. 29 (2), 203–206. Available from: https://doi.org/10.1002/mds.25773.

Hulley, S., Grady, D., Bush, T., Furberg, C., Herrington, D., Riggs, B., et al., 1998. Heart and Estrogen/progestin Replacement Study (HERS) Research Group: randomized trial of estrogen plus progestin for secondary prevention of coronary heart disease in postmenopausal women. J. Am. Med. Assoc. 280, 605–613.

Hume, D., 1748. An enquiry concerning human understanding. A Letter from a Gentleman to His Friend in Edinburgh. In: Steinberg, E., (Ed.), An Abstract of a Treatise of Human Nature. Hackett Publishing Company, Indianapolis, Indiana, US.

Hutchison, W.L., Lang, A.E., Dostrovsky, J.O., Lozano, A.M., 2003. Pallidal neuronal activity: implications for models of dystonia. Ann. Neurol. 53, 480–488.

Hutson, M., 2018. Has artificial intelligence become alchemy. Science 360, 478.

Institute of Medicine [US] Committee on the Public Health Effectiveness of the FDA 510(k) Clearance Process. 2011, National Academies Press (US); 2010, Washington, DC.

Ioannidis, J.P., 2005. Contradicted and initially stronger effects in highly cited clinical research. J. Am. Med. Assoc. 294 (2), 218–228.

Jaeger, H.M., Liu, A.J., 2007. Far-from-equilibrium physics: an overview. Condensed-Matter and Materials Physics: The Science of the World Around Us, Condensed-Matter and Materials Physics: The Science of the World Around Us. National Academies Press, Washington, DC.

James, W., 1981. The Principles of Psychology. Harvard University Press, Cambridge, MA, p. 462, originally published in 1890.

Jankovic, J., Gelineau-Kattner, R., Davidson, A., 2010. Tourette's syndrome in adults. Mov. Disord. 25 (13), 2171–2175.

Jesseph, D., 1996. Hobbes and the method of natural science. In: Sorell, T. (Ed.), The Cambridge Companion to Hobbes. Cambridge University Press, p. 96. (Chapter 4).

Johnston, J.D., 2005. The contribution of Dr. Mary Walker towards myasthenia gravis and periodic paralysis whilst working in poor law hospitals in London. J. Hist. Neurosci. 14 (2), 121–137.

Johnson-Laird, P., 2008. How We Reason. Oxford University Press, New York, US.

Jordan, D., 2017. The King's City: London under Charles II: A City that Transformed a Nation and Created Modern Britain. Little Brown, London.

Joyner, M.J., 2011. Giant sucking sound: can physiology fill the intellectual void left by the reductionists? J. Appl. Physiol. (1985) 111 (2), 335–342. Available from: https://doi.org/10.1152/japplphysiol.00565.2011.

Joyner, M.J., Proctor, D.N., 1999. Muscle blood flow during exercise: the limits of reductionism. Med. Sci. Sports Exerc. 31 (7), 1036–1040. Review.

Kandela, I., Aird, F., 2017. Reproducibility project: cancer biology. Replication study: discovery and preclinical validation of drug indications using compendia of public gene expression data. eLife 6, e17044. Available from: https://doi.org/10.7554/eLife.17044.

Kaneko, K., 1990. Globally coupled chaos violates the law of large numbers but not the central-limit theorem. Phys. Rev. Lett. 65 (12), 1391–1394. Erratum Phys. Rev. Lett. 1991, 66, 243.

Kant, I. 1781. A Critique of Pure Reason N.K. Smith, Trans. St. Martin's Press, New York.

Kelemen, D., Rosset, E., 2009. The human function compunction: teleological explanation in adults. Cognition 111 (1), 138–143. Available from: https://doi.org/10.1016/j.cognition.2009.01.001.

Koch, M.W., Korngut, L., Patry, D.G., Agha-Khani, Y., White, C., Sarna, J.R., et al., 2015. The promise of futility trials in neurological diseases. Nat. Rev. Neurol. 11 (5), 300–305. Available from: https://doi.org/10.1038/nrneurol.2015.34.

Kordower, J.H., Goetz, C.G., Chu, Y., Halliday, G.M., Nicholson, D.A., Musial, T.F., et al., 2017. Robust graft survival and normalized dopaminergic innervation do not obligate recovery in a Parkinson disease patient. Ann. Neurol. 81 (1), 46–57. Available from: https://doi.org/10.1002/ana.24820.

Kraus, L.M., 2012. A Universe from Nothing: Why There Is Something Rather Than Nothing. Simon and Schuster, New York.

Kuhn, T., 1962. The Structure of Scientific Revolutions. University of Chicago Press, Chicago, IL.

Kurkela, K.A., Dennis, N.A., 2016. Event-related fMRI studies of false memory: an activation likelihood estimation meta-analysis. Neuropsychologia 81, 149–167. Available from: https://doi.org/10.1016/j.neuropsychologia.2015.12.006.

Landis, S., Insel, T.R., 2008. The "neuro" in neurogenetics. Science 322 (5903), 821.

Landis, S.C., Amara, S.G., Asadullah, K., Austin, C.P., Blumenstein, R., et al., 2012. A call for transparent reporting to optimize the predictive value of preclinical research. Nature 490, 187–191.

Laudan, L., 1984. Science and Values: The Aims of Science and Their Role in Scientific Debate. University of California Press, Berkely, Los Angles, London.

Le Fanuref, J., 1999. The Rise and Fall of Modern Medicine. Basic Books, New York, NY.

Leckman, J.F., Riddle, M.A., Hardin, M.T., Ort, S.I., Swartz, K.L., Stevenson, J., et al., 1989. The Yale Global Tic Severity Scale: initial testing of a clinician-rated scale of tic severity. J. Am. Acad. Child Adolesc. Psychiatry 28 (4), 566–573.

Leckman, J.F., Zhang, H., Vitale, A., et al., 1998. Course of tic severity in Tourette syndrome: the first two decades. Pediatr. 102 (1 Pt 1), 14–19.

Lee, A.S., Hubona, G.S., 2009. A scientific basis for rigor in information systems research. MIS Q. 33 (2), 237–262.

Legenstein, R., Maass, W., 2007. Edge of chaos and prediction of computational performance for neural circuit models. Neural Netw. 20 (3), 323–334.

Lei, X., Yang, T., Wu, T., 2015. Functional neuroimaging of extraversion-introversion. Neurosci. Bull. 31 (6), 663–675. Available from: https://doi.org/10.1007/s12264-015-1565-1.

Leroi, A.M., 2014. The Lagoon: How Aristotle Invented Science. Penguin Books, New York.

Lewis, J., Breeze, C.E., Charlesworth, J., Maclaren, O.J., Cooper, J., 2016. Where next for the reproducibility agenda in computational biology? BMC Syst. Biol. 10 (1), 52. Available from: https://doi.org/10.1186/s12918-016-0288-x.

LeWitt, P.A., Fahn, S., 2016. Levodopa therapy for Parkinson disease: a look backward and forward. Neurology 86 (14 Suppl 1). Available from: https://doi.org/10.1212/WNL.0000000000002509. S3-12.

Li, X., Yu, M., Zhu, M., 2009. Innate immune signaling pathways in animals: beyond reductionism. Int. Rev. Immunol. 28 (3-4), 207–238. Available from: https://doi.org/10.1080/08830180902839777. Review.

Lipovetsky, S., 2010. Supercritical pitchfork bifurcation in implicit regression modeling. Int. J. Artif. Life Res. 1, 4. Available from: https://doi.org/10.4018/jalr.2010100101.

Lipton, P., 2000. Inference to the best explanation. In: Newton-Smith, W.H. (Ed.), A Companion to the Philosophy of Science. Blackwell, Malden, Massachusetts, US, pp. 184–193.

Ludbrook, J., 1994. Repeated measurements and multiple comparisons in cardiovascular research. Cardiovasc. Res. 28 (3), 303–311. Review.

Luppi, P.H., Billwiller, F., Fort, P., 2017. Selective activation of a few limbic structures during paradoxical (REM) sleep by the claustrum and the supramammillary nucleus: evidence and function. Curr. Opin. Neurobiol. 44, 59–64. Available from: https://doi.org/10.1016/j.conb.2017.03.002.

MacDougall-Shackleton, S.A., 2011. The levels of analysis revisited. Philos. Trans. R. Soc. Lond. B Biol. Sci. 366 (1574), 2076–2085. Available from: https://doi.org/10.1098/rstb.2010.0363. Review.

Machamer, P., Wolters, G., 2004. Science, Values and Objectivity. University of Pittsburg Press, Pittsburgh, Pennsylvania, US.

Mantis, C., Kandela, I., Aird, F., 2017. reproducibility project: cancer biology. Replication study. Coadministration of a tumor-penetrating peptide enhances the efficacy of cancer drugs. eLife 6, e17584. Available from: https://doi.org/10.7554/eLife.17584.

Mariadason, J.G., Wang, W.N., Wallack, M.K., Belmonte, A., Matari, H., 2012. Negative appendicectomy rate as a quality metric in the management of appendicitis: impact of computed tomography, Alvarado score and the definition of negative appendicectomy. Ann. R. Coll. Surg. Engl. 94 (6), 395–401. Available from: https://doi.org/10.1308/003588412X13171221592131.

Marsden, C.D., 1982. The mysterious motor function of the basal ganglia: the Robert Wartenberg Lecture. Neurology 32 (5), 514–539.

Marsden, C.D., Obeso, J.A., 1994. The functions of the basal ganglia and the paradox of stereotaxic surgery in Parkinson's disease. Brain 117 (Pt 4), 877–897. Review.

Martinez-Lavin, M., Infante, O., Lerma, C., 2008. Hypothesis: the chaos and complexity theory may help our understanding of fibromyalgia and similar maladies. Semin. Arthritis Rheum. 37 (4), 260–264.

Matosin, N., Frank, E., Engel, M., Lum, J.S., Newell, K.A., 2014. Negativity towards negative results: a discussion of the disconnect between scientific worth and scientific culture. Dis. Model Mech. 7 (2), 171–173. Available from: https://doi.org/10.1242/dmm.015123.

Matrosov, V.V., Kazantsev, V.B., 2011. Bifurcation mechanisms of regular and chaotic network signaling in brain astrocytes. Chaos 21 (2), 023103. Available from: https://doi.org/10.1063/1.3574031.

Matthews, J.R., 1995. Quantification and the Quest for Medical Certainty. Princeton University Press, Princeton, NJ.

May, R.M., 1972. Limit cycles in predator-prey communities. Science 177 (4052), 900–902.

McComas, A., 2011. Galvani's Spark: The Story of the Nerve Impulse. Oxford University Press, New York.

McGough, J.J., Faraone, S.V., 2009. Estimating the size of treatment effects: moving beyond p values. Psychiatry (Edgmont) 6 (10), 21–29.

McNutt, M., 2014. Journals unite for reproducibility. Science 346 (6210), 679.

Michelson, A.A., Morley, E.W., 1887. On the relative motion of the earth and the luminiferous ether. Am. J. Sci. 34, 333–345. Available from: https://doi.org/10.2475/ajs.s3-34.203.333.

Mikkola, T.S., Savolainen-Peltonen, H., Venetkoski, M., Ylikorkala, O., 2017. New evidence for cardiac benefit of postmenopausal hormone therapy. Climacteric 20 (1), 5–10. Available from: https://doi.org/10.1080/13697137.2016.1262839.

Miller, D.W., Abercrombie, E.D., 1999. Role of high-affinity dopamine uptake and impulse activity in the appearance of extracellular dopamine in striatum after administration of exogenous L-DOPA: studies in intact and 6-hydroxydopamine-treated rats. J. Neurochem. 72 (4), 1516–1522.

Minto, C.F., Schnider, T.W., Shafer, S.L., 1997. Pharmacokinetics and pharmacodynamics of remifentanil. II. Model application. Anesthesiology 86 (1), 24–33.

Moalem, S., Pèrcy, M.E., 2002. The quandary of reductionism: relevance to Alzheimer disease research. J. Alzheimers Dis. 4 (6), 531–537. Review.

Moeller, S.J., Bederson, L., Alia-Klein, N., Goldstein, R.Z., 2016. Neuroscience of inhibition for addiction medicine: from prediction of initiation to prediction of relapse. Prog. Brain Res. 223, 165–188. Available from: https://doi.org/10.1016/bs.pbr.2015.07.007.

Montgomery Jr., E.B., 1992. Pharmacokinetics and pharmacodynamics of levodopa. Neurology 42 (1 Suppl 1), 17–22. discussion 57-60. Review.

Montgomery Jr., E.B., 2017. Deep brain stimulation programming: mechanisms, Principles, and Practice, second ed. Oxford University Press, New York.

Montgomery Jr., E.B., 2018. Medical Reasoning: The Nature and Use of Medical Knowledge. Oxford University Press, New York.

Montgomery Jr., E.B., 2007. Basal ganglia physiology and pathophysiology: a reappraisal. Parkinsonism Relat. Disord. 13 (8), 455–465.

Montgomery Jr., E.B., 2008. Subthalamic nucleus neuronal activity in Parkinson's disease and epilepsy subjects. Parkinsonism Relat. Disord. 14 (2), 120–125.

Montgomery Jr., E.B., 2012. The epistemology of deep brain stimulation and neuronal pathophysiology. Front. Integr. Neurosci. 6, 78.

Montgomery Jr., E.B., 2015. Twenty Things to Know about Deep Brain Stimulation. Oxford University Press, New York.

Montgomery Jr., E.B., 2016. Modeling and theories of pathophysiology and physiology of the basal ganglia-thalamic-cortical system: critical analysis. Front. Hum. Neurosci. 10, 469.

Montgomery Jr., E.B., Gale, J.T., 2008. Mechanisms of action of deep brain stimulation (DBS). Neurosci. Biobehav. Rev. 32 (3), 388–407.

Montgomery Jr., E.B., Turkstra, L.S., 2003. Evidenced based medicine: let's be reasonable. J. Med. Speech Language Pathol. 11, ix–xii.

Montgomery Jr., E.B., Buchholz, S.R., Delitto, A., Collins, R.C., 1986. Alterations in basal ganglia physiology following MPTP in monkeys. In: Markey, S.P., Castagnoli, N., Trevor, A.J., Kopin, I.J. (Eds.), A Neurotoxin Producing a Parkinsonian Syndrome. Academic Press, Inc, Cambridge, Massachusetts, US, pp. 679–682.

Montgomery Jr., E.B., Huang, H., Assadi, A., 2005. Unsupervised clustering algorithm for N-dimensional data. J. Neurosci. Methods 144 (1), 19–24.

Muralidharan, A., Jensen, A.L., Connolly, A., Hendrix, C.M., Johnson, M.D., Baker, K.B., et al., 2016. Physiological changes in the pallidum in a progressive model of Parkinson's disease: are oscillations enough? Exp. Neurol. 279, 187–196. Available from: https://doi.org/10.1016/j.expneurol.2016.03.002.

Musser, G., 2015. Spooky Action at a Distance. Scientific American/Farrar, Straus and Giroux, New York, NY.

National Commission for the Protection of Human Subjects of Biomedical and Behavioral Research, Department of Health, Education and Welfare (DHEW), 1978. The Belmont Report (PDF). United States Government Printing Office, Washington, DC.

Neugebauer, E.A., Willy, C., Sauerland, S., 2001. Complexity and non-linearity in shock research: reductionism or synthesis? Shock 16 (4), 252–258. Review.

Nguyen, D.H., Widrow, B., 1990. Neural networks for self-learning control systems. IEEE Contr. Syst. Mag. 10 (3), 18–23.

Nilles, K.M., London, B., 2007. Knockin animal models of inherited arrhythmogenic diseases: what have we learned from them? J. Cardiovasc. Electrophysiol. 18 (10), 1117–1125.

Norrelykke, S.F., 2011. Oscillator in heat bath: exact simulation of time-lapse-recorded data. Exact analytical benchmark statistics. Phys. Rev. E 83, 041103. Available from: https://doi:10.1103/PhysRevE.83.041103.

Northoff, G., 2016. How do resting state changes in depression translate into psychopathological symptoms? From 'Spatiotemporal correspondence' to 'Spatiotemporal

Psychopathology'. Curr. Opin. Psychiatry 29 (1), 18–24. Available from: https://doi. org/10.1097/YCO.0000000000000222.

Olanow, C.W., Stocchi, F., 2017. Levodopa: a new look at an old friend. Mov. Disord. 27. Available from: https://doi.org/10.1002/mds.27216.

Olanow, C.W., Goetz, C.G., Kordower, J.H., Stoessl, A.J., Sossi, V., Brin, M.F., et al., 2003. A double-blind controlled trial of bilateral fetal nigral transplantation in Parkinson's disease. Ann. Neurol. 54 (3), 403–414.

Olson, R., 2015. Houston, We Have a Narrative: Why Science Needs a Story. University of Chicago Press, Chicago, IL.

Orcutt, G.H., Watts, H.W., Edwards, J.B., 1968. Data aggregation and information loss. Am. Econ. Rev. 58 (4), 773–787.

Pappert, E.J., Goetz, C.G., Louis, E.D., Blasucci, L., Leurgans, S., 2003. Objective assessments of longitudinal outcome in Gilles de la Tourette's syndrome. Neurology 61 (7), 936–940.

Pham, T.D., Ichikawa, K., 2013. Spatial chaos and complexity in the intracellular space of cancer and normal cells. Theor. Biol. Med. Model. 10, 62. Available from: https://doi. org/10.1186/1742-4682-10-62.

Picciotto, M.R., 1999. Knock-out mouse models used to study neurobiological systems. Crit. Rev. Neurobiol. 13 (2), 103–149. Review.

Pizzuti, C., Rombo, S.E., 2014. Algorithms and tools for protein-protein interaction networks clustering, with a special focus on population-based stochastic methods. Bioinformatics 30 (10), 1343–1352. Available from: https://doi.org/10.1093/bioinformatics/btu034.

Poirier, L.J., 1971. The development of animal models for studies of Parkinson's disease. In: McDowell, F.H., Markham, C.H. (Eds.), Recent Advances in Parkinson's Disease. F. A. Davis Company, Philadelphia, PA (Chapter 4).

Poldrack, R.A., 2008. The role of fMRI in cognitive neuroscience: where do we stand? Curr. Opin. Neurobiol. 18 (2), 223–227. Available from: https://doi.org/10.1016/j. conb.2008.07.006.

Popper, K., 2002. The Logic of Scientific Discovery, originally Logik der Forschung, 1934 (Mohr Siebeck, in German), 1959 (in English), Routledge edition, 2002.

Prinz, A.A., Bucher, D., Marder, E., 2004. Similar network activity from disparate circuit parameters. Nat. Neurosci. 7 (12), 1345–1352.

Prinz, F., Schlange, T., Asadullah, K., 2011. Believe it or not: how much can we rely on published data on potential drug targets? Nat. Rev. Drug Discov. 10 (9), 712. Available from: https://doi.org/10.1038/nrd3439-c1.

Quine, W.V.O., Ullian, J.S., 1978. The Web of Belief, second ed. Random House, Inc, New York.

Ramón y Cajal, S., 1999. Advice for a Young Investigator. MIT Press, Cambridge, MA, p. 1.

Rose, S., 1998. What is wrong with reductionist explanations of behaviour? Novartis Found. Symp. 213, 176–186. discussion 186-192, 218-221. Review.

Rossouw, J.E., Anderson, G.L., Prentice, R.L., LaCroix, A.Z., Kooperberg, C., Stefanick, M. L., et al., 2002. Risks and benefits of estrogen plus progestin in healthy postmenopausal women: principal results from the Women's Health Initiative randomized controlled trial. J. Am. Med. Assoc. 288, 321–333.

Russell, B., 1945. The History of Western Philosophy. Simon & Schuster, New York, NY, p. 472.

Russell, J.F., 2013. If a job is worth doing, it is worth doing twice. Nature 496 (7443). Available from: https://doi.org/10.1038/496007a.

Saks, V., Monge, C., Guzun, R., 2009. Philosophical basis and some historical aspects of systems biology: from Hegel to Noble—applications for bioenergetic research. Int. J. Mol. Sci. 10 (3), 1161–1192. Available from: https://doi.org/10.3390/ijms10031161.

Saltelli, A., Funtowicz, S., 2017. What is science's crisis really about? Futures 91, 5—11.

Sandve, G.K., Nekrutenko, A., Taylor, J., Hovig, E., 2013. Ten simple rules for reproducible computational research. PLoS Comput. Biol. 9 (10), e1003285. Available from: https://doi.org/10.1371/journal.pcbi.1003285.

Schadt, E.E., Sachs, A., Friend, S., 2005. Embracing complexity, inching closer to reality. Sci. STKE 2005 (295), pe40. Review.

Schaffer, S., Shapin, S., 1985. Leviathan and the Air-Pump: Hobbes, Boyle, and the Experimental Life. Princeton University Press, Princeton, NJ.

Schierbeck, L., 2015. Primary prevention of cardiovascular disease with hormone replacement therapy. Climacteric 18 (4), 492—497. Available from: https://doi.org/10.3109/13697137.2015.1034098.

Schmidt, P.F., 1956. Some merits and misinterpretations of scientific method. Sci. Month. 82 (1), 20—24.

Schork, N.J., 2015. Personalized medicine: time for one-person trials. Nature 520 (7549), 609—611. Available from: https://doi.org/10.1038/520609a.

Schultz, W., 1986. Responses of midbrain dopamine neurons to behavioral trigger stimuli in the monkey. J. Neurophysiol. 56 (5), 1439—1461.

Scott, C., Wasserman, T., 1997. When is a negative study not negative? Int. J. Radiat. Oncol. Biol. Phys. 39 (4), 859—861. Review.

Scott, S., Kranz, J.E., Cole, J., Lincecum, J.M., Thompson, K., Bostrom, A., et al., 2008. Design, power, and interpretation of studies in the standard murine model of ALS. Amyotroph. Lateral Scler. 9, 4—15.

Ségalat, L., 2007. Loss-of-function genetic diseases and the concept of pharmaceutical targets. Orphanet. J. Rare Dis. 2, 30. Available from: https://doi.org/10.1186/1750-1172-2-30.

Shapin, S., Schaffer, S., 1985. Leviathan and the Air-Pump: Hobbes, Boyle, and the Experimental Life. Princeton University Press, Princeton, NJ.

Shen, P., Larter, R., 1995. Chaos in intracellular Ca^{2+} oscillations in a new model for non-excitable cells. Cell Calcium 17 (3), 225—232.

Sigalov, A.B., 2011. Cells diversify transmembrane signaling through the controlled chaos of protein disorder. Self Nonself 2 (2), 75—79.

Sjoberg, E.A., 2017. Logical fallacies in animal model research. Behav. Brain Funct. 13 (1), 3. Available from: https://doi.org/10.1186/s12993-017-0121-8. Review.

Snapinn, S., Chen, M.G., Jiang, Q., Koutsoukos, T., 2006. Assessment of futility in clinical trials. Pharm. Stat. 5 (4), 273—281.

Sokal, A., Bricmont, J., 1998. Fashionable Nonsense. Picador, New York, NY.

Sokal, A., Bricmont, J., 1999. Fashionable nonsense: postmodern intellectuals' abuse of science, French: Impostures intellectuelles, Odile Jacob, in French, 1997, Picador USA, in English.

Spirov, A., Holloway, D., 2013. Using evolutionary computations to understand the design and evolution of gene and cell regulatory networks. Methods 62 (1), 39—55. Available from: https://doi.org/10.1016/j.ymeth.2013.05.013.

Stahel, W.A., 2016. Statistical issues in reproducibility. In: Atamspacher, H., Massen, S. (Eds.), Reproducibility: Principles, Problems, Practices, and Prospects. John Wiley & Sons, New York, NY (Chapter 5).

Stampfer, M.J., Colditz, G.A., Willett, W.C., Manson, J.E., Rosner, B., Speizer, F.E., et al., 1991. Postmenopausal estrogen therapy and cardiovascular disease: ten-year follow-up from the Nurses' Health Study. N. Engl. J. Med. 325, 756—762.

Steen, R.G., 2011. Retractions in the scientific literature: is the incidence of research fraud increasing? Research ethics. J. Med. Ethics 37 (4), 249—253. Available from: https://doi.org/10.1136/jme.2010.040923.

Steinle, F., 2016. Stability and replication of experimental results: a historical perspective. In: Atamspacher, H., Massen, S. (Eds.), Reproducibility: Principles, Problems, Practices, and Prospects. John Wiley & Sons, New York, NY (Chapter 3).

Steriade, M., 2001. The Intact and Sliced Brain. MIT Press, Cambridge, Massachusetts, US.

Steward, O., Balice-Gordon, R., 2014. Rigor or mortis: best practices for preclinical research in neuroscience. Neuron 84 (3), 572–581. Available from: https://doi.org/10.1016/j.neuron.2014.10.042.

Stewart, I., 2007. Why Beauty Is Truth: A History of Symmetry. Basic Books, New York.

Stigler, S.M., 2016. The Seven Pillars of Statistical Wisdom. Harvard University Press, Cambridge Massachusetts, US.

Strange, K., 2005. The end of "naive reductionism": rise of systems biology or renaissance of physiology? Am. J. Physiol. Cell Physiol. 288 (5), C968–C974. Review.

Street, M.E., Buscema, M., Smerieri, A., Montanini, L., Grossi, E., 2013. Artificial neural networks, and evolutionary algorithms as a systems biology approach to a data-base on fetal growth restriction. Prog. Biophys. Mol. Biol. 113 (3), 433–438. Available from: https://doi.org/10.1016/j.pbiomolbio.2013.06.003.

Strogatz, S.H., 1994. Non-linear Dynamics and Chaos. Perseus Books Publishing, Cambridge Massachusetts, US.

Tang, J.M., Moro, E., Lozano, A.M., Lang, A.E., Hutchison, W.D., Mahant, N., et al., 2005. Firing rates of pallidal neurons are similar in Huntington's and Parkinson's disease patients. Exp. Brain Res. 166, 230–236.

Thompson, P.D., Berardelli, A., Rothwell, J.C., Day, B.L., Dick, J.P., Benecke, R., et al., 1988. The coexistence of bradykinesia and chorea in Huntington's disease and its implications for theories of basal ganglia control of movement. Brain 111 (Pt 2), 223–244.

Tragante, V., Moore, J.H., Asselbergs, F.W., 2014. The ENCODE project and perspectives on pathways. Genet. Epidemiol. 38 (4), 275–280. Available from: https://doi.org/10.1002/gepi.21802.

Tsang, R., Colley, L., Lynd, L.D., 2009. Inadequate statistical power to detect clinically significant differences in adverse event rates in randomized controlled trials. J. Clin. Epidemiol. 62 (6), 609–616. Available from: https://doi.org/10.1016/j.jclinepi.2008.08.005.

Turing, A.M., 1937. On computable numbers, with an application to the entscheidungsproblem. Proc. Lond. Math. Soc. 2 (42), 230–265. Available from: https://doi.org/10.1112/plms/s2-42.1.230.

Turing, A.M., 1938. On computable numbers, with an application to the entscheidungsproblem: a correction. Proc. Lond. Math. Soc. 2 (43), 544–546. Available from: https://doi.org/10.1112/plms/s2-43.6.544 (published 1937).

Underwood, R., Kumari, V., Peters, E., 2016. Cognitive and neural models of threat appraisal in psychosis: a theoretical integration. Psychiatry Res. 239, 131–138. Available from: https://doi.org/10.1016/j.psychres.2016.03.016.

Van Regenmortel, M.H., 2001. Pitfalls of reductionism in the design of peptide-based vaccines. Vaccine 19 (17–19), 2369–2374. Review.

Verma, I.M., 2002. Hypothesis-driven science. Mol. Ther. 6 (1), 1.

Villoslada, P., Steinman, L., Baranzini, S.E., 2009. Systems biology and its application to the understanding of neurological diseases. Ann. Neurol. 65 (2), 124–139. Available from: https://doi.org/10.1002/ana.21634. Review.

Walker, E., Nowacki, A.S., 2011. Understanding equivalence and noninferiority testing. J. Gen. Intern. Med. 26 (2), 192–196. Available from: https://doi.org/10.1007/s11606-010-1513-8.

Walker, E., Hernandez, A.V., Kattan, M.W., 2008. Meta-analysis: its strengths and limitations. Cleve. Clin. J. Med. 75 (6), 431–439. Review.

Warner, J.H., 1991. Ideals of science and their discontents in late nineteenth-century American medicine. Isis 82, 454–478.

Wolkenhauer, O., Green, S., 2013. The search for organizing principles as a cure against reductionism in systems medicine. FEBS J. 280 (23), 5938–5948. Available from: https://doi.org/10.1111/febs.12311.

Worzel, W.P., Yu, J., Almal, A.A., Chinnaiyan, A.M., 2009. Applications of genetic programming in cancer research. Int. J. Biochem. Cell Biol. 41 (2), 405–413. Available from: https://doi.org/10.1016/j.biocel.2008.09.025.

Yahr, M.D., Duvoisin, R.C., Schear, M.J., Barrett, R.E., Hoehn, M.M., 1969. Treatment of parkinsonism with levodopa. Arch. Neurol. 21 (4), 343–354.

Zaner, R.M., 1998. Ethics and the Clinical Encounter. Prentice Hall, Englewood Cliffs, NJ.

Zhu, C.L., Jia, Y., Liu, Q., Yang, L.J., Zhan, X., 2007. A mesoscopic stochastic mechanism of cytosolic calcium oscillations. Biophys. Chem. 125 (1), 201–212.

Zimmerli, W.Ch, 2016. Reproducibility between production and prognosis. In: Atamspacher, H., Massen, S. (Eds.), Reproducibility: Principles, Problems, Practices, and Prospects. John Wiley & Sons, New York, NJ (Chapter 2).

Zimmerman, D.W., 2004. Conditional probabilities of rejecting H_o by pooled and separate-variances t tests given heterogeneity of sample variances. Commun. Stat. Simul. Comput. 33, 69–81.

Zuckerman, D.M., Brown, P., Nissen, S.E., 2011. Medical device recalls and the FDA approval process. Arch. Intern. Med. 171 (11), 1006–1011. Available from: https://doi.org/10.1001/archinternmed.2011.30.

Index